# Lecture Notes in Physics

## Volume 856



For further volumes:
www.springer.com/series/5304

## The Lecture Notes in Physics

The series Lecture Notes in Physics (LNP), founded in 1969, reports new developments in physics research and teaching—quickly and informally, but with a high quality and the explicit aim to summarize and communicate current knowledge in an accessible way. Books published in this series are conceived as bridging material between advanced graduate textbooks and the forefront of research and to serve three purposes:

- to be a compact and modern up-to-date source of reference on a well-defined topic
- to serve as an accessible introduction to the field to postgraduate students and nonspecialist researchers from related areas
- to be a source of advanced teaching material for specialized seminars, courses and schools

Both monographs and multi-author volumes will be considered for publication. Edited volumes should, however, consist of a very limited number of contributions only. Proceedings will not be considered for LNP.

Volumes published in LNP are disseminated both in print and in electronic formats, the electronic archive being available at springerlink.com. The series content is indexed, abstracted and referenced by many abstracting and information services, bibliographic networks, subscription agencies, library networks, and consortia.

Proposals should be sent to a member of the Editorial Board, or directly to the managing editor at Springer:

Christian Caron
Springer Heidelberg
Physics Editorial Department I
Tiergartenstrasse 17
69121 Heidelberg/Germany
christian.caron@springer.com

A.P.J. Jansen

# An Introduction to Kinetic Monte Carlo Simulations of Surface Reactions

A.P.J. Jansen
ST/SKA
Eindhoven University of Technology
Eindhoven, Netherlands

ISSN 0075-8450 ISSN 1616-6361 (electronic)
Lecture Notes in Physics
ISBN 978-3-642-29487-7 ISBN 978-3-642-29488-4 (eBook)
DOI 10.1007/978-3-642-29488-4
Springer Heidelberg New York Dordrecht London

Library of Congress Control Number: 2012940387



Printed on acid-free paper

Springer is part of Springer Science+Business Media (www.springer.com)

*For Esther.*

# Preface

Kinetic Monte Carlo (kMC) simulations form still a quite new area of research. Figure 1 shows the number of publications (articles or reviews) with "kinetic Monte Carlo" in the title or abstract according to the abstract and citation database Scopus. There are two things to note. On the one hand it is not a very extensive area of research yet. A very diligent researcher can still keep track of all publications that appear. On the other hand, the number of publications is rapidly growing.

Figure 1 shows that there were no publications before 1993 that used the term kMC. This does not mean that there have been no kMC simulations before that year. There have been some but the term was not used yet. In fact, there are still people, who do what we will call kMC simulations here, but who do not use the term. One mundane reason for that is probably that they use an algorithm that they regard as one of many possible algorithms for doing Monte Carlo (MC) simulations. Why give it a special name? Another reason may be historical. Instead of kMC, people have used and still use the term dynamic MC. This is a term introduced by D.T. Gillespie for his algorithms that use MC to solve macroscopic rate equations. These algorithms are often almost identical to the ones we will describe in Chap. 3, and it seems reasonable to use the same term even when the algorithms are used for different problems. There has been a tendency to be more strict in the terminology however. For example, the term Stochastic Simulation Algorithm is now often used when using MC for rate equations. There are even people that restrict the term kMC to one particular algorithm, the Variable Step Size Method in our terminology (see Sect. 3.2), even though all other algorithms in Chap. 3 give exactly the same results. But the term kMC has also been used for rate equations. So the situation concerning terminology is still fluent.

So what do we mean when we use the term kMC? There are always two aspects to kMC as we will discuss it here. We will regard a system as a set of minima of a potential-energy surface (PES). The evolution of a system in real time will then be regarded as hops from one minimum to a neighboring one. These are the elementary events of kMC. The second aspect concerns the algorithms. The hops in kMC will be seen to be stochastic processes and the algorithms use random numbers to determine at which times the hops occur and to which neighboring minimum they go. This is

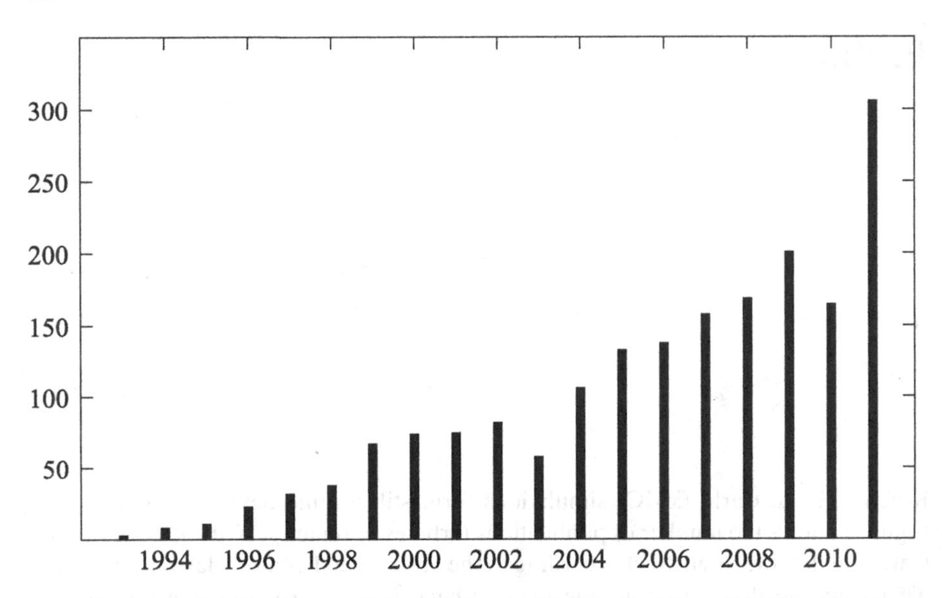


**Fig. 1** Number of publications (articles or reviews) with "kinetic Monte Carlo" in the title or abstract as a function of the year of publication

our general definition of kMC. We will use it however only in Chap. 2 and Sect. 8.4. In the rest of this book we will make an additional assumption. This is where the surface reactions in the title of this book come in. The surface on which the reactions take place is often periodic and has translational symmetry in two directions. The minima of the PES are related to the adsorption sites of the surface. The latter form a lattice and the reactions can be modeled with a lattice-gas model. We will see that this is even possible if the periodicity of the surface is not perfect. So kMC in this book stands for a lattice-gas model that describes the evolution of the system in real time and with elementary events that are stochastic and that correspond to reactions and other processes.

This book has two objectives. First, it is about the kMC method. A derivation of the method will be given from first principles, and we will discuss various algorithms that can be used to do actual simulations. This means that much of the book is also supposed to be useful to people who use kMC for other systems than surface reactions. For example, the derivation of the master equation in Chap. 2, which forms the basis of our theory of kMC, does not use any information particular to surface reactions. It only assumes that you have a system that can be described by a single-valued PES. This includes a very large majority of all systems one encounters in chemical physics. Chapter 8 also has a section that discusses kMC for when this is all one knows about a system.

Most of the book does however assume that a lattice-gas model is used, because this simplifies the applicability of kMC enormously. However, this still does not restrict the usefulness only to surface reactions. In fact, most publications using lattice-gas kMC are not about surface reactions. There are many applications of kMC in crystalline solids, polymers, crystal growth, chemical vapor deposition, molecular-beam epitaxy, ion implantation, etching, nanoparticles, and non-reactive processes. The discussions of algorithms in Chap. 3 and the way processes can be modeled in Chaps. 5, 6, 7, and 8 are just as useful for those applications as for surface reactions. However, the second objective of this book is to show what kMC simulations can teach us about the kinetics of surface reactions that one finds in catalysis and surface science. The book was mainly written with this in mind. This means that there are aspects that are relevant for the application of kMC to other areas that will not be found here, whereas some aspects that are discussed here may not be relevant for these areas.

The book is called an introduction because it is meant to give all information on kMC simulations of surface reactions that you need if you want to start from scratch. A lot of space is devoted to the basics, which are discussed in detail. The term "introduction" is not meant to imply that everything in this book is low level or easy. Some things are but others definitely are not. It is for example quite easy to implement the algorithms of Chap. 3 for a simple system of surface reactions, and the resulting code will probably yield very useful and interesting information on the kinetics of the system. Writing a general-purpose code however is much harder. Also the theoretical derivation of the master equation on which we base kMC, advanced aspects of the algorithms, and certain new developments in Chap. 8 are anything but easy.

The structure of this book is as follows. Chapter 1 discusses why one would want to do kMC simulations. The kinetics of surface reactions is normally described with macroscopic rate equations. There are different ways in which these equations can be used, but it is shown that they all have substantial drawbacks.

Chapter 2 deals with the basic theory. It introduces the lattice gas as the model for the systems in this book, and it gives the derivation of the master equation. This is the central equation for kMC. It forms the basis of all kMC algorithms, it relates quantum chemical calculations of rate constants to kMC, and it relates kMC to other kinetic theories like microkinetics.

Chapter 3 discusses kMC algorithms. kMC generates a sequence of configurations and times when the transitions between these configurations occur. This solves the master equation. There are many algorithms that yield such a sequence of configurations and which are statistically equivalent. We discuss a few in detail because they are the ones that are efficient for models of surface reactions. Time-dependent rate constants are discusses separately as the determination of when processes take place pose special problems. Parallelization is discussed as well as some older algorithms. Some guidelines are given of how to choose an algorithm for a simulation.

Chapter 4 shows how the rate constants that are needed for kMC simulations can be obtained. It shows how rate constants can either be calculated or be derived from experimental results. Calculating rate constants involves determining the initial and the transition state of a process, the energies of these states, and their partition functions. The phenomenological or macroscopic equation is the essential equation to get rate constants from experiments. Lateral interactions can affect rate constants substantially, but because they are relatively weak and special attention needs to be given to the reliability of calculations of these interactions.

Chapters 5 and 6 discuss ways to model surface processes. These chapters deal with the same topic, but approach it from different angles. Chapter 5 shows the tools that we can use in modeling. For simple systems there is a lattice corresponding to the adsorption sites and the labels of the lattice points describe the occupation of the sites. The labels can however also be used to model steps and other defects and sites on bimetallic substrates. The lattice points don't need to correspond to sites however, but can also be used to store other information like the presence of certain structures in the adlayer. Processes need not always to correspond to reactions or other actual processes, but when they have an infinite rate constant they can be used in a general-purpose code to handle exceptional situations that are normally hard-coded in special-purpose codes.

Chapter 6 discusses typical surface processes and how each of them can be modeled in different ways using the tools from Chap. 5. The way to model many processes for kMC simulations is straightforward. There are however also processes that one encounters regularly and for which there are more modeling options and for which it is not always clear which the best. We discuss several of them.

Chapter 7 shows how the modeling of various surface processes can be integrated. We discuss a number of complete surface reaction systems and show the benefits of kMC simulations for them. Chapter 8 finally discusses some aspects of kMC that one might want to improve and some likely new developments. kMC is a very versatile and powerful method to study the kinetics of surface reactions, but there are nevertheless some systems and phenomena for which one would like it to be even more efficient or one would like to extend it.

Geldrop, Netherlands Tonek Jansen

# Acknowledgements

This book would never have been written without the support of many people. I would like to thank Rutger van Santen, Risto Nieminen, Juha-Pekka Hovi, Hans Niemantsverdriet, Vladimir Kuzovkov, Mark Koper, and Rasmita Raval for many discussions on the kinetics of surface reactions. Peter Hilbers and John Segers contributed a lot to my understanding of the algorithms of kinetic Monte Carlo. The Ph.D. students and postdocs Ronald Gelten, Rafael Salazar, Silvia Nedea, Cristina Popa, Sander van Bavel, Joris Hagelaar, Maarten Jansen, Zhang Xueqing, and Minhaj Ghouri have taught me a lot about how to model surface reactions. In particular they have taught me to trust our computer codes also for very complicated systems. I am grateful to Ivo Filot for his critical comments on Chap. 4. Special thanks go to Chrétien Hermse and Johan Lukkien for endless discussions, unfailing support and enthusiasm even after they stopped working on kinetic Monte Carlo themselves. Chrétien has come up with many of the modeling tricks without which Chaps. 5, 6, and 7 would have looked very different. Johan has been crucial in the development of the algorithms that can be found in Chap. 3. He has also written the Carlos program that has been essential for almost all of the work in kinetic Monte Carlo that I have done. Finally I would like to thank friends and family for bearing with me when I was preoccupied with matters kinetic Monte Carlo when I shouldn't have been.

I have to give credit for many of the good ideas in this book to the people mentioned above, but I do claim that the bad ideas, errors, and other shortcomings are all mine.

# Contents


**1 Introduction** . . . 1
- 1.1 Why Do Kinetic Monte Carlo Simulations? . . . 1
- 1.2 Some Comparisons . . . 4
- 1.3 Length and Time Scales . . . 11
- References . . . 12

**2 A Stochastic Model for the Description of Surface Reaction Systems** 13
- 2.1 The Lattice Gas . . . 13
  - 2.1.1 Lattices, Sublattices, and Unit Cells . . . 14
  - 2.1.2 Examples of Lattices . . . 15
  - 2.1.3 Labels and Configurations . . . 18
  - 2.1.4 Examples of Using Labels . . . 18
  - 2.1.5 Shortcomings of Lattice-Gas Models . . . 20
  - 2.1.6 Boundary Conditions . . . 22
- 2.2 The Master Equation . . . 22
  - 2.2.1 The Definition and Some Properties of the Master Equation 22
  - 2.2.2 The Derivation of the Master Equation . . . 26
  - 2.2.3 The Master Equation for Lattice-Gas Models . . . 32
- References . . . 35

**3 Kinetic Monte Carlo Algorithms** . . . 37
- 3.1 Introduction . . . 37
- 3.2 The Variable Step Size Method . . . 38
  - 3.2.1 The Integral Form of the Master Equation . . . 38
  - 3.2.2 The Concept of the Variable Step Size Method . . . 39
  - 3.2.3 Enabled and Disabled Processes . . . 41
- 3.3 Some General Techniques . . . 43
  - 3.3.1 Selection Methods . . . 43
  - 3.3.2 Using Disabled Processes . . . 46
  - 3.3.3 Reducing Memory Requirements . . . 49
  - 3.3.4 Supertypes . . . 50

3.4 The Random Selection Method . . . . . 51
3.5 The First Reaction Method . . . . . 53
3.6 Time-Dependent Rate Constants . . . . . 55
3.7 A Comparison with Other Methods . . . . . 58
3.7.1 The Fixed Time Step Method . . . . . 59
3.7.2 Algorithmic Approach . . . . . 59
3.7.3 The Original Kinetic Monte Carlo . . . . . 60
3.7.4 Cellular Automata . . . . . 61
3.8 Parallel Algorithms . . . . . 61
3.9 Practical Considerations Concerning Algorithms . . . . . 65
References . . . . . 69

**4 How to Get Kinetic Parameters** . . . . . 73
4.1 Introductory Remarks on Kinetic Parameters . . . . . 73
4.2 Two Expressions for Rate Constants . . . . . 74
4.2.1 The General Expression . . . . . 75
4.2.2 The Arrhenius Form . . . . . 76
4.3 Partition Functions . . . . . 78
4.3.1 Classical and Quantum Partition Functions . . . . . 78
4.3.2 Zero-Point Energy . . . . . 78
4.3.3 Types of Partition Function . . . . . 79
4.3.4 Vibrations . . . . . 80
4.3.5 Rotations . . . . . 81
4.3.6 Hindered Rotations . . . . . 83
4.3.7 Translations . . . . . 84
4.3.8 Floppy Molecules . . . . . 84
4.4 The Practice of Calculating Rate Constants . . . . . 85
4.4.1 Langmuir–Hinshelwood Reactions . . . . . 85
4.4.2 Desorption . . . . . 86
4.4.3 Adsorption . . . . . 87
4.4.4 Eley–Rideal Reactions . . . . . 91
4.4.5 Diffusion . . . . . 91
4.4.6 Examples . . . . . 91
4.4.7 Summary . . . . . 93
4.5 Lateral Interactions . . . . . 94
4.5.1 The Cluster Expansion . . . . . 94
4.5.2 Linear Regression . . . . . 96
4.5.3 Cross Validation . . . . . 97
4.5.4 Bayesian Model Selection . . . . . 98
4.5.5 The Effect of Lateral Interactions on Transition States . . . . . 103
4.5.6 Other Models for Lateral Interactions . . . . . 103
4.6 Rate Constants from Experiments . . . . . 104
4.6.1 Relating Macroscopic Properties to Microscopic Processes 105
4.6.2 Simple Desorption . . . . . 106
4.6.3 Simple Adsorption . . . . . 108
4.6.4 Unimolecular Reactions . . . . . 110

4.6.5 Diffusion . . . 110
4.6.6 Bimolecular Reactions . . . 111
4.6.7 Dissociative Adsorption . . . 114
4.6.8 A Brute-Force Approach . . . 115
References . . . 117

**5 Modeling Surface Reactions I** . . . 121
5.1 Introduction . . . 121
5.2 Reducing Noise . . . 122
5.3 A Modeling Framework . . . 125
5.4 Modeling the Occupation of Sites . . . 128
5.4.1 Simple Adsorption, Desorption, and Unimolecular Conversion . . . 128
5.4.2 Bimolecular Reactions and Diffusion . . . 130
5.5 Modeling Adsorption Sites . . . 133
5.5.1 Using the Sublattice Index . . . 133
5.5.2 Using Labels to Distinguish Sublattices . . . 135
5.5.3 Systems Without Translational Symmetry . . . 137
5.5.4 Bookkeeping Sites . . . 141
5.6 Using Immediate Processes . . . 142
5.6.1 Very Fast Processes . . . 142
5.6.2 Flagging Structural Elements . . . 143
5.6.3 Counting . . . 146
5.6.4 Decomposing the Implementation of Processes . . . 148
5.6.5 Implementing Procedures . . . 150
References . . . 153

**6 Modeling Surface Reactions II** . . . 155
6.1 Introduction . . . 155
6.2 Large Adsorbates and Strong Repulsion . . . 156
6.3 Lateral Interactions . . . 159
6.4 Diffusion and Fast Reversible Reactions . . . 162
6.5 Combining Processes . . . 163
6.6 Isotope Experiments and Diffusion . . . 165
6.7 Simulating Nanoparticles and Facets . . . 170
6.8 Making the Initial Configuration . . . 177
References . . . 179

**7 Examples** . . . 181
7.1 Introduction . . . 181
7.2 NO Reduction on Rh(111) . . . 181
7.3 $NH_3$ Induced Reconstruction of (111) Steps on Pt(111) . . . 189
7.4 Phase Transitions and Symmetry Breaking . . . 193
7.4.1 TPD with Strong Repulsive Interactions . . . 194
7.4.2 Voltammetry and the Butterfly . . . 197
7.4.3 The Ziff–Gulari–Barshad Model . . . 200
7.5 Non-linear Kinetics . . . 203

7.5.1 The Lotka Model . . . . . . . . . . 204
7.5.2 Oscillations of CO Oxidation on Pt Surfaces . . . . . . . . . 206
References . . . . . . . . . . 208

**8 New Developments** . . . . . . . . . . 211
8.1 Longer Time Scales and Fast Processes . . . . . . . . . . 211
8.1.1 When Are Fast Processes a Problem? . . . . . . . . . . 212
8.1.2 A Simple Solution . . . . . . . . . . 212
8.1.3 Reduced Master Equations . . . . . . . . . . 213
8.1.4 Dealing with Slightly Slower Reactions . . . . . . . . . . 215
8.1.5 Two Other Approaches . . . . . . . . . . 224
8.2 Larger Length Scales . . . . . . . . . . 226
8.3 Embedding kMC in Larger Simulations . . . . . . . . . . 231
8.4 Off-lattice kMC . . . . . . . . . . 234
References . . . . . . . . . . 240

**Glossary** . . . . . . . . . . 243

**Index** . . . . . . . . . . 251

# Acronyms

| | |
|---|---|
| DES | Discrete Event Simulation |
| DFT | Density-Functional Theory |
| DMC | Dynamic Monte Carlo |
| FRM | First Reaction Method |
| kMC | kinetic Monte Carlo |
| MC | Monte Carlo |
| MD | Molecular Dynamics |
| MEP | Maximum Entropy Principle |
| MF | Mean Field |
| MFA | Mean-Field Approximation |
| PES | potential-energy surface |
| RSM | Random Selection Method |
| SSA | Stochastic Simulation Algorithm |
| TPD | Temperature-Programmed Desorption |
| TPR | Temperature-Programmed Reaction |
| TST | Transition-State Theory |
| VSSM | Variable Step Size Method |
| VTST | Variational Transition-State Theory |
| ZGB | Ziff–Gulari–Barshad |

# Chapter 1
# Introduction

**Abstract** The kinetics of surface reactions is normally described with macroscopic rate equations. There are different ways in which these equations can be used, but it is shown that they all have substantial drawbacks, which is the reason why we want to do kinetic Monte Carlo simulations. These simulations allow us to bridge the gap of many orders of magnitude in length and time scales between the processes on the atomic scale and the macroscopic kinetics.

## 1.1 Why Do Kinetic Monte Carlo Simulations?

Kinetics of surface reactions is generally described using macroscopic rate equations, which are also just called rate equations, mass balance equations, Mean Field equations, or phenomenological equations. If we have adsorbates A, B, C, et cetera, then these equations can be written as [1, 2]

$$\frac{d\theta_{\mathrm{X}}}{dt} = \sum_n I_{\mathrm{X}}^{(n)} k^{(n)} f^{(n)}(\theta_{\mathrm{A}}, \theta_{\mathrm{B}}, \ldots) \tag{1.1}$$

where $\theta_{\mathrm{X}}$ is the coverage of adsorbate X, $d\theta_{\mathrm{X}}/dt$ is the rate with which the coverage of X changes, the sum is over all reactions, $k^{(n)}$ is the rate constant of reaction $n$, and $f^{(n)}$ is a factor indicating how the rate of reaction $n$ depends on the coverages. $I_{\mathrm{X}}^{(n)}$ is the change in the number of X's in reaction $n$ with $I_{\mathrm{X}}^{(n)} > 0$ if X's are formed, $I_{\mathrm{X}}^{(n)} < 0$ if X's react away, and $I_{\mathrm{X}}^{(n)} = 0$ in all other cases (usually X doesn't participate in the reaction, but it might also be for example a catalyst).

There are two extreme ways in which the rate Eq. (1.1) can be interpreted. The first takes the different terms on the right-hand-side to reflect the rates of the actual processes taking place on the atomic scale. This is sometimes called microkinetics [3]. We will see that with this interpretation the rate equations can be derived from first principles (see Sect. 4.6), provided some assumptions are made. In this interpretation the constants $k$ are the rate constants of the individual processes and the $f$'s indicate the number of ways a reaction can take place normalized with respect to the area of the surface.

It is important to have a good understanding of this interpretation of the rate equations, because one objective of this book is to relate the processes on the atomic

A.P.J. Jansen, *An Introduction to Kinetic Monte Carlo Simulations of Surface Reactions*, Lecture Notes in Physics 856,
DOI 10.1007/978-3-642-29488-4_1, © Springer-Verlag Berlin Heidelberg 2012

scale to the macroscopic kinetics. Let's therefore look at a simple example. Suppose we have two different adsorbates A and B that can react when they are at neighboring sites to form a product AB: i.e., we have the reaction $A + B \to AB$. A specific example that we will encounter various times in this book is CO oxidation. In that case A is CO and B an oxygen atom. Let's also assume that all adsorbates prefer to adsorb on the same sites at all coverages, and that these sites form a square lattice. If we have only the reaction then

$$\frac{d\theta_A}{dt} = \frac{d\theta_B}{dt} = -kf(\theta_A, \theta_B). \quad (1.2)$$

We see that $I_A = I_B = -1$, because each reaction removes one A and one B. If we have a surface with $S$ sites and there are $N_A$ A's and $N_B$ B's, then $\theta_A = N_A/S$ and $\theta_B = N_B/S$.

What does $f(\theta_A, \theta_B)$ look like? To answer this question it is convenient to multiply Eq. (1.2) by $S$ to get

$$\frac{dN_A}{dt} = \frac{dN_B}{dt} = -kSf(\theta_A, \theta_B). \quad (1.3)$$

The right-hand-side now stands for the rate with which the total numbers of A's and B's change. The rate constant $k$ stands for the rate constant of an individual reaction, so $Sf(\theta_A, \theta_B)$ must stand for the number of individual reactions that can take place. Because A and B can only react when they are at neighboring sites $Sf(\theta_A, \theta_B)$ must equal the number of neighboring A–B pairs. In general this number depends on how the A's and B's are distributed over the sites: i.e., the adlayer structure. The normal assumption for rate equations is that the adsorbates are distributed randomly over the sites. In that case each A has a probability $N_B/(S-1)$ that a particular neighboring site is occupied by a B. Each A has four neighboring sites when we have a square lattice, so the number of neighboring A–B pairs is then $4N_A N_B/(S-1)$. From this we get $f(\theta_A, \theta_B) = 4N_A N_B/[S(S-1)]$ which equals $4\theta_A\theta_B$ if $S$ is large. So our rate equations becomes

$$\frac{d\theta_A}{dt} = \frac{d\theta_B}{dt} = -4k\theta_A\theta_B. \quad (1.4)$$

Textbooks on the kinetics of surface reactions generally simply pose equations like (1.4) with little justification and generally leave out the factor 4 which derives from the structure of the substrate. The problem with this equation is however that the assumption that the adsorbates are randomly distributed over the sites is rarely correct. The main reason is that there are interactions between the adsorbates. At low temperature they lead to correlation in the occupation of neighboring sites and at very low temperature may even result in island formation or ordered adlayers. This correlation will become negligible only at very high temperatures. Calculations show that for a transition metal surface, small molecular adsorbates like CO and NO or atoms, and adsorption at neighboring sites of the same type (e.g., two adsorbates at neighboring top sites) there may be repulsive interactions of 20 kJ/mol or more [4]. With such an interaction there will be correlation in the occupation of neighboring sites at any temperature relevant to surface science or catalysis as it is substantially higher than the thermal energy $k_B T$.

Although interactions between adsorbates are the main reason that they will not be randomly distributed over the sites, they are not the only reason. Another one is that the sites might differ because of defects in the substrate. Less obvious is that reactions themselves may also lead to correlation. The reaction $A + B \rightarrow AB$ in the example removes A's and B's when they are neighbors. This will make it less likely that a neighboring A–B pair is found. How strong this effect is depends on the rate constant of the reaction and how fast the adsorbates diffuse. In extreme cases (see Sect. 7.4.3) the reaction may lead to ordered adlayers.

One might think that the problem is really the particular forms that we have used above for the coverage dependence $f$. If the adsorbates are not randomly distributed, then we might try to derive a form for $f$ that reflects the way the adsorbates are actually found on the surface. For example, if one adsorbate forms islands then the coverage dependence in $f$ of the adsorbate should reflect that only the adsorbates on the edge of the island can react with other adsorbates.

There are two problems with this idea. The lesser is that it is generally not clear in which way the adsorbates are distributed over the sites. In fact, this is part of the kinetic problem. A kinetic theory that does not have an answer to the question of what is the structure of the adlayer is at best incomplete. The bigger problem is that there are situations where the macroscopic rate equations can never be correct. It is quite possible to have two situations for a system with exactly the same coverage but with a very different coverage dependence of the rates. Many systems form ordered adlayers at low temperatures and disordered adlayers at high temperatures. Of course, at different temperatures the rate constants are different, but there will also be a different coverage dependence $f$ at low and high temperature because the correlation in the occupation of neighboring sites is different. This means that it is not even in principle possible to describe the kinetics at low and high temperature for such a system with only one macroscopic rate equation.

One should also be aware that when there are interactions between the adsorbates the coverage dependence of the rate $d\theta_X/dt$ is not only described by the factors $f^{(n)}(\theta_A, \theta_B, \ldots)$. A consequence of these interactions is that the rate constants also become dependent on the coverage or the adlayer structure. One should not mix up these coverage dependences, because they have a different origin. The factor $f^{(n)}$ stands for the number of possible occurrences of reaction $n$ because of the way the adsorbates are distributed over the surface. The coverage dependence of the rate constants is a consequence of how interactions between the adsorbates change these rate constants.

The arguments above show that the macroscopic rate equations have severe limitations when we want to interpret them as describing the processes on the atomic scale. This however should not be interpreted to mean that the rate equations are always wrong and useless. In fact, they have been and still are extensively used in chemical engineering with great success. There the rate equations are interpreted in a different way. They are used to fit kinetic experiments and then to use the results to predict the kinetics at other reaction conditions. This is possible provided these other conditions do not differ too much from to ones that were used to fit the rate equations. The reason for this is that the rate equations used in this way generally

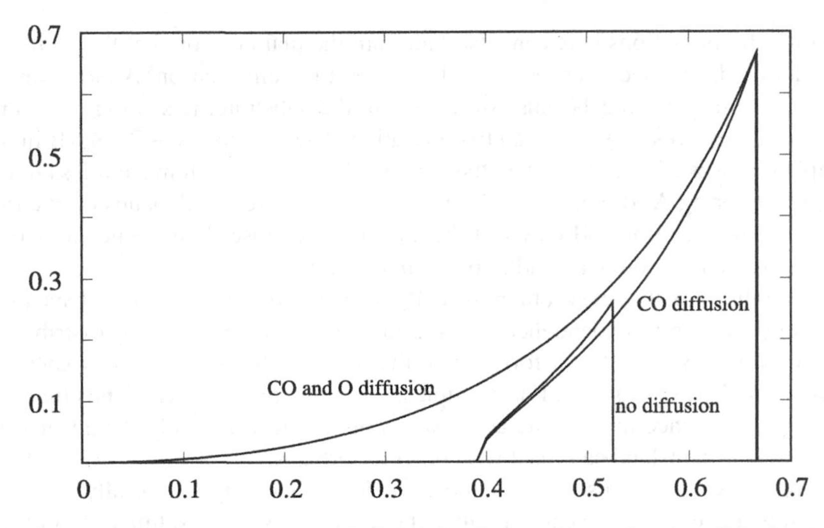


**Fig. 1.1** The turnover number in the Ziff–Gulari–Barshad model as a function of the fraction of CO molecules in the gas phase

do not really capture the chemistry of the system. Instead they just form mathematical expressions yielding reasonable numerical values for reaction rates. This is most clearly when one looks at the rate constants. They can get highly unphysical values in the fitting procedure, and should therefore not be interpreted as rate constants of the microscopic processes. In this book we will not look at this descriptive or phenomenological interpretation of the rate equations.

The discussion above shows the shortcomings of the rate equations and the need for a more sophisticated approach. Such an approach is kinetic Monte Carlo (kMC), which simulates individual processes on the microscopic scale, can include, among other things, interactions between adsorbates and incorporates the dependence on the structure of the adlayers properly.

## 1.2 Some Comparisons

To give some idea of the difference that one might expect between kMC simulations and macroscopic rate equations, we discuss a few examples. Figure 1.1 shows the reactivity in the Ziff–Gulari–Barshad (ZGB) model with and without diffusion [5]. The ZGB model is a simple model for CO oxidation. The original model has only CO adsorption, dissociative adsorption of oxygen, and formation of $CO_2$ immediately followed by desorption. The $CO_2$ formation is assumed to by infinitely fast. The steady state of the model can therefore be characterized by only one parameter: the fraction of molecules in the gas phase that are CO. (For details see Sect. 7.4.3.)

In spite of its simplicity the model shows some very interesting behavior. There are three phases. There is one reactive phase and two phases in which the surface is poisoned: one with CO and one with oxygen. There are two kinetic phase transitions: one in which CO poisoning takes place when we are in the reactive phase and then increase the fraction of CO molecules in the gas phase, and one in which oxygen poisoning takes place when we are in the reactive phase and then decrease the fraction of CO molecules in the gas phase. Figure 1.1 shows the reactive window. If there is too little CO in the gas phase, then the surface will become completely covered by oxygen. If there is too much CO in the gas phase, then the surface will become completely covered by CO.

One point of critique on the ZGB model that one can have is that there is no diffusion of the adsorbates, and that it is only to be expected that macroscopic rate equations will be bad, because there is no mechanism that randomizes the adsorbates over the surface. Indeed, without diffusion islands of CO and islands of oxygen are formed, and if both adsorbates diffuse kMC simulations give the same results as macroscopic rate equations. With diffusion the reactive window becomes much wider. There is still CO poisoning, but only when the fraction of CO molecules in the gas phase is substantially higher, and there is no oxygen poisoning anymore.

If we want to use the ZGB model to represent a real system, then the question will be how much diffusion should be included. If both adsorbates diffuse, then macroscopic rate equations can be used. It seems however more likely that only CO diffuses substantially, whereas oxygen atoms are bound to tightly to diffuse easily. So what if there is only CO diffusion. Figure 1.1 shows that macroscopic rate equations then do not work. The phase transition to CO poisoning is then correctly described by the macroscopic rate equations, but oxygen poisoning is still possible and occurs at the same point as when there is no diffusion at all. The reason for this behavior is that oxygen still forms islands. The islands are small just below the point where CO poisoning occurs, and the adsorbates can be regarded as well mixed. Near the point where oxygen poisoning occurs, the oxygen islands are large however, and the adlayer is anything but well mixed.

Slow or no diffusion of some adsorbates is an important cause of incorrect results from macroscopic rate equations. In the case of the ZGB model one might regard them as only quantitative errors. In Sect. 7.5.1 the Lotka model will be discussed. That model also has no diffusion, but for that model there is even a qualitative difference between kMC and rate equations. The kMC simulations show very well-defined oscillations (see Fig. 7.16), but the rate equations only give a steady state.

Even if diffusion is fast, there may still be substantial differences between kMC results and results obtained with macroscopic rate equations. Section 5.4.2 shows that the diffusion manages to randomize adsorbates only on certain length and time scales, and that the system shows structure at larger lengths and shorter times. More common for the difference between kMC and rate equations are interactions between adsorbates (see Sect. 7.4.1). Figure 1.2 shows a Temperature-Programmed Desorption (TPD) spectrum for CO desorption from a Rh(100) surface assuming that CO prefers the top site at all coverages [6]. Although all sites are equivalent, the spectrum obtained from kMC shows two peaks. There is a symmetry breaking due to strong repulsive interactions between the CO molecules. At low temperatures

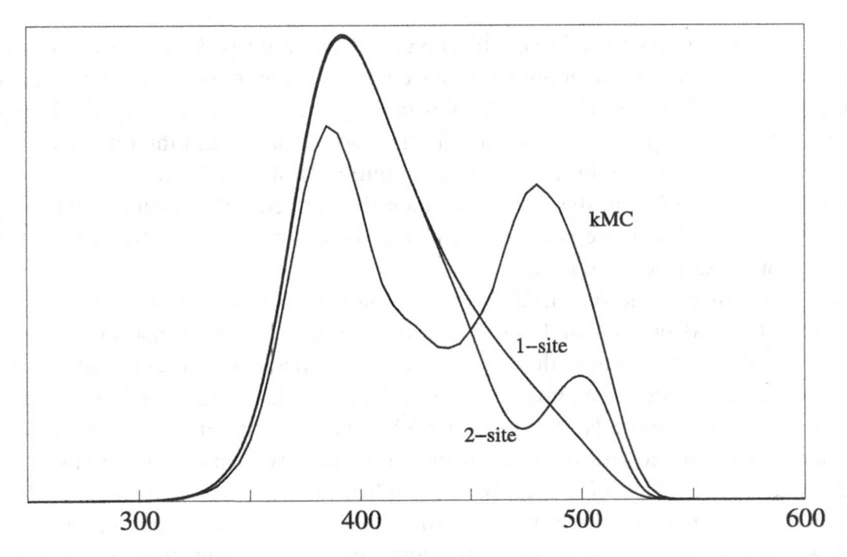


**Fig. 1.2** Temperature-Programmed Desorption spectrum (desorption rate versus temperature in Kelvin) of adsorbates repelling each other obtained with kMC and with macroscopic rate equations. Activation energy for desorption is $E_{\mathrm{act}} = 121.3$ kJ/mol and the prefactor is $\nu = 1.435 \cdot 10^{12}\ \mathrm{s}^{-1}$. These numbers were taken from CO desorption from Rh(100) at low coverage [6]. For diffusion (i.e., hops from one site to a neighboring one) we have used the same activation energy but a prefactor that is a factor ten higher. The repulsion between two adsorbates is 6.65 kJ/mol. The heating rate is 5 K/s and the initial coverage 1.0 ML

the coverage is high. Each CO molecule has four neighbors that reduce the effective adsorption energy. As a result we find a desorption peak at the relatively low temperature of about 385 K. After half the CO's have desorbed, the remaining ones form a checkerboard structure. None of the CO's in that structure has a neighbor, and as a consequence the remaining molecules have a much high adsorption energy. They therefore only desorb at the much higher temperature of about 480 K.

We can derive a macroscopic rate equation that includes the repulsion between the CO's when we assume that they are randomly distributed over the surface in spite of the strong repulsion. The derivation is given in Sect. 7.2 and the result is

$$\frac{d\theta}{dt} = -W_{\mathrm{des}}^{(0)} \theta \left[\theta e^{\varphi/k_{\mathrm{B}}T} + (1-\theta)\right]^4 \tag{1.5}$$

with

$$W_{\mathrm{des}}^{(0)} = \nu e^{-E_{\mathrm{act}}/k_{\mathrm{B}}T} \tag{1.6}$$

the rate constant for desorption of an isolated CO molecule, $E_{\mathrm{act}}$ the activation energy, $\nu$ the prefactor, and $\varphi$ the interaction energy of neighboring CO's. This expression gives a TPD spectrum with just one peak: in Fig. 1.2 it is marked "1-site". Although the effective adsorption energy increases with the decreasing coverage, this occurs in the same way for all adsorbates. A symmetry breaking with some sites becoming vacant and others retaining a CO molecule is not possible.

One might try to extend the rate equations in a way that does allow for symmetry breaking. We can partition all sites in two groups. Because the CO molecules form a checkerboard structure at coverage 0.5 ML, we divide all sites into white and black sites as for a checkerboard. We also allow the probability of these sites to be occupied to differ. The rate equations then become

$$\begin{aligned} \frac{d\theta_{\mathrm{W}}}{dt} = & -W^{(0)}_{\mathrm{des}}\theta_{\mathrm{W}}\left[\theta_{\mathrm{B}}e^{\varphi/k_{\mathrm{B}}T} + (1-\theta_{\mathrm{B}})\right]^4 \\ & - 4W^{(0)}_{\mathrm{diff}}\theta_{\mathrm{W}}(1-\theta_{\mathrm{B}})\left[\theta_{\mathrm{B}}e^{\varphi/2k_{\mathrm{B}}T} + (1-\theta_{\mathrm{B}})\right]^3 \\ & \times \left[\theta_{\mathrm{W}}e^{-\varphi/2k_{\mathrm{B}}T} + (1-\theta_{\mathrm{W}})\right]^3 \\ & + 4W^{(0)}_{\mathrm{diff}}\theta_{\mathrm{B}}(1-\theta_{\mathrm{W}})\left[\theta_{\mathrm{W}}e^{\varphi/2k_{\mathrm{B}}T} + (1-\theta_{\mathrm{W}})\right]^3 \\ & \times \left[\theta_{\mathrm{B}}e^{-\varphi/2k_{\mathrm{B}}T} + (1-\theta_{\mathrm{B}})\right]^3 \end{aligned} \tag{1.7}$$

with $\theta_{\mathrm{W}}$ the probability that a white site is occupied and $\theta_{\mathrm{B}}$ a black site. There is also an equation for $d\theta_{\mathrm{B}}/dt$ that looks the same except with $\theta_{\mathrm{W}}$ and $\theta_{\mathrm{B}}$ interchanged. Because $\theta_{\mathrm{W}}$ and $\theta_{\mathrm{B}}$ need not be the same, there might be hops of CO from black to white sites and back. These hops are represented by the terms with $W^{(0)}_{\mathrm{diff}}$. This is the rate constant for an isolated CO molecule hopping from one site to a neighboring one. Note that the way in which $\theta_{\mathrm{W}}$ and $\theta_{\mathrm{B}}$ appear in the equations reveal the origin. The factor $\theta_{\mathrm{W}}$ in the desorption term on the right-hand-side corresponds to a factor $f$ in Eq. (1.1). The factor in square brackets with $\theta_{\mathrm{B}}$ derives from the coverage dependence of the desorption rate constant. For the first diffusion term the factor $f$ equals $4\theta_{\mathrm{W}}(1-\theta_{\mathrm{B}})$, and the factors in the square brackets derive from the coverage dependence of the hopping rate constant.

Equation (1.7) by itself does not necessarily result in a two-peak spectrum. If we take initial coverages $\theta_{\mathrm{W}} = \theta_{\mathrm{B}} = 1$, then we get the same result as with Eq. (1.5). However, the solution is unstable and the coverages want to diverge. This is what we want for the symmetry breaking. We therefore start with $\theta_{\mathrm{W}} = 1$ and $\theta_{\mathrm{B}} = 0.99$. This gives the curve marked "2-site" in Fig. 1.2. We see that we indeed get a second peak, but it is much too small. So in spite of using an ad hoc assumption we still do not get a correct spectrum.

The ZGB model and the model for CO/Rh(100) show phase transitions. The macroscopic rate equations are based on a Mean Field Approximation (MFA) as will be shown in Sect. 4.6.6. It is well known that MFA is worse for lower dimensional systems. The Ising model on a square lattice is a prototype model to study phase transitions. It shows an order-disorder phase transition that can be solved analytically [7]. The exact temperature of the phase transition is a factor $2\ln(1+\sqrt{2}) \approx 1.763$ lower than the temperature that is obtained from MFA. Such a large discrepancy is typical, and forms a good reason for doing simulations.

These simple models above may appear contrived. Realistic models are more complicated, and one might think that when there are more factors affecting the kinetics these factors may cause some averaging out or cancellation of errors and that the net result can be described by rate equations after all. That turns out not to be the case. The complexity may hide errors, but it doesn't remove them. One reason is that there are large interactions between the adsorbates in complicated reactions

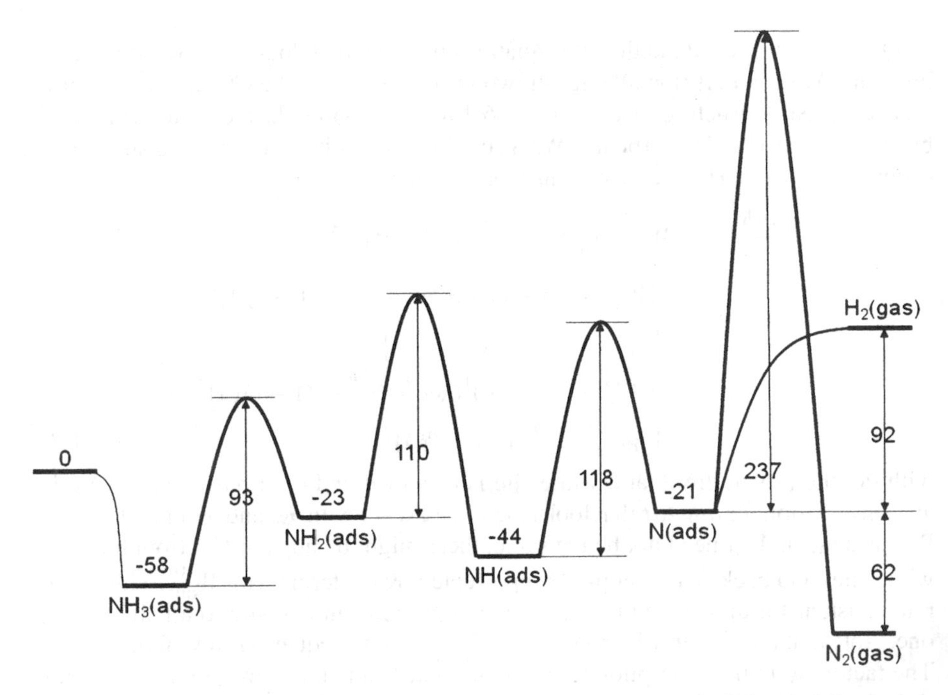


**Fig. 1.3** The reaction profile of the reduction of $NH_3$ on Pt(111). The energies are in kJ/mol

systems. This is because such systems will have adsorbates that prefer to adsorb on different sites. These sites will be so close together, that two neighboring ones can not be occupied simultaneously because there will be a very strong repulsion between the adsorbates. This causes a very strong correlation in the occupation of the sites, which can not be described properly by MFA.

As an example we look at a system for which all site preferences, reactions, and their rate constants have been computed using Density-Functional Theory (DFT). The system is the reduction of $NH_3$ to molecular nitrogen and hydrogen on Pt(111). Figure 1.3 shows the reaction profile [8]. $NH_3$ prefers top sites, $NH_2$ prefers bridge sites, NH and atomic nitrogen prefer fcc hollow sites, and atomic hydrogen does not really have a site preference. The DFT calculations also showed that there is a strong repulsion between adsorbates at a distance equal to the Pt-Pt distance or closer.

Table 1.1 shows the coverages of the adsorbates obtained using kMC, microkinetics, and a Boltzmann distribution based on adsorption energies. The reaction profile in Fig. 1.3 shows that ammonia is the most stable species on the surface. The Boltzmann distribution therefore has that adsorbate as the one with the highest coverage. (The distribution is normalized so that the total coverage equals that of the kMC simulation.) The system is however not at equilibrium but at steady state. The desorption of molecular nitrogen and hydrogen make it irreversible. The consequence is that there is no simple relation between the stabilities of the adsorbates and their coverages, and an equilibrium approach using for example a grand canonical ensemble will not give good results [9].

**Table 1.1** Coverages for the reduction of ammonia on Pt(111) at $T = 1000$ K and $P_{NH_3} = 1.18$ atm

| | kMC | Microkinetics | Boltzmann |
|---|---|---|---|
| $NH_3$ | 0.002 | 0.0002 | 0.16 |
| $NH_2$ | 0.0002 | 0.0002 | 0.002 |
| NH | 0.004 | 0.17 | 0.03 |
| N | 0.28 | 0.81 | 0.006 |
| H | 0.0005 | 0.0002 | 0.10 |

The desorption of molecular nitrogen has by far the highest activation energy. However, at steady state the rate of this process must be equal to half the net rates of the dissociations of $NH_3$, $NH_2$, and NH. With net rates we mean the difference between rates of the forward and reverse reactions. The factor of one half stems from the fact that nitrogen desorbs associatively. Because the rate constant for nitrogen desorption is extremely small due to the high activation energy, the rate can only be high enough if there is a high coverage of nitrogen. This is indeed what is obtained from kMC and microkinetics.

Although microkinetics is qualitatively correct, the coverages are predicted quite wrong. The reason is that we made the usual assumption for microkinetics: one site per unit cell and no interactions between the adsorbates. In this case it would not have made sense to use equations like (1.5) or (1.7), because the interactions are simply too strong. In fact, in the kMC simulations they were even assumed to be infinite. Because of the absence of the interactions the total coverage according to microkinetics is too high. The pressure is high and almost all sites are occupied.

In the kMC simulations this is simply not possible because of the repulsion between the adsorbates. As a consequence the coverages are lower. The difference with microkinetics is not however just a matter of scaling or normalization. For example, the ratio between the N and NH coverage is about 70 in kMC but less than 5 in microkinetics. Moreover, even if it would be matter of normalization, how would we be able to determine the normalization constant? So we see that also for such a realistic system it is necessary to do kMC simulations to get good kinetics.

The reason for the failure of the macroscopic rate equations is that they assume that the system is homogeneous and the adsorbates are randomly distributed over the surface. We have seen in the examples above that this is not the case. For the ZGB model the fast formation of $CO_2$ caused island formation, and for the TPD of CO/Rh(100) and the reduction of $NH_3$ on ammonia the interactions between the adsorbates lead to a strong correlation in the occupation of neighboring sites. Another reason for a non-random distribution of the adsorbates can be the substrate. The substrates in the examples above have been perfect. In reality substrates have defects.

Figure 1.4 shows simulated TPD spectra of a model of $N_2$ desorption from Ru(0001). It was hypothesized that the experimental spectra showed not desorption from a flat surface, but from steps on the surface. Gold was added to the surface to block the steps sites and to test this hypothesis [10]. Indeed this shifted the peak to

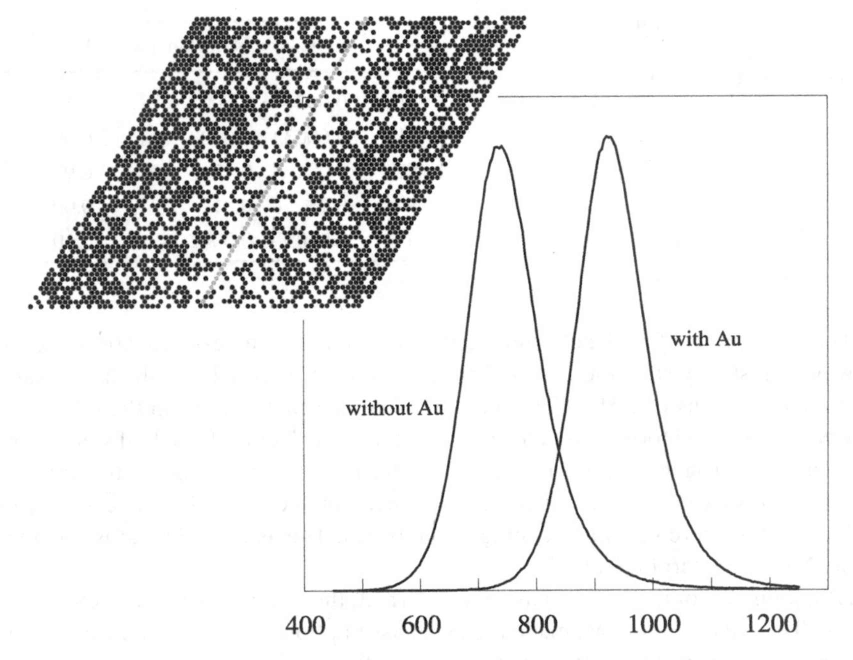


**Fig. 1.4** Simulated TPD spectra (desorption rate versus temperature in Kelvin) of $N_2$ desorption from Ru(0001). The surface without Au has steps and terrace of four rows of Ru atoms wide. The Au atoms on the surface with Au blocks the steps sites. The *inset* shows a snapshot of a simulation and how the nitrogen atoms are distributed near a step. *Black circles* are nitrogen atoms and *gray circles* indicate the step

much higher temperatures. The figure shows this effect, but from simulations. The figure also shows a snapshot of a simulation with just a single step. Note that there is a gradient in the coverage of nitrogen. The coverage is low near the step where the nitrogen desorbs, and high on the terraces. Nitrogen on the terraces diffuses to the steps.

Because of this gradient, macroscopic rate equations with a single coverage will not be able to describe the desorption properly in this case, although it is not solely the presence of the step why this is the case. Section 7.3 shows how for a simple adsorption-desorption equilibrium with a step one can use macroscopic rate equations and even get exact results. The same procedure does not work here, because the desorption is associative. We will see in Sect. 4.6.6 that the macroscopic rate equations are then an approximation to the correct kinetics.

A step is just one example of a substrate that will induce a structure on the adlayer. More examples can be found in Chap. 7. There you will find an adsorbate-induced step reconstruction (Sect. 7.3), how the width of a terrace can affect a phase transition (Sect. 7.4.2), a synergistic effect of a bimetallic surface (Sect. 7.3), and how surface reconstruction can lead to oscillations and chaos in the kinetics (Sect. 7.5). All these phenomena can be studied in detail with kMC, but only at best approximately and with great difficulty with macroscopic rate equations.

## 1.3 Length and Time Scales

The discussion above has shown that we need to know the structure of the adlayer on an atomic scale to understand the kinetics. To get the rate of the A + B reaction it was necessary to know the number of A–B pairs. On the other hand kinetics is generally studied on meso- or macroscopic scales. Atomic scales are of the order of Ångstrøm and femtoseconds. Typical length scales in laboratory experiments vary between micrometers to centimeters. The difference in time scales is often even larger. Vibrations of molecules have periods in the order of femtoseconds. Some reactions take only nanoseconds, but others seconds to many hours. This means that there are many orders of difference in length and time scales between the individual reactions and the resulting kinetics.

The length gap is not always a problem. Many systems are homogeneous, and the kinetics of a macroscopic system can be reduced to the kinetics of a few reacting molecules. This is generally the case for reactions in the gas phase and in solutions. For reactions on the surface of a catalyst it is not always clear when this is the case. It is certainly the case that in the overwhelming number of studies on the kinetics in heterogeneous catalysis it is implicitly assumed that the adsorbates are well-mixed, and that macroscopic rate equations (1.1) can be used. We have already seen examples of systems that show correlation in the occupation of sites that are close. There may however also be ordering on a larger scale. For example, there are systems that show pattern formation with a characteristic length scale of micro- to centimeters [11]. For such systems the macroscopic rate equations can be extended by making the coverages position dependent: $\theta = \theta(\mathbf{r}, t)$. One generally then also adds a diffusion term. The resulting expressions are called reaction-diffusion equations [12]. The problem with these equations is however the same as the one with the macroscopic rate equations. On the atomic length scale the adsorbates are assumed to be randomly distributed, but it is just this assumption that is rarely correct.

The real problem however is the time gap. The typical atomic time scale is given by the period of a molecular vibration. The fastest vibrations have a reciprocal wavelength of up to 4000 $\text{cm}^{-1}$, and a period of about 8 fs. Reactions in catalysis take place in seconds or more. It is important to be aware of the origin of these fifteen orders of magnitude difference. A reaction can be regarded as a movement of the system from one local minimum on a potential-energy surface (PES) to another. In such a move a so-called activation barrier has to be overcome. Most of the time the system moves around one local minimum. This movement is fast, takes place in the order of femtoseconds, and corresponds to a superposition of all possible vibrations. Every time that the system moves in the direction of the activation barrier can be regarded as an attempt to react. The probability that the reaction actually succeeds can be estimated by calculating a Boltzmann factor that gives the relative probability of finding the system at a local minimum or on top of the activation barrier. This Boltzmann factor is given by $\exp[-E_{\text{bar}}/k_B T]$, where $E_{\text{bar}}$ is the height of the barrier. A barrier of $E_{\text{bar}} = 100$ kJ/mol at room temperature gives a Boltzmann factor of about $10^{-18}$. Hence we see that the very large difference in time scales is due to the very small probability that the system overcomes activations barriers.

The standard method to study the evolution of a system on an atomic scale is Molecular Dynamics (MD) [13–15]. In MD a reaction with a high activation barrier is called a rare event, and various techniques have been developed to get a reaction even when a standard simulation would never show it. These techniques, however, work for one reacting molecule or two molecules that react together, but not when one is interested in the combination of thousands or more reacting molecules that one has when studying kinetics. The objective of this book is to show how one deals with such a collection of reacting molecules. It turns out that one has to sacrifices some of the detailed information that one has in MD simulations. One can still work on atomic length scales, but one cannot work with the exact position of all atoms in a system. Instead one only specifies near which minimum of the PES the system is. One does not work with the atomic time scale. Instead one has the reactions as elementary events: i.e., one specifies at which moment the system moves from one minimum of the PES to another. Moreover, because one doesn't know where the atoms are exactly and how they are moving, one cannot determine the times for the reactions exactly either. Instead one can only give probabilities for the times of the reactions. It turns out, however, that this information is more than sufficient for studying kinetics. The resulting method is kMC: the topic of this book.

## References


1. M. Boudart, *Kinetics of Chemical Processes* (Prentice Hall, Englewood Cliffs, 1968)
2. R.A. van Santen, J.W. Niemantsverdriet, *Chemical Kinetics and Catalysis* (Plenum, New York, 1995)
3. J.A. Dumesic, D.F. Rudd, L.M. Aparicio, *The Microkinetics of Heterogeneous Catalysis* (Am. Chem. Soc., Washington, 1993)
4. C.G.M. Hermse, A.P.J. Jansen, in *Catalysis*, vol. 19, ed. by J.J. Spivey, K.M. Dooley (Royal Society of Chemistry, London, 2006)
5. R.M. Ziff, E. Gulari, Y. Barshad, Phys. Rev. Lett. **56**, 2553 (1986)
6. A.P.J. Jansen, Phys. Rev. B **69**, 035414 (2004)
7. R.J. Baxter, *Exactly Solved Models in Statistical Mechanics* (Academic Press, London, 1982)
8. W.K. Offermans, A.P.J. Jansen, R.A. van Santen, Surf. Sci. **600**, 1714 (2006)
9. C. Wu, D.J. Schmidt, C. Wolverton, W.F. Schneider, J. Catal. **286**, 88 (2012)
10. S. Dahl, A. Logadottir, R.C. Egeberg, J.H. Larsen, I. Chorkendorff, E. Törnqvist, J.K. Nørskov, Phys. Rev. Lett. **83**, 1814 (1999)
11. R. Imbihl, G. Ertl, Chem. Rev. **95**, 697 (1995)
12. P. Grindrod, *The Theory and Applications of Reaction-Diffusion Equations: Patterns and Waves* (Clarendon, Oxford, 1996)
13. M.P. Allen, D.J. Tildesley, *Computer Simulation of Liquids* (Clarendon, Oxford, 1987)
14. D. Frenkel, B. Smit, *Understanding Molecular Simulation: From Algorithms to Applications* (Academic Press, London, 2001)
15. D.C. Rapaport, *The Art of Molecular Dynamics Simulation* (Cambridge University Press, Cambridge, 2004)

# Chapter 2
# A Stochastic Model for the Description of Surface Reaction Systems

**Abstract** The most important concept for surface reactions is the adsorption site. For simple crystal surfaces the adsorption sites form a lattice. Lattices form the basis for the description of surface reactions in kinetic Monte Carlo. We give the definition of a lattice and discuss related concepts like translational symmetry, primitive vectors, unit cells, sublattices, and simple and composite lattices. Labels are introduced to describe the occupation of the adsorption sites. This leads to lattice-gas models. We show how these labels can be used to describe reactions and other surfaces processes and we make a start with showing how they can also be used to model surfaces that are much more complicated than simple crystal surfaces. Kinetic Monte Carlo simulates how the occupation of the sites changes over time. We derive a master equation that gives us probability distributions for what processes can occur and when these processes occur. The derivation is from first principles. Some general mathematical properties of the master equation are discussed and we show how a lattice-gas model simplifies the master equation so that it becomes feasible to use it as a basis for kinetic Monte Carlo simulations.

## 2.1 The Lattice Gas

We start the discussion of the way how we will model surface reactions by specifying how we will describe our systems. We want an atomic scale description of our systems and relate this to the macroscopic kinetics: i.e., we want to be able to talk about individual atoms and molecules reacting on a surface, and then link this to global changes and reaction rates of the layer of adsorbates. It turns out that the proper way to described a system is related to the different time scales with which things change on the atomic and on the macroscopic scale. We will see that we need to do some coarse-graining on the atomic length scale to bridge the gap in time scales.

If we regard the evolution of a layer of atoms and molecules adsorbed on a surface on an atomic scale, we will notice that there is a huge difference in time scale of the motion of individual atoms and molecules on the one and of the macroscopic properties on the other hand. For most systems of interest in catalysis, for example, the latter typically vary over a period of seconds or even longer. Motions of atoms occur typically on a time scale of femtoseconds. This enormous gap in time scales

A.P.J. Jansen, *An Introduction to Kinetic Monte Carlo Simulations of Surface Reactions*, Lecture Notes in Physics 856,
DOI 10.1007/978-3-642-29488-4_2, 

poses a large problem if we want to predict or even explain the kinetics (i.e., reaction rates) in terms of the processes that take place on the atomic scale.

The conventional method to simulate the motions of atoms and molecules is Molecular Dynamics (MD) [1–3]. This method generally discretizes time in intervals of equal lengths. The size of this so-called time step, and with it the computational costs, is determined by the fast vibrations of chemical bonds [1]. A stretch vibration of a C-H bond has a typical frequency of around 3000 $cm^{-1}$. This corresponds to a period of about 10 fs. If one wants to study the kinetics of surface reactions, then one needs a method that does away with these fast motions.

The kinetic Monte Carlo (kMC) method that we present here does this by using the concept of sites. The forces working on an atom or a molecule that adsorbs on a surface move it to well-defined positions on the surface [4, 5]. These positions are called sites. They correspond to minima on the potential-energy surface (PES) for the adsorbate. Most of the time adsorbates stay very close to these minima. If we would take a snapshot of a layer of adsorbates at normal temperatures, only about 1 in $10^{13}$ of them would not be near a minimum at normal reaction conditions. Only when they diffuse from one site to another or during a reaction they will not be near such a minima, but only for a very short time. Now instead of specifying the precise positions, orientations, configurations, and motions of the adsorbates we will only specify for each sites its occupation. A reaction and a diffusion from one site to another will be modeled as a sudden change in the occupation of the sites. These changes are the elementary events in a kMC simulation. The vibrations of the adsorbates do not change the occupations of the sites. So they are not simulated in kMC, and hence they do not determine the time scale of a kMC simulation. Reactions and diffusion take place on a much longer time scale. Thus by taking a slightly larger length scale, we can simulate a much longer time scale.

If the surface has two-dimensional translational symmetry, or when it can be modeled as such, the sites form a regular grid or a lattice. Our model is then a so-called lattice-gas model. This chapter shows how this model can be used to describe a large variety of problems in the kinetics of surface reactions.

### *2.1.1 Lattices, Sublattices, and Unit Cells*

If the surface has two-dimensional translational symmetry then there are two linearly independent vectors, $\mathbf{a}_1$ and $\mathbf{a}_2$, with the property that when the surface is translated over any of these vectors the result is indistinguishable from the situation before the translation. It is said that the system is invariant under translation over these vectors. In fact the surface is then invariant under translations for any vector of the form

$$n_1\mathbf{a}_1 + n_2\mathbf{a}_2 \tag{2.1}$$

where $n_1$ and $n_2$ are integers. If all translations that leave the surface invariant can be written as (2.1), then $\mathbf{a}_1$ and $\mathbf{a}_2$ are so-called primitive vectors or primitive translations, and the vectors of the form (2.1) are the lattice vectors. Primitive vectors are not uniquely defined. For example a (111) surface of a fcc metal is translationally invariant for $\mathbf{a}_1 = a(1,0)$ and $\mathbf{a}_2 = a(1/2, \sqrt{3}/2)$, where $a$ is the lattice spacing. But one can just as well choose $\mathbf{a}_1 = a(1,0)$ and $\mathbf{a}_2 = a(-1/2, \sqrt{3}/2)$. The area defined by

$$x_1\mathbf{a}_1 + x_2\mathbf{a}_2 \tag{2.2}$$

with $x_1, x_2 \in [0, 1\rangle$ is called the unit cell. The whole system is obtained by tiling the plane with the contents of a unit cell.

Expression (2.1) defines a simple lattice, Bravais lattice, or net. Simple lattices have just one lattice point, or grid point, per unit cell. It is also possible to have more than one lattice point per unit cell. The lattice is then given by all points

$$\mathbf{s}^{(i)} + n_1\mathbf{a}_1 + n_2\mathbf{a}_2 \tag{2.3}$$

with $i = 0, 1, \ldots, N_{\text{sub}} - 1$ and $N_{\text{sub}}$ the number of lattice points in the unit cell. Each $\mathbf{s}^{(i)}$ is a different vector in the unit cell. The set $\mathbf{s}^{(i)} + n_1\mathbf{a}_1 + n_2\mathbf{a}_2$ for a particular vector $i$ forms a sublattice, which is itself a simple lattice. There are $N_{\text{sub}}$ sublattices, and they are all equivalent: they are only translated with respect to each other. (For more information on lattices, also for a discussion of their symmetry, see for example references [4] and [6].) All points of the form (2.3) from a composite lattice.

The sites of a simple crystal surface form a lattice. The description so far suggests that the different lattice points in a unit cell, corresponding to sites, are all in the some plane, but that does not need to be the case. As we will see in Sect. 4.6.3, that different lattice points may also correspond to positions for adsorbates in different layers that are stacked on top of each other. Lattices can also be used to model surfaces that are much more complicated than simple crystal surfaces (see Sects. 5.5.2 and 5.5.3). In fact, we will see that sometimes lattice points do not correspond to physical adsorption sites at all (see Sect. 5.5.4).

### *2.1.2 Examples of Lattices*

Figure 2.1 shows top and hollow sites of the (100) surface of an fcc metal. Such a surface has $\mathbf{a}_1 = a(1,0)$ and $\mathbf{a}_2 = a(0,1)$ as primitive translations with $a$ the distance between the surface atoms. CO for example prefers the top sites on such surface if the metal is rhodium [7–10]. We have $N_{\text{sub}} = 1$ if we would only include these top sites. We can choose the origin of our reference frame any way we want so we take $\mathbf{s}^{(0)} = (0,0)$ for simplicity. If we would want to include the hollow sites as well then $N_{\text{sub}} = 2$ and $\mathbf{s}^{(1)} = a(1/2, 1/2)$.

Figure 2.2 shows bridge sites of the same surface. Some CO moves to these bridge sites at high coverages [7–10]. If we would include the top and bridge sites to describe all adsorption sites for CO/Rh(100), then we would have $N_{\text{sub}} = 3$ and $\mathbf{s}^{(0)} = (0,0)$, $\mathbf{s}^{(1)} = a(1/2, 0)$, and $\mathbf{s}^{(2)} = a(0, 1/2)$ for the top and the two types of bridge site, respectively.

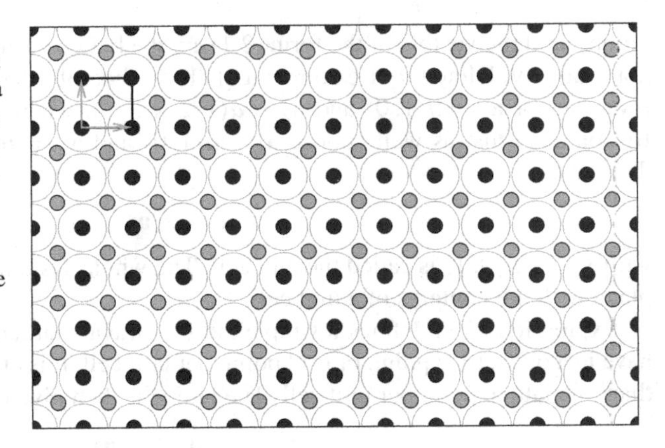

**Fig. 2.1** The large *white circles with gray edges* depict the atoms of the top layer of a (100) surface of an fcc metal. The *black circles* indicate the positions of the top sites, and the *gray circles with black edges* the positions of the hollow sites. The top sites form a simple lattice as do the hollow sites. In the *top-left* corner the unit cell and the primitive translations of the surface are shown

Figure 2.3 shows a (111) surface of an fcc metal. CO on Pt prefers to adsorb on this surface on the top sites [4]. We can therefore model CO on this surface with a simple lattice with the lattice points corresponding to the top sites. We have $\mathbf{a}_1 = a(1,0)$ and $\mathbf{a}_2 = a(1/2, \sqrt{3}/2)$. As $N_{\text{sub}} = 1$ we choose the origin of our reference frame so that $\mathbf{s}^{(0)} = (0,0)$ for simplicity. Each lattice point corresponds to a site that is either vacant or occupied by CO.

NO on Rh(111) forms a $(2 \times 2)$-3NO structure in which equal numbers of NO molecules occupy top, fcc hollow, and hcp hollow sites [11, 12]. Figure 2.3 shows all the sites that are involved. We now have three sublattices with $\mathbf{s}^{(0)} = (0,0)$ (top

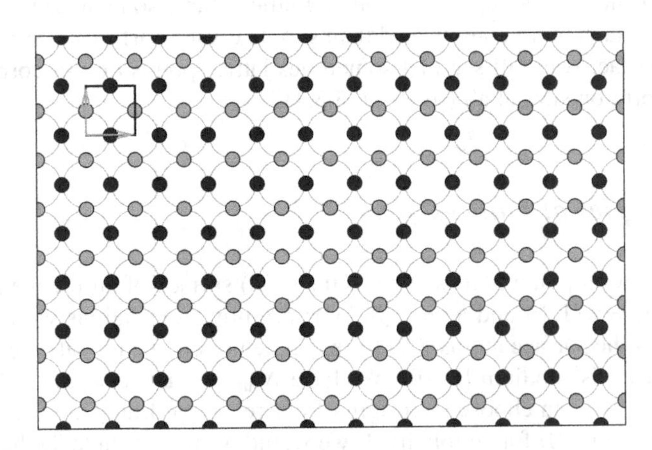

**Fig. 2.2** The *large white circles with gray edges* depict the atoms of the top layer of a (100) surface of an fcc metal. The *black circles* and the *gray circles with black edges* indicate the positions of the bridge sites. Although all bridge sites have the same adsorption properties, together they do not form a simple lattice, but a composite lattice. This is because the relative positions of the surface atoms with respect to the "black" bridge sites is different from those of the "gray" bridge sites. However, if we ignore the surface atoms, then all bridge sites together from a square simple lattice. In the *top-left* corner the unit cell and the primitive translations of the surface are shown

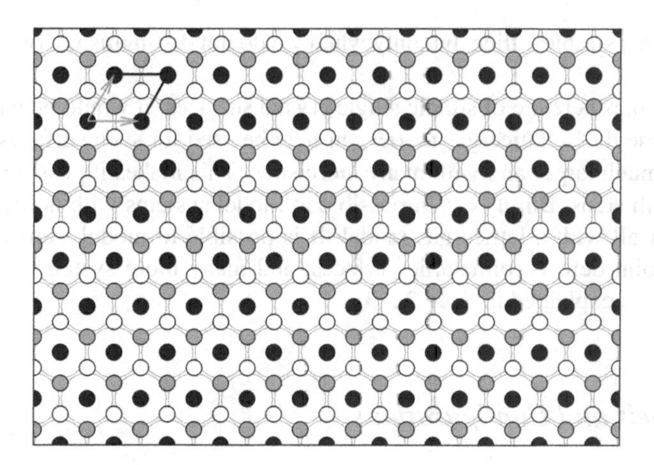

**Fig. 2.3** The *large white circles with gray edges* depict the atoms of the top layer of a (111) surface of an fcc metal. The *black circles* indicate the positions of the top sites, and the *gray circles with black edges* the positions of one type of hollow site, say fcc, and the *small white circle with black edges* the positions of the other type, say hcp, of hollow site. The top sites form a simple lattice as do the fcc sites and the hcp sites separately. The top and hollow sites together also form a simple lattice if we disregard the different adsorption properties of the sites and the different positions with respect to the surface atoms. Otherwise they form a composite lattice with three sublattices. In the *top-left* corner the unit cell and the primitive translations of the surface are shown

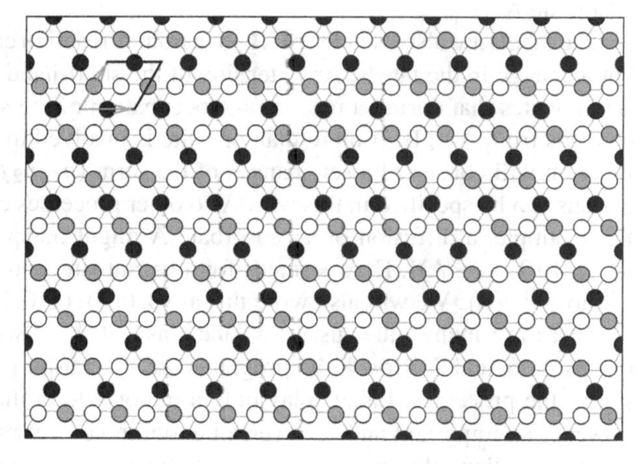

**Fig. 2.4** The *large white circles with gray edges* depict the atoms of the top layer of a (111) surface of an fcc metal. The *black circles*, the *gray circles with black edges*, and the *small white circle with black edges* indicate the positions of the bridge sites. Together they form a composite lattice even though they have the same adsorption properties. In the *top-left* corner the unit cell and the primitive translations of the surface are shown

sites), $\mathbf{s}^{(1)} = a(1/2, \sqrt{3}/6)$ (fcc hollow sites), and $\mathbf{s}^{(2)} = a(1, \sqrt{3}/3)$ (hcp hollow sites).

At high coverages the repulsion between the CO molecules on Pt(111) forces some of them again to bridge sites [13]. Figure 2.4 shows the bridge sites. We have now four sublattices with $\mathbf{s}^{(0)} = (0, 0)$, $\mathbf{s}^{(1)} = a(1/2, 0)$, $\mathbf{s}^{(2)} = a(1/4, \sqrt{3}/4)$, $\mathbf{s}^{(3)} = a(3/4, \sqrt{3}/4)$. The first one is for the top sites (not shown in the figure, but see Fig. 2.3). The others are for the three sublattices of bridge sites. The four sublattices together form a simple lattice, but only when we do not distinguish between top and bridge sites.

The examples here are of simple single crystal surfaces. It would be wrong however to assume that a lattice-gas model can only be used for such surfaces. The unit cell can be much larger and with many more sites. This makes it possible to model a surface with steps. But it is even possible to model systems with no translational symmetry at all with a lattice-gas model. It is possible to model steps at variable distances, point defects, bimetallic surfaces, and many more systems through the use of labels as explained in Sect. 2.1.3.

### *2.1.3 Labels and Configurations*

The sites are the positions where the adsorbates are found on the surface, but for each site we need something to indicate if it is occupied or not, and if it is occupied with which adsorbate. We use labels for this.

We assign a label to each lattice point. The lattice points correspond to the sites, and the labels specify properties of the sites. A particular labeling of all lattice points together we call a configuration. The most common property that one wants to describe with the label is the occupation of the site. We use the short-hand notation $(n_1, n_2/s : \mathrm{A})$ to mean that the site at position $\mathbf{s}^{(s)} + n_1\mathbf{a}_1 + n_2\mathbf{a}_2$ is occupied by an adsorbate A.

The labels are also used to specify reaction. A reaction can be regarded as nothing but a change in the labels. An extension of the short-hand notation $(n_1, n_2/s : \mathrm{A} \to \mathrm{B})$ indicates that during a reaction the occupation of the site at $\mathbf{s}^{(s)} + n_1\mathbf{a}_1 + n_2\mathbf{a}_2$ changes from A to B. If more than one site is involved in a reaction then the specification will consist of a set changes of the form $(n_1, n_2/s : \mathrm{A} \to \mathrm{B})$. Not only reactions can be specified in this way. Also other processes can be described like this. For example, a diffusion of an adsorbate A might be specified by $\{(0, 0/0 : \mathrm{A} \to *), (1, 0/0 : * \to \mathrm{A})\}$. Here $*$ stands for a vacant site, and the diffusion is from site $\mathbf{s}^{(0)}$ to $\mathbf{s}^{(0)} + \mathbf{a}_1$. We will also write this as $(0, 0/0), (1, 0/0) : \mathrm{A}* \to *\mathrm{A}$.

There are many other uses for labels as will be discussed in Sect. 2.1.4 and Chaps. 5, 6, and 7. Most kMC programs are special-purpose codes with hard coding of the processes. Labels play only a minor role in these programs. Labels are however an important and very versatile tool in general-purpose kMC codes. They allow great flexibility in creating models for reaction systems, and a clever use of them can greatly enhance the speed of simulations.

### *2.1.4 Examples of Using Labels*

Desorption of CO from Pt(111) can be written as $(0, 0 : \mathrm{CO} \to *)$ when we use a model of the top sites shown in Fig. 2.3. We have left out the index of the sublattice,

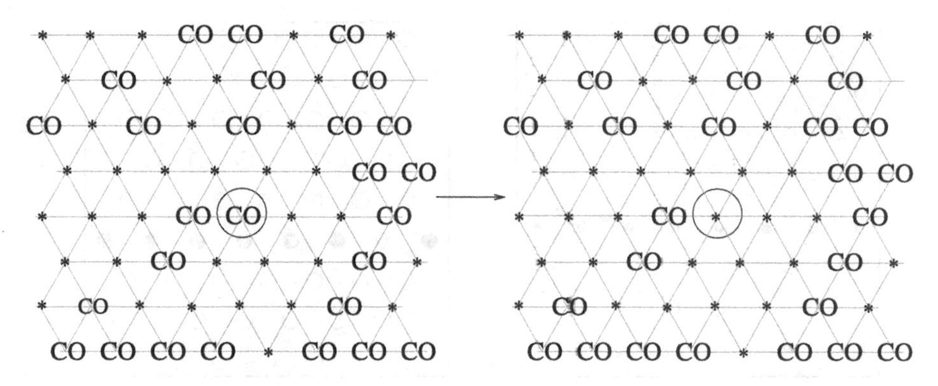


**Fig. 2.5** The change of labels for CO desorption from a Pt(111) surface. The *encircled* CO molecule on the *left* desorbs and the label becomes $*$ indicating a vacant site. The *lines* are guides for the eyes

because, as there is only one, it is clear on which sublattice the reaction takes place (see Fig. 2.5). Desorption on other sites can be obtained by translations over lattice vectors: i.e., $(0,0\!:\!\mathrm{CO} \to *)$ is really representative for $(n_1,n_2\!:\!\mathrm{CO} \to *)$ with $n_1$ and $n_2$ arbitrary integers. Diffusion of CO can be modeled as hops from one site to a neighboring site. We can write that as $\{(0,0\!:\!\mathrm{CO} \to *),(1,0\!:\!* \to \mathrm{CO})\}$ or $(0,0),(1,0)\!:\!\mathrm{CO}* \to *\mathrm{CO}$. Hops on other sites can again be obtained from these descriptions by translations over lattice vectors, but also by rotations that leave the surface is invariant.

Specifying adsorbates is the most obvious and most frequent use of labels, but other properties can be modeled by labels to great effect. Note that in the case of NO on Rh(111) (Fig. 2.3) the lattice is a composite one, but if we ignore the difference between the sites we get a simple lattice with $\mathbf{a}_1 = a(1/2,\sqrt{3}/6)$ and $\mathbf{a}_2 = a(\sqrt{3}/3,0)$. It is possible to use the simple lattice and at the same time retain the difference between the sites. The trick is to use labels not just for the occupation, but also for indicating the type of site. So instead of labels NO and $*$ indicating the occupation, we use NOt, NOf, NOh, $*$t, $*$f, and $*$h. The last letter indicates the type of site (t stands for top, f for fcc hollow, and h for hcp hollow) and the rest for the occupation. Instead of $(0,0/0\!:\!\mathrm{NO})$ and $(0,0/1\!:\!*)$ we have $(0,0\!:\!\mathrm{NOt})$ and $(1,0\!:\!*\mathrm{f})$, respectively. It depends very much on the processes that we want to simulate which way of describing the system is more convenient and computationally more efficient.

Because a lattice is used to represent the adsorption sites, one might think that only systems with translational symmetry can be modeled. That is not the case however. Figure 2.6 shows how to model a step [14, 15]. The difference in the top sites can be modeled with different labels just as for the NO/Rh(111) example above. If the terraces are small then it might also be possible to work with a unit cell spanning the width of a terrace, but when the terraces become large this will be inconvenient as there will be many sublattices. If the width of the terraces varies this is even impossible. In a similar way as in Fig. 2.6 bimetallic surfaces can be modeled [16]. Notice that some distances between the sites on the left in Fig. 2.6 are different

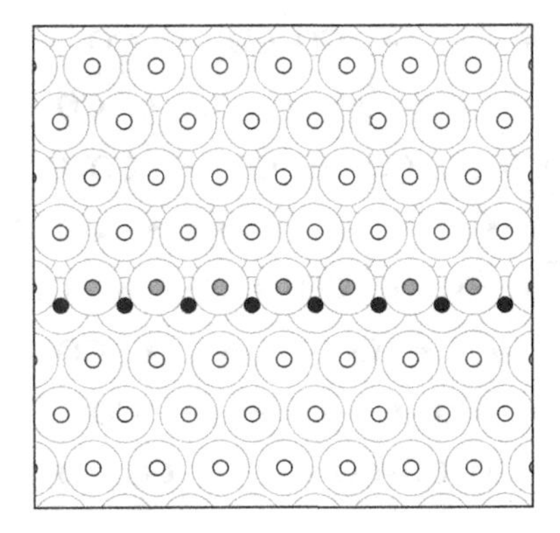

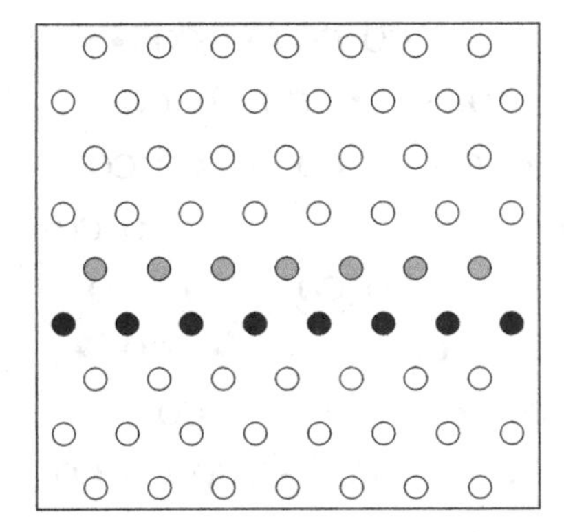

**Fig. 2.6** A Ru(0001) surface with a step with the top sites indicated on the *left*. On the *right* is shown the lattice. The *large open circles* are the atoms. The *small open circles* indicate top sites on the terraces, the *small black circles* top sites at the bottom of the step, and the *small gray circles* top sites at the top of the step. Notice the difference in distance between the top sites at the step on the *left* and on the *right*

from those on the right. The distance between the sites on the top and bottom of a step is smaller on the left than on the right. On the right this distance is increased so that the sites form a lattice. Such a distortion of the system is quite acceptable in kMC simulations. The elementary events (reactions, diffusion, and possibly other processes) are described in terms of changes of the labels of sites. We only need to know which sites and how the labels change. Distances between sites are not part of the description of events.

Site properties like the sublattice of which the site is part of and if it is a step site or not are static properties. The occupation of a site is a dynamic property. There are also other properties of sites that are dynamic. Bare Pt(100) reconstructs into a quasi-hexagonal structure [17]. CO oxidation on Pt(100) is substantially influenced by this reconstruction because oxygen adsorbs much less readily on the reconstructed surfaces than on the unreconstructed one. This can lead to oscillations, chaos, and pattern formation [17, 18]. It is possible to model the effect of the reconstruction on the CO oxidation by using a label that specifies whether the surface is locally reconstructed or not [19–21].

Chapters 5, 6, and 7 will show more esoteric uses of labels. Labels can be used as flags or as counters as well. Very often this use is combined with fictitious processes and fictitious sites, which can either be a way of modeling actual physical and chemical processes, or a way to get kinetic information.

### *2.1.5 Shortcomings of Lattice-Gas Models*

The lattice-gas model is simple yet very powerful, as it allows us to model a large variety of systems and phenomena. Still, not everything can be modeled with it.

Let's look again at CO oxidation on Pt(100). As stated above this system shows reconstruction which can be modeled with a label indicating that the surface is reconstructed or not. This way of modeling has shown to be very successful [19–21], but it does neglect some aspects of the reconstruction. The reconstructed and the unreconstructed surface have very different unit cells, and the adsorption sites are also different [22, 23]. In fact, the unit cell of the reconstructed surface is very large, and there are a large number of adsorption sites with slightly different properties. These aspects have been neglected in the kinetic simulations so far. As these simulations have been quite successful, it seems that these aspects are not very relevant in this case, but that need not always be so. Another example would be catalytic partial oxidation (CPO), which takes place at high temperatures at which the surface is so dynamic that all translational symmetry is lost. In this case using a lattice to model the kinetics seems inappropriate.

The example of CO on Pt(111) has shown that at high coverage the position at which the molecules adsorb change. The reason for this is that these positions are not only determined by the interactions between the adsorbates and the substrate, but also by the interactions between the adsorbates themselves. At low coverages the former dominate, but at high coverages the latter may be more important. This may lead to adlayer structures that are incommensurate with the substrate [4]. Examples are formed by the nobles gases. These are weakly physisorbed, whereas at high coverages the packing onto the substrate is determined by the steric repulsion between them. At low and high coverages different lattices are needed to describe the positions of the adsorbates, but a single lattice describing both the low and the high coverage sites is not possible. Simulations in which the coverages change in such a way from low to high coverage and/or vice versa then cannot be based on a lattice-gas model except by making substantial approximations.

Although not all systems can be modeled well by a lattice gas, it is a much more flexible model than might initially appear. Figure 2.6 already shows some of this flexibility. Note that the sites in the system in the figure have only translational symmetry in the direction parallel to the step, whereas in the corresponding lattice-gas model there is the usual two-dimensional periodicity. This is accomplished by displacing the sites at the step somewhat from their real positions. As has been explained above, this is perfectly acceptable. Similarly, it is possible to describe with a lattice-gas model a layer of adsorbates that have been displaced from their normal site positions by the interactions between them, and that have formed an adlayer with a structure that is incommensurate with that of the preferred adsorption sites. The reason for this flexibility is the labels that we attach to each lattice point. It will be shown in Chaps. 5, 6, and 7 that these labels make it possible to model a very large variety of systems with a lattice gas. Whether or not a system can be described by a lattice-gas model depends very much on one's ingenuity.

### 2.1.6 Boundary Conditions

The surface of a real catalyst will very often contain many more sites than we can include in a kMC simulation. In fact, such a surface is generally regarded as infinite in two directions. In a kMC simulation we need to restrict ourselves to a much more limited number of sites. It is possible to do kMC simulations with all sites in a small part of the catalyst's surface. This gives an acceptable description except for the sites at the edge. It is more customary to use periodic boundary conditions. In that case all sites $\mathbf{s}^{(i)}+n_1\mathbf{a}_1+n_2\mathbf{a}_2$ with $n_1=0,1,\ldots,N_1-1$ and $n_2=0,1,\ldots,N_2-1$ are explicitly included in the simulation. Sites with values of $n_1$ and/or $n_2$ outside this range are thought to have the same label as those of $n_1 \bmod N_1$ and $n_2 \bmod N_2$. The system can be thought as if being rolled up on a torus. The values of $N_1$ and $N_2$ in real simulations vary. Sometimes they can be smaller than 100, but simulations with $N_1=N_2=8192$ have been reported as well [24].

## 2.2 The Master Equation

### 2.2.1 The Definition and Some Properties of the Master Equation

The derivation of the algorithms and a large part of the interpretation of the results of kMC simulations are based on a master equation

$$\frac{dP_\alpha}{dt}=\sum_\beta[W_{\alpha\beta}P_\beta-W_{\beta\alpha}P_\alpha]. \tag{2.4}$$

In this equation $t$ is time, $\alpha$ and $\beta$ are configurations of the adlayer (i.e., different ways in which adsorbates are distributed over the sites, or more generally ways in which labels can be assigned to lattice points), $P_\alpha$ and $P_\beta$ are their probabilities, and $W_{\alpha\beta}$ and $W_{\beta\alpha}$ are so-called transition probabilities per unit time that specify the rate with which the adlayer changes due to reactions and other processes. The master equation is the single most important equation in kMC. It relates everything that we do in kinetics to each other as will be shown below. Here we start with looking at some of its mathematical properties. The master equation is a loss-gain equation. The first term on the right stands for increases in $P_\alpha$ because of processes that change other configurations into $\alpha$. The second term stands for decreases because of processes in $\alpha$. From

$$\frac{d}{dt}\sum_\alpha P_\alpha=\sum_\alpha\frac{dP_\alpha}{dt}=\sum_{\alpha\beta}[W_{\alpha\beta}P_\beta-W_{\beta\alpha}P_\alpha]=0 \tag{2.5}$$

we see that the total probability is conserved. (The last equality can be seen by swapping the summation indices in one of the terms.)

The master equation can be derived from first principles as will be shown below, and hence forms a solid basis for all subsequent work. There are other advantages

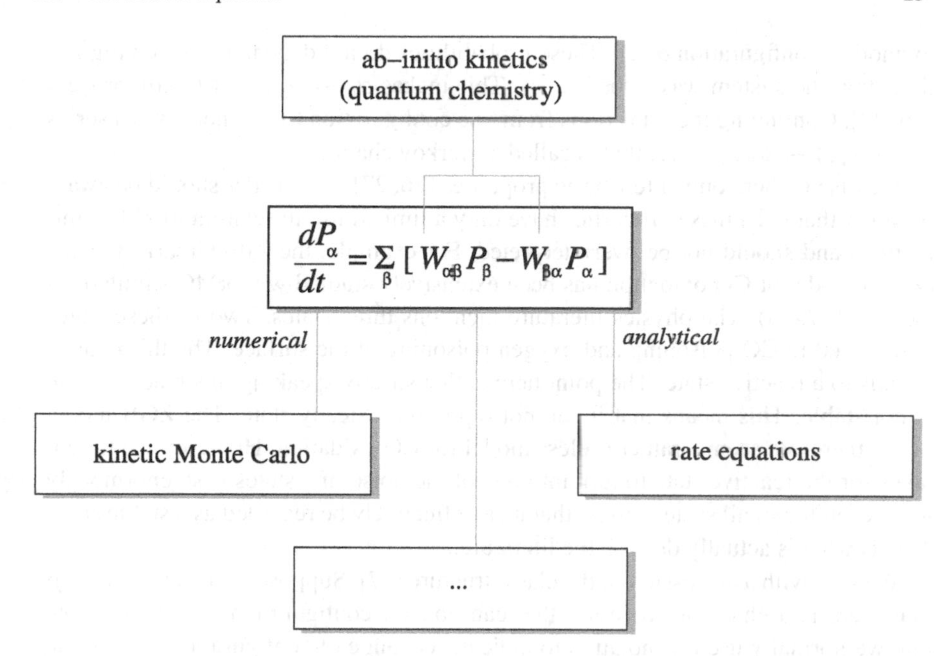


**Fig. 2.7** Scheme showing the central position of the master equation for kinetics. Quantum chemical calculations yield the rate constants of the master equation, and kMC, rate equations, and other kinetic theories, methods, and approaches can be regarded as ways to solve the master equation

as well. First, the derivation of the master equation yields expressions for the transition probabilities that can be computed with quantum chemical methods [25]. This makes ab-initio kinetics for catalytic processes possible. Such calculations generally use the term rate constants instead of transition probabilities and we will use that term for the $W$'s in the master equation as well. We will show however that these rate constants are generally not the same as the rate constants in macroscopic rate equations (see Sect. 4.6). Second, there are many different algorithms for kMC simulations. Those that are derived from the master equation all give necessarily results that are statistically identical. Those that cannot be derived from the master equation conflict with first principles. Third, kMC is a way to solve the master equation, but it is not the only one. The master equation can, for example, be used to derive the normal macroscopic rate equation (see Sect. 4.6), although this generally involves the introduction of approximations. In general, it forms a good basis to compare different theories of kinetics quantitatively, and also to compare these theories with simulations. Figure 2.7 shows that the master equation can be regarded as the central equation of the kinetics of surface reactions and that it relates the quantum chemical calculation of rate constants, kMC, and other kinetic theories to each other.

There is an extensive mathematical literature on the master equation. This literature also often talks about continuous-time Markov chains. If a system is in configuration $\alpha_n$ then the master equation gives the probabilities that the system will move to another configuration $\alpha_{n+1}$. These probabilities do not depend on the configurations that the system was in before $\alpha_n$. This makes $\alpha_n \rightarrow \alpha_{n+1}$ a Markov process [26, 27]. Continuing the transitions from one configuration to another gives a series $\alpha_n \rightarrow \alpha_{n+1} \rightarrow \alpha_{n+2} \rightarrow \ldots$ that is called a Markov chain.

We discuss here only a few basic properties [26, 27]. The reader should be aware however that even these properties have only a limited use in relation to kMC simulations and should not be overinterpreted. For example, the Ziff–Gulari–Barshad (ZGB) model of CO oxidation has been extensively studied with kMC simulations (see Sect. 7.4.3). The physics literature mentions three states. Two of these states correspond to CO poisoning and oxygen poisoning of the surface. The third corresponds to a reactive state. The point here is that strictly speaking this reactive state is not stable. This means that it can not represent a steady state. The ZGB model might therefore seem a rather useless model for CO oxidation. However, the time it takes for the reactive state to turn into one of the poisoning states is so enormously long, even for small system sizes, that it can effectively be regarded as a stable state. This is what is actually done in the literature.

We start with a discussion of the class structure [27]. Suppose we mean by $\alpha \rightarrow \beta$ that there is a chain of processes that can convert configuration $\alpha$ into $\beta$. (Note that we normally use this notation to indicate a change of configuration by a single process.) We then define $\alpha \leftrightarrow \beta$ to mean $\alpha \rightarrow \beta$ and $\beta \rightarrow \alpha$. This relation $\leftrightarrow$ defines an equivalence relation [28]. This means $\alpha \leftrightarrow \alpha$ for all $\alpha$, if $\alpha \leftrightarrow \beta$ then $\beta \leftrightarrow \alpha$, and if $\alpha \leftrightarrow \beta$ and $\beta \leftrightarrow \gamma$ then $\alpha \leftrightarrow \gamma$. Because of these properties we can partition all configurations into equivalence classes: i.e., $\alpha$ and $\beta$ belong to the same class if and only if $\alpha \leftrightarrow \beta$.

A class is closed if $\alpha$ is in the class and $\alpha \rightarrow \beta$ implies that $\beta$ is also in the same class as $\alpha$: i.e., $\beta \rightarrow \alpha$ also holds. If the system ends up in a closed class, then it will never leave it. Such a situation may for example arise when no reactions are possible in a system anymore, and the adsorbates can only hop from one site to another. A system in a closed class visits each of the configurations in the class an infinite number of times. This does not hold for configurations in a class that is not closed. If a closed class consists of a single configuration, then the configuration is called an absorbing state. The poisoned states mentioned above for the ZGB model are examples of such absorbing states.

If all configurations can be partitioned into two or more sets such that neither $\alpha \rightarrow \beta$ nor $\beta \rightarrow \alpha$ holds if $\alpha$ and $\beta$ are in different sets, then the system is said to be reducible or decomposable. This terminology is related to the way in which the master equation can be written in matrix-vector form. With the vector $\mathbf{P}$ with components $\mathbf{P}_\alpha = P_\alpha$, and the matrix $\mathbf{W}$ with components

$$\mathbf{W}_{\alpha\beta} = \begin{cases} W_{\alpha\beta}, & \text{if } \alpha \neq \beta, \text{ and} \\ -\sum_\gamma W_{\gamma\beta}, & \text{if } \alpha = \beta, \end{cases} \tag{2.6}$$

we can write the master equation as

$$\dot{\mathbf{P}} = \mathbf{W}\mathbf{P}. \tag{2.7}$$

For a completely reducible or decomposable system the matrix $\mathbf{W}$ can be written in the form

$$\begin{pmatrix} \mathbf{A} & \mathbf{0} \\ \mathbf{0} & \mathbf{B} \end{pmatrix} \tag{2.8}$$

with $\mathbf{A}$ and $\mathbf{B}$ square matrices by appropriately ordering the configurations. The configurations in the different sets can be treated completely independently from each other. A system that is not reducible is called irreducible.

If there is a closed class, but the system is irreducible, then we can write $\mathbf{W}$ in the form

$$\begin{pmatrix} \mathbf{A} & \mathbf{D} \\ \mathbf{0} & \mathbf{B} \end{pmatrix} \tag{2.9}$$

with $\mathbf{D}$ a non-zero matrix, $\mathbf{A}$ and $\mathbf{B}$ square matrices, and the rows and columns of $\mathbf{A}$ indexed by the configurations in the closed class. Such a system is called (incompletely) reducible. If there are two closed classes, then $\mathbf{W}$ can be written as

$$\begin{pmatrix} \mathbf{A} & \mathbf{0} & \mathbf{D} \\ \mathbf{0} & \mathbf{B} & \mathbf{E} \\ \mathbf{0} & \mathbf{0} & \mathbf{C} \end{pmatrix} \tag{2.10}$$

with $\mathbf{A}$, $\mathbf{B}$, and $\mathbf{C}$ square matrices, $\mathbf{D}$ and $\mathbf{E}$ non-zero matrices, and the rows and columns of $\mathbf{A}$ and $\mathbf{B}$ being indexed by the configurations in the two closed classes. Such a system is called splitting, because it ends up in one closed class or the other.

The definition of $\mathbf{W}$ immediately gives

$$\sum_{\alpha} \mathbf{W}_{\alpha\beta} = 0. \tag{2.11}$$

Matrices with this property can be shown to have at least one right eigenvector with eigenvalue equal to zero [26]. The components of such eigenvector are all non-negative and by proper normalization can be identified with the probabilities to find the system in a configuration when the system is in a steady state. It can also be shown that the system will evolve toward such an eigenvector.

There are various ways to prove this [26]. One way is to write the solution of the master equation as

$$\mathbf{P}(t) = e^{\mathbf{W}t}\mathbf{P}(0), \tag{2.12}$$

where $\mathbf{P}(0)$ are the probabilities of the configurations at $t = 0$. If we can diagonalize $\mathbf{W}$ and write

$$\mathbf{WU} = \mathbf{UV}, \tag{2.13}$$

then this becomes

$$\mathbf{P}(t) = \mathbf{U}e^{\mathbf{V}t}\mathbf{U}^{-1}\mathbf{P}(0). \tag{2.14}$$

The matrix $\mathbf{V}$ is diagonal and all matrix elements on the diagonal are non-positive. This means that for large $t$, all components of $\exp[\mathbf{V}t]$ vanish, except those with zero on the diagonal of $\mathbf{V}$. $\mathbf{P}$ evolves to the corresponding eigenvector in $\mathbf{U}$.

Note that an eigenvector with eigenvalue equal to zero does not exclude the existence of an oscillation. Suppose we have configurations $\alpha_1, \alpha_2, \ldots, \alpha_N$ with $N \geq 2$ and the system always moves from $\alpha_n$ directly to $\alpha_{n+1}$ for $n = 1, 2, \ldots, N-1$ and from $\alpha_N$ directly to $\alpha_1$. This means that we have a cycle. The eigenvector with zero eigenvalue has $W_{\alpha_{n+1}\alpha_n} P_{\alpha_n} = W_{\alpha_1\alpha_N} P_N$ for $n = 1, 2, \ldots, N-1$. Although the system has a cycle, this can still be regarded as a stationary state because the time it takes the system to move through the cycle will not always be exactly the same. It can be shown (see Sect. 3.7.3) that on average this will take a time equal to

$$T = \sum_{n=1}^{N-1} W^{-1}_{\alpha_{n+1}\alpha_n} + W^{-1}_{\alpha_1\alpha_N}. \tag{2.15}$$

If we would start the system in $\alpha_1$ at time $t = 0$ and follow the system, then after a time $T$ we would find it most likely again in $\alpha_1$, but also with some small probability in $\alpha_N$ and $\alpha_2$, with an even smaller probability in $\alpha_{N-1}$ and $\alpha_3$, et cetera. After another period $T$ the probability of finding the system in $\alpha_1$ would be reduced and the probabilities for configurations before and after $\alpha_1$ would be increased. After many periods $T$ we would find the system in one of the configurations with a probability corresponding to the one given by the eigenvector with zero eigenvalue.

### 2.2.2 The Derivation of the Master Equation

The master equation can be derived by looking at the surface and its adsorbates in phase space.[1] This is, of course, a classical mechanics concept, and one might wonder if it is correct to look at the processes on an atomic scale and use classical mechanics. The situation here is the same as for the derivation of the rate equations for gas phase reactions. The usual derivations there also use classical mechanics [29–33]. Although it is possible to give a completely quantum mechanical derivation formalism [34–37], the mathematical complexity hides much of the important parts of the chemistry. We therefore give only a classical derivation. It is possible at the end to replace the classical expressions by quantum mechanical ones, in exactly the same way as for gas phase reactions. The new expressions will depend on the type of motion (vibration, rotation, et cetera). This will be shown in detail in Chap. 4.

The derivation of the master equation is usually based on the observation that there is a separation between the time scale on which reactions take place and the time scale of much faster motions like vibrations [38, 39]. The longer time scale of reactions defines states, in which the system is localized in configuration space, and the transitions between them can be described by a master equation. The rates of the individual transitions can each be computed separately by one of the methods

[1] Parts of Sect. 2.2.2 have been reprinted with permission from X.Q. Zhang, A.P.J. Jansen, Kinetic Monte Carlo method for simulation reactions in solutions, Phys. Rev. E **82**, 046704 (2010). Copyright 2010, American Physical Society.

of chemical kinetics: e.g., Transition-State Theory (TST) [38–40]. We present here a different derivation that incorporates all process at the same time [41–43]. It is a generalization of the derivation of Variational TST (VTST): i.e., we partition phase space in many regions [29, 32, 33], and it is an alternative to the derivation using projection operators [44, 45]. It has the advantage that the result is somewhat more general. We will use this in Sects. 2.2.3 and 8.2. The derivation does not use the fact that we are interested in surface reactions, and has a much more general validity. We will show in Sect. 2.2.3 how the results simplify for surface reactions.

A point in phase space completely specifies the positions and momenta of all atoms in the system. In MD simulations one uses these positions and momenta at some starting point to compute them at later times. One thus obtains a trajectory of the system in phase space. We are not interested in that amount of detail, however. In fact, as was stated before, too much detail is detrimental if one is interested in simulating many processes. The time interval that one can simulate a system using MD is typically of the order of nanoseconds. Reactions in catalysis have a characteristic time scale that is many orders of magnitude longer. To overcome this large difference we need an approach that removes the fast processes (vibrations) that determine the time scale of MD, and leaves us with the slow processes (reactions). This approach looks as follows.

Instead of the precise position of each atom, we only want to know how the different adsorbates are distributed over the sites of a surface. So our physical model is a lattice. Each lattice point corresponds to one site, and has a label that specifies which adsorbate is adsorbed. (A vacant site is simply a special label.) This gives us a configuration. As each point in phase space is a precise specification of the position of each atom, we also know which adsorbates are at which sites: i.e., we know the corresponding configuration. Different points in phase space may, however, correspond to the same configuration. These points differ only in slight variations of the positions and momenta of the atoms. This means that we can partition phase space in many regions, each of which corresponds to one configuration. The processes are then nothing but the motion of the system in phase space from one region to another.

More generally and in line with the idea of different time scales mentioned above, we can start with identifying the regions in configuration space where the fast motions take place. Figure 2.8 shows a sketch of a PES of an arbitrary system. We assume that only the electronic ground state is relevant, so that the PES is a single-valued function of the positions of all the atoms in the system. The points in the figure indicate the minima of the PES. Each minimum of the PES has a catchment region. This is the set of all points that lead to the minimum if one follows the gradient of the PES downhill [46].

We now partition phase space into these catchment regions and then extend each catchment region with the momenta. Let's call $\mathcal{C}$ the configuration space of a system and $\mathcal{P}$ its phase space [47, 48]. The minima of the PES are points in configuration space. We define $\mathcal{C}_\alpha$ to be the catchment region of minimum $\alpha$. This catchment

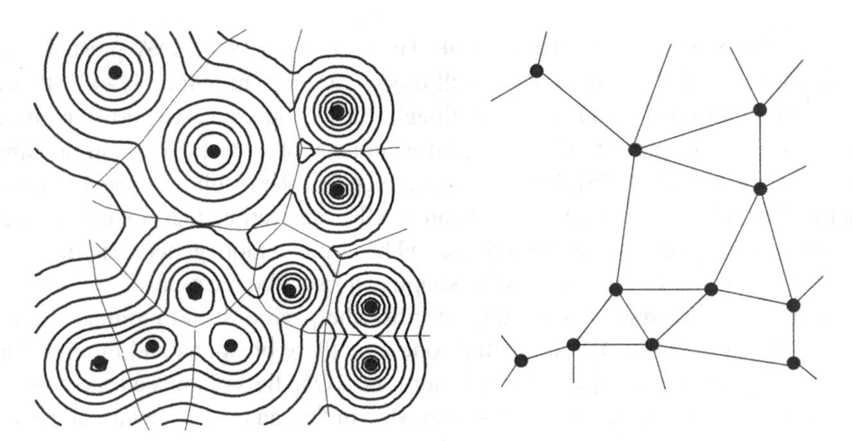

**Fig. 2.8** A sketch of a potential-energy surface of an arbitrary system and its corresponding graph. The *points* are minima of the potential-energy surface. The *edges* in the graph connect minima that have catchment regions that border on each other. They correspond to reactions or other activated processes

region is a subspace of configuration space $\mathcal{C}$, and all catchment regions form a partitioning of the configuration space.

$$\mathcal{C} = \bigcup_{\alpha} C_{\alpha}. \tag{2.16}$$

(There is a small difficulty with those points of configuration space that do not lead to minima, but to saddle points, and with maxima. These points are irrelevant because the number of such points are vanishing small with respect to the other points. They are found where two or more catchment regions meet, and we can arbitrarily assign them to one of these catchment regions.) With $\mathbf{q}$ the set of all coordinates and $\mathbf{p}$ the set of all conjugate momenta we can extend the catchment region $C_{\alpha}$ to a corresponding region in phase space $R_{\alpha}$ as follows.

$$R_{\alpha} = \left\{ (\mathbf{q}, \mathbf{p}) \in \mathcal{P} | \mathbf{q} \in C_{\alpha} \right\}. \tag{2.17}$$

We then have for phase space

$$\mathcal{P} = \bigcup_{\alpha} R_{\alpha}. \tag{2.18}$$

If we use the regions $R_{\alpha}$, we can derive the master equation exactly as for the lattice-gas model. This starting point based on the PES for the derivation of the master equation is more general than the one that defines the regions in phase space in terms of configurations of adlayers. It applies in principle to any molecular system. However, for an adlayer both lead to the same partitioning of phase space.

The probability to find the system in region $R_{\alpha}$ is given by

$$P_{\alpha}(t) = \int_{R_{\alpha}} \frac{d\mathbf{q}\, d\mathbf{p}}{h^D} \rho(\mathbf{q}, \mathbf{p}, t), \tag{2.19}$$

where $h$ is Planck's constant, $D$ is the number of degrees of freedom, and $\rho$ is the phase space density. The denominator $h^D$ is not needed for a purely classical description of the kinetics. However, it makes the transition later on from a classical to a quantum mechanical description easier [47].

The master equation tells us how these probabilities $P_\alpha$ change in time. Differentiating Eq. (2.19) yields

$$\frac{dP_\alpha}{dt} = \int_{R_\alpha} \frac{d\mathbf{q}d\mathbf{p}}{h^D} \frac{\partial \rho}{\partial t}(\mathbf{q}, \mathbf{p}, t). \tag{2.20}$$

This can be transformed using the Liouville-equation [48]

$$\frac{\partial \rho}{\partial t} = -\sum_{i=1}^{D} \left[ \frac{\partial \rho}{\partial q_i} \frac{\partial H}{\partial p_i} - \frac{\partial \rho}{\partial p_i} \frac{\partial H}{\partial q_i} \right] \tag{2.21}$$

into

$$\frac{dP_\alpha}{dt} = \int_{R_\alpha} \frac{d\mathbf{q}d\mathbf{p}}{h^D} \sum_{i=1}^{D} \left[ \frac{\partial \rho}{\partial p_i} \frac{\partial H}{\partial q_i} - \frac{\partial \rho}{\partial q_i} \frac{\partial H}{\partial p_i} \right], \tag{2.22}$$

where $H$ is the system's classical Hamiltonian. To simplify the mathematics, we will assume that the coordinates are Cartesian and the Hamiltonian has the usual form

$$H = \sum_{i=1}^{D} \frac{p_i^2}{2m_i} + V(\mathbf{q}), \tag{2.23}$$

where $m_i$ is the mass corresponding to coordinate $i$. The area $R_\alpha$ has been defined above by coordinates only, and the limits of integration for the momenta are $\pm\infty$. Although $R_\alpha$ can be defined more generally (we would like to mention reference [49] for a more general derivation), the definition here allows us to go from phase space to configuration space. The first term on the right-hand-side of Eq. (2.22) now becomes

$$\begin{aligned}
\int_{R_\alpha} \frac{d\mathbf{q}d\mathbf{p}}{h^D} \sum_{i=1}^{D} \frac{\partial \rho}{\partial p_i} \frac{\partial H}{\partial q_i} &= \sum_{i=1}^{D} \int_{R_\alpha} d\mathbf{q} \frac{\partial V}{\partial q_i} \int_{-\infty}^{\infty} \frac{d\mathbf{p}}{h^D} \frac{\partial \rho}{\partial p_i} \\
&= \sum_{i=1}^{D} \int_{R_\alpha} d\mathbf{q} \frac{\partial V}{\partial q_i} \int_{-\infty}^{\infty} \frac{dp_1 \ldots dp_{i-1} dp_{i+1} \ldots dp_D}{h^D} \\
&\quad \times \left[ \rho(p_i = \infty) - \rho(p_i = -\infty) \right] \\
&= 0,
\end{aligned} \tag{2.24}$$

because $\rho$ has to go to zero for any of its variables going to $\pm\infty$ to be integrable. The second term becomes

$$-\int_{R_\alpha} \frac{d\mathbf{q}d\mathbf{p}}{h^D} \sum_{i=1}^{D} \frac{\partial \rho}{\partial q_i} \frac{\partial H}{\partial p_i} = -\int_{R_\alpha} \frac{d\mathbf{q}d\mathbf{p}}{h^D} \sum_{i=1}^{D} \frac{\partial}{\partial q_i} \left( \frac{p_i}{m_i} \rho \right). \tag{2.25}$$

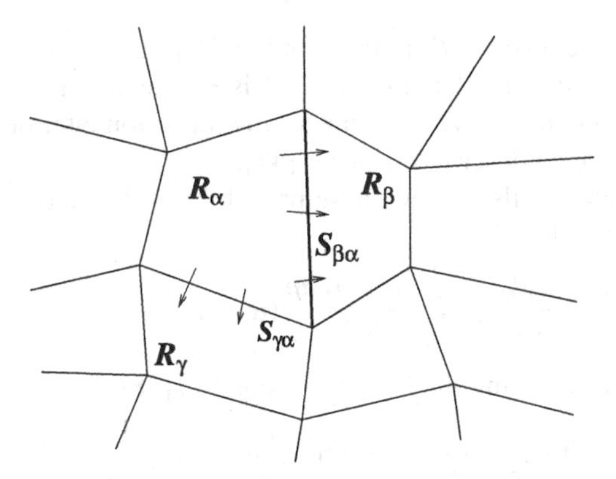


**Fig. 2.9** Schematic drawing of the partitioning of configuration space into regions $R$, each of which corresponds to some particular configuration of the adlayer or catchment region of a minimum of the potential-energy surface. The process that changes $\alpha$ into $\beta$ corresponds to a flow from $R_\alpha$ to $R_\beta$. The transition probability $W_{\beta\alpha}$ for this process equals the flux through the surface $S_{\beta\alpha}$, separating $R_\alpha$ from $R_\beta$, divided by the probability to find the system in $R_\alpha$

This particular form suggests using the divergence theorem for the integration over the coordinates [50]. The final result is then

$$\frac{dP_\alpha}{dt} = -\int_{S_\alpha} dS \int_{-\infty}^{\infty} \frac{d\mathbf{p}}{h^D} \sum_{i=1}^{D} n_i \frac{p_i}{m_i} \rho, \tag{2.26}$$

where the first integration is a surface integral over the surface of $R_\alpha$, and $n_i$ are the components of the outward pointing normal of that surface. Both the area $R_\alpha$ and the so-called dividing surface $S_\alpha$ are now regarded as parts of the configuration space of the system [32, 33]. As $p_i/m_i = \dot{q}_i$, we see that the summation in the last expression is the flux through $S_\alpha$ in the direction of the outward pointing normal (see Fig. 2.9).

The final step is now to decompose this flux in two ways. First, we split the dividing surface $S_\alpha$ into sections $S_\alpha = \bigcup_\beta S_{\beta\alpha}$, where $S_{\beta\alpha}$ separates $R_\alpha$ from $R_\beta$. Second, we distinguish between an outward flux, $\sum_i n_i p_i/m_i > 0$, and an inward flux, $\sum_i n_i p_i/m_i < 0$. Equation (2.26) can then be rewritten as

$$\begin{aligned} \frac{dP_\alpha}{dt} =& \sum_\beta \int_{S_{\alpha\beta}} dS \int_{-\infty}^{\infty} \frac{d\mathbf{p}}{h^D} \left(\sum_{i=1}^{D} n_i \frac{p_i}{m_i}\right) \Theta\left(\sum_{i=1}^{D} n_i \frac{p_i}{m_i}\right) \rho \\ &- \sum_\beta \int_{S_{\beta\alpha}} dS \int_{-\infty}^{\infty} \frac{d\mathbf{p}}{h^D} \left(\sum_{i=1}^{D} n_i \frac{p_i}{m_i}\right) \Theta\left(\sum_{i=1}^{D} n_i \frac{p_i}{m_i}\right) \rho, \end{aligned} \tag{2.27}$$

where in the first term $S_{\alpha\beta}$ ($= S_{\beta\alpha}$) is regarded as part of the surface of $R_\beta$, and the $n_i$ are components of the outward pointing normal of $R_\beta$. The function $\Theta$ is the Heaviside step function [51]

$$\Theta(x) = \begin{cases} 1, & \text{if } x \geq 0, \text{ and} \\ 0, & \text{if } x < 0, \end{cases} \tag{2.28}$$

Equation (2.27) can be cast in the form of the master equation

$$\frac{dP_\alpha}{dt} = \sum_\beta [W_{\alpha\beta} P_\beta - W_{\beta\alpha} P_\alpha], \tag{2.29}$$

if we define the transition probabilities as

$$W_{\beta\alpha} = \frac{\int_{S_{\beta\alpha}} dS \int_{-\infty}^{\infty} d\mathbf{p}/h^D (\sum_{i=1}^{D} n_i p_i/m_i) \Theta(\sum_{i=1}^{D} n_i p_i/m_i) \rho}{\int_{R_\alpha} d\mathbf{q} \int_{-\infty}^{\infty} d\mathbf{p}/h^D \rho}. \tag{2.30}$$

The expression for the transition probabilities can be cast in a more familiar form by using a few additional assumptions. We assume that $\rho$ can locally (i.e., in $R_\alpha$ and on $S_{\beta\alpha}$) be approximated by a Boltzmann-distribution

$$\rho = N \exp\left[-\frac{H}{k_B T}\right], \tag{2.31}$$

where $T$ is the temperature, $k_B$ is the Boltzmann-constant, and $N$ is a normalizing constant. We also assume that we can define $S_{\beta\alpha}$ and the coordinates in such a way that $n_i = 0$, except for one coordinate $i$, called the reaction coordinate, for which $n_i = 1$. The integral of the momentum corresponding to the reaction coordinate can then be done and the result is

$$W_{\beta\alpha} = \frac{k_B T}{h} \frac{Q^\ddagger}{Q}, \tag{2.32}$$

with

$$Q^\ddagger = \int_{S_{\beta\alpha}} dS \int_{-\infty}^{\infty} \frac{dp_1 \dots dp_{i-1} dp_{i+1} \dots dp_D}{h^{D-1}} \exp\left[-\frac{H}{k_B T}\right], \tag{2.33}$$

$$Q = \int_{R_\alpha} d\mathbf{q} \int_{-\infty}^{\infty} \frac{d\mathbf{p}}{h^D} \exp\left[-\frac{H}{k_B T}\right]. \tag{2.34}$$

We see that this is an expression that is formally identical to the TST expression for rate constants [52]. There are differences in the definition of the partition functions $Q$ and $Q^\ddagger$, but they can generally be neglected. For example, it is quite common that the PES has a well-defined minimum in $R_\alpha$ and on $S_{\beta\alpha}$, and that it can be replaced by a quadratic form in the integrals above. The borders of the integrals can then be extended to infinity and the normal partition functions for vibrations are obtained. This is sometimes called harmonic TST (see Chap. 4) [53]. The dividing surface $S_{\beta\alpha}$ is then chosen so that it contains the transition state of the process, which is then also where the PES has its minimum on $S_{\beta\alpha}$.

The $W$'s indicate how fast the system moves from (the catchment region of) one minimum to another. We will often call them therefore rate constants. The system can only move from minimum $\alpha$ to minimum $\beta$ if the catchment region of these minima border on each other. Only in such a case we have $W_{\beta\alpha} \neq 0$. The right-hand-side of Fig. 2.8 shows the minima of the PES as points. Two minima are connected if their catchment regions border on each other, and the system can move from one to the other without having to go through a third catchment region. The result is the graph in Fig. 2.8. The vertices of the graph are the minima of the PES and the edges indicate how the system can move from one minimum to another.

Although we have presented the partitioning of phase space based on the catchment regions of the PES, this is actually not required. In fact, we have not used this particular partitioning in the derivation up to Eq. (2.30) anywhere. One can in principle partition phase space in any way one likes and derive a master equation. It is the partitioning that defines the processes that the master equation describes. Of course, most partitionings lead to processes that are hard to interpret physically, but there are variations in the partitioning above that are useful.

The dividing surface $S_{\beta\alpha}$ was split to distinguish fluxes in opposite directions. If there is a trajectory of the system that crosses the surface and then recrosses it, then effectively no process has occurred, but both crossings contribute to the rate constants of $\alpha \to \beta$ and $\beta \to \alpha$. For surface reactions such a recrossing is well known for adsorption and leads to the definition of a sticking coefficient (see Sect. 4.4.3). The idea of VTST is to move $S_{\beta\alpha}$ to remove recrossings and to minimize the rate constants [29, 32, 33, 40]. It can be shown that when we have a canonical ensemble, this is equivalent to locating $S_{\beta\alpha}$ at a maximum of the Gibbs energy along the reaction coordinate [54, 55]. The transition state need then not be on $S_{\beta\alpha}$. The transition state need then generally not be on $S_{\beta\alpha}$. As our derivation is a generalization of VTST, it has the same limitations and possible ways to deal with them. We refer to Chap. 4 of [40] for a fuller discussion.

### 2.2.3 The Master Equation for Lattice-Gas Models

In the derivation of the master equation above the subscripts $\alpha$ and $\beta$ refer to minima of the PES. We want however a master equation with subscripts referring to configurations. We have already stated that using configurations to derive the master equation gives the same result, but there are some subtleties. So we take a closer look at how minima of the PES and configurations relate to each other. As the coordinates of all atoms have well-defined values for a minimum of a PES, we also know which adsorbate is at each site: i.e., we know the corresponding configuration. So it is easy to go from the minima of the PES to the configurations. The reverse is not true in general.

The first problem is that not all configurations need to correspond to a minimum of the PES. For example, suppose that we have an adsorbate that is so large that there is a very high repulsion when another adsorbate is at a neighboring site. A configuration with two adsorbates at such relative positions may then not correspond to a minimum. The forces between them may push them farther apart and to other sites.

This however does not prevent us from identifying the subscripts in the master equation with configurations. We only need to make sure that the rate constant $W_{\alpha\beta}$ for the process $\beta \to \alpha$ equals zero when the configuration $\alpha$ can not be identified with a minimum of the PES.

The other problem is that there may be more than one minimum of the PES that leads to the same configuration. This is the case when an adsorbate has different (meta)stable structures or adsorption modes. For example, an NO molecule on Rh(100) may be adsorbed perpendicular to the surface with the N end down or it may be adsorbed parallel to the surface [56, 57]. In this case we have a choice. We may regard the different geometries as different adsorbates. This solves the problem, because we then do get a 1-to-1 correspondence between configurations and minima of the PES. Alternatively we may ignore these differences in geometry. We can deal with this by redefining the areas $R_\alpha$ in the partitioning of configuration space. Instead of the catchment region of one minimum, we define them as the union of the catchment regions of all the minima leading to the same configuration. We can also first regard the different geometries as different adsorbates as before and then do a coarse-graining as explained in Sect. 8.2. The result is the same. The advantage of disregarding these differences is that it leads to a simpler model and faster kMC simulations. We need to point out however that disregarding the difference in geometry may constitute an approximation that needs to be justified.

Now that we have established that the master equation can be regarded in terms of configurations of lattice-gas models, we can discuss the advantage this gives us. The number of rate constants $W_{\alpha\beta}$ is enormous even if we use only a small model. Suppose that we do a simulation with a modest $100 \times 100$ lattice. Also suppose that we have only one type of adsorbate so that each site can either be occupied or vacant. This gives us $2^{100\times 100}$ different configurations and $2^{2\times 100\times 100} \approx 10^{6000}$ rate constants $W_{\alpha\beta}$. In general, if we have a number of sites $S$ and each site can be occupied in $A$ ways then the number of configurations equals $A^S$ and the number of rate constants is $A^{2S}$. A lattice model allows us to reduce this number by the same order of magnitude, because what matters is only the number of different non-zero values that the rate constants can have.

First we note that $W_{\alpha\beta} = 0$ unless the change $\beta \to \alpha$ corresponds to an actual physical or chemical process like a reaction or diffusion of an adsorbate. This means that for a configuration $\beta$ the number of configurations $\alpha$ with values $W_{\alpha\beta} \neq 0$ does not equal the number of possible configurations, but only the number of processes that can actually take place in $\beta$. This number is only proportional to the number of sites, and not to an exponential function of the number of sites. So the number of rate constants is thereby reduced to $cSA^S$ with $c$ a constant depending on the number of types of process but not on $S$.

This is still a huge number of values for the rate constants. Using translational symmetry only reduces this number by a factor $S$. Point-group symmetry only reduces it by a factor of the order of unity. To reduce it to a workable number of rate constants, we need to make some assumptions. Fortunately, such assumptions are almost always valid and easy to find. The reason why the number of values for the rate constants depends exponentially on the number of sites is that so far we have assumed that the occupation of each and every site in the system can affect these values. This will often be unlikely however. If we have a process involving one site or a pair of neighboring sites, then the occupation of a site well away from this site or these sites will not be relevant. The extreme case is where we only need to look at the sites that change occupation, and which define the process, to determine the rate constant. This means that we have just one value for a desorption of a particular adsorbate, one value for the reaction of two adsorbates, et cetera. Adsorbates that do not desorb, react, et cetera do not affect the value of the rate constant. In such a case the number of different values of the rate constants reduces to $c$, the proportionality constant introduced above. A large majority of kMC simulations done so far have assumed that this case is valid.

Suppose that the occupation of $S_{\text{env}}$ sites in the environment of a process does not change but does affect the value of the rate constant because of interactions with the adsorbates that change in the process. These interactions are called lateral interactions. The number of values of the rate constants is then $cA^{S_{\text{env}}}$. This number need not be large. Suppose we look at CO desorption from a Rh(100) surface. At low coverage we can model this with a square lattice representing the top sites which CO prefers [10]. We also assume that only interactions between CO molecules at nearest-neighbor positions need to be included. We then have $A = 2$. For desorption we have $S_{\text{env}} = 4$ so that there are at most $2^4 = 16$ possibly different values for the rate constants for desorption. For diffusion modeled as a CO hopping from one site to a neighboring one we have $S_{\text{env}} = 6$ and $2^6 = 64$ possibly different values. We can use point-group symmetry to reduce the number of values further.

This example shows that there are situations in which the number of values of rate constants is relatively small even with lateral interactions, but the exponential dependence on $S_{\text{env}}$ will often necessitate another approach. We want to reduce the number of values for the rate constants, because these values are generally hard to determine as will be shown in Chap. 4. For a lattice-gas model it is often possible to split the effect of the lateral interactions from the determination of a rate constants without lateral interactions. The determination of rate constants even without lateral interactions is difficult, but need only be done for a few processes. The determination of lateral interactions is also difficult, but here too often only few values need to be determined. It is the large number of combinations of lateral interactions that leads to a very large number of possible values for rate constants with lateral interactions. These combinations can however often be determined quite easily. This is shown explicitly in Chap. 4 and in particular Sect. 4.5. However, the number of values of the rate constants does also determine which kMC algorithm is the most efficient. If the number of values is small, faster algorithms can be used than when we have a large number of values of the rate constants even if these values can be determined easily and fast. See Chap. 3.

To summarize. The reason why kMC simulations of surface reactions can be done much more efficiently than simulations of systems without translational symmetry has to do with the number of rate constants. For surface reactions this number is either very limited when there are no lateral interactions, or can often be computed easily from a limited set of parameters when there are lateral interactions. In both cases only a relatively small number of values need to be determined, although they require costly calculations or time-consuming experiments (see Chap. 4). This can however be done before and separately from the kMC simulations. This is not the case for simulations of other systems.

## References


1. M.P. Allen, D.J. Tildesley, *Computer Simulation of Liquids* (Clarendon, Oxford, 1987)
2. D. Frenkel, B. Smit, *Understanding Molecular Simulation: From Algorithms to Applications* (Academic Press, London, 2001)
3. D.C. Rapaport, *The Art of Molecular Dynamics Simulation* (Cambridge University Press, Cambridge, 2004)
4. A. Zangwill, *Physics at Surfaces* (Cambridge University Press, Cambridge, 1988)
5. J.M. Thomas, W.J. Thomas, *Principles and Practice of Heterogeneous Catalysis* (VCH, Weinheim, 1997)
6. N.W. Ashcroft, N.D. Mermin, *Solid State Physics* (Holt, Rinehart & Winston, New York, 1976)
7. A.M. de Jong, J.W. Niemantsverdriet, J. Chem. Phys. **101**, 10126 (1994)
8. R. Kose, W.A. Brown, D.A. King, J. Phys. Chem. B **103**, 8722 (1999)
9. M.J.P. Hopstaken, J.W. Niemantsverdriet, J. Chem. Phys. **113**, 5457 (2000)
10. M.M.M. Jansen, C.G.M. Hermse, A.P.J. Jansen, Phys. Chem. Chem. Phys. **12**, 8053 (2010)
11. C.T. Kao, G.S. Blackman, M.A.V. Hove, G.A. Somorjai, C.M. Chan, Surf. Sci. **224**, 77 (1989)
12. M. van Hardeveld, Elementary reactions in the catalytic reduction of NO on rhodium surfaces. Ph.D. thesis, Eindhoven University of Technology, Eindhoven (1997)
13. N.V. Petrova, I.N. Yakovkin, Surf. Sci. **519**, 90 (2002)
14. S. Dahl, A. Logadottir, R.C. Egeberg, J.H. Larsen, I. Chorkendorff, E. Törnqvist, J.K. Nørskov, Phys. Rev. Lett. **83**, 1814 (1999)
15. B. Hammer, Phys. Rev. Lett. **83**, 3681 (2000)
16. M.T.M. Koper, J.J. Lukkien, A.P.J. Jansen, R.A. van Santen, J. Phys. Chem. B **103**, 5522 (1999)
17. R. Imbihl, G. Ertl, Chem. Rev. **95**, 697 (1995)
18. M.M. Slin'ko, N.I. Jaeger, *Oscillating Heterogeneous Catalytic Systems*. Studies in Surface Science and Catalysis, vol. 86 (Elsevier, Amsterdam, 1994)
19. R.J. Gelten, A.P.J. Jansen, R.A. van Santen, J.J. Lukkien, J.P.L. Segers, P.A.J. Hilbers, J. Chem. Phys. **108**, 5921 (1998)
20. V.N. Kuzovkov, O. Kortlüke, W. von Niessen, Phys. Rev. Lett. **83**, 1636 (1999)
21. O. Kortlüke, V.N. Kuzovkov, W. von Niessen, Phys. Rev. Lett. **83**, 3089 (1999)
22. M. Gruyters, T. Ali, D.A. King, Chem. Phys. Lett. **232**, 1 (1995)
23. M. Gruyters, T. Ali, D.A. King, J. Phys. Chem. **100**, 14417 (1996)
24. R. Salazar, A.P.J. Jansen, V.N. Kuzovkov, Phys. Rev. E **69**, 031604 (2004)
25. A.R. Leach, *Molecular Modelling. Principles and Applications* (Longman, Singapore, 1996)
26. N.G. van Kampen, *Stochastic Processes in Physics and Chemistry* (North-Holland, Amsterdam, 1981)
27. J.R. Norris, *Markov Chains* (Cambridge University Press, Cambridge, 1998)
28. D.E. Knuth, *The Art of Computer Programming, Volume I: Fundamental Algorithms* (Addison-Wesley, Reading, 1973)
29. J.C. Keck, J. Chem. Phys. **32**, 1035 (1960)
30. J.C. Keck, Discuss. Faraday Soc. **33**, 173 (1962)
31. J.C. Keck, Adv. Chem. Phys. **13**, 85 (1967)
32. P. Pechukas, in *Dynamics of Molecular Collisions, Part B*, ed. by W. Miller (Plenum, New York, 1976), pp. 269–322

33. D.G. Truhlar, A.D. Isaacson, B.C. Garrett, in *Theory of Chemical Reaction Dynamics, Part IV*, ed. by M. Baer (CRC Press, Boca Raton, 1985), pp. 65–138
34. W.H. Miller, J. Chem. Phys. **61**, 1823 (1974)
35. W.H. Miller, J. Chem. Phys. **62**, 1899 (1975)
36. G.A. Voth, J. Phys. Chem. **97**, 8365 (1993)
37. V.A. Benderskii, D.E. Makarov, C.A. Wight, Adv. Chem. Phys. **88**, 1 (1994)
38. R.S. Berry, R. Breitengraser-Kunz, Phys. Rev. Lett. **74**, 3951 (1995)
39. R.E. Kunz, R.S. Berry, J. Chem. Phys. **103**, 1904 (1995)
40. K.J. Laidler, *Chemical Kinetics* (Harper & Row, New York, 1987)
41. A.P.J. Jansen, Comput. Phys. Comm. **86**, 1 (1995)
42. R.J. Gelten, R.A. van Santen, A.P.J. Jansen, in *Molecular Dynamics: From Classical to Quantum Methods*, ed. by P.B. Balbuena, J.M. Seminario (Elsevier, Amsterdam, 1999), pp. 737–784
43. A.P.J. Jansen, An introduction to Monte Carlo simulations of surface reactions. http://arxiv.org/abs/cond-mat/0303028 (2003)
44. I. Prigogine, *Introduction to Thermodynamics of Irreversible Processes* (Interscience, New York, 1968)
45. C. van Vliet, *Equilibrium and Non-equilibrium Statistical Mechanics* (World Scientific, Singapore, 2008)
46. P.G. Mezey, *Potential Energy Hypersurfaces* (Elsevier, Amsterdam, 1987)
47. D.A. McQuarrie, *Statistical Mechanics* (Harper, New York, 1976)
48. R. Becker, *Theorie der Wärme* (Springer, Berlin, 1985)
49. A.P.J. Jansen, J. Chem. Phys. **94**, 8444 (1991)
50. E. Kreyszig, *Advanced Engineering Mathematics* (Wiley, New York, 1993)
51. A.H. Zemanian, *Distribution Theory and Transform Analysis* (Dover, New York, 1987)
52. R.A. van Santen, J.W. Niemantsverdriet, *Chemical Kinetics and Catalysis* (Plenum, New York, 1995)
53. G. Henkelman, G. Jóhannesson, H. Jónsson, in *Progress in Theoretical Chemistry and Physics*, ed. by S.D. Schwarts (Kluwer Academic, London, 2000)
54. B.C. Garrett, D.G. Truhlar, J. Am. Chem. Soc. **101**, 5207 (1979)
55. B.C. Garrett, D.G. Truhlar, J. Am. Chem. Soc. **102**, 2559 (1980)
56. A.P. van Bavel, Understanding and quantifying interactions between adsorbates: CO, NO, and N- and O-atoms on Rh(100). Ph.D. thesis, Eindhoven University of Technology, Eindhoven (2005)
57. C. Popa, A.P. van Bavel, R.A. van Santen, C.F.J. Flipse, A.P.J. Jansen, Surf. Sci. **602**, 2189 (2008)

# Chapter 3
# Kinetic Monte Carlo Algorithms

**Abstract** Kinetic Monte Carlo generates a sequence of configurations and times when the transitions between these configurations occur. This solves the master equation in the sense that a configuration $\alpha$ is obtained at time $t$ with a probability $P_\alpha(t)$ that is a solution of the master equation. There are many algorithms that yield such a sequence of configurations and which are statistically equivalent. They all need to determine repeatedly the time that the next process will occur, the type of process that will occur, and the position on the surface where the process will occur. Each of these can be determined in a number of ways, which can be combined in even more ways. This results in many algorithms. Few of them are however efficient. We discuss in detail the Variable Step Size Method, the Random Selection Method, and the First Reaction Method. We use the Variable Step Size Method to show how to handle lists of processes, different ways to make selections of processes and process types, and how computer time and memory scales with system size. Time-dependent rate constants are discusses separately as the determination of when processes take place pose special problems. Parallelization is discussed as well as some older algorithms. Some guidelines are given of how to choose an algorithm for a simulation.

## 3.1 Introduction

Deriving analytical results from the master equation is not possible for most systems of interest. Approximations can be used, but they may not be satisfactory. In such cases one can resort to kinetic Monte Carlo (kMC) simulations.

Monte Carlo methods have been known already for several decades for the master equation [1]. Following Gillespie they have become quite popular to simulate reactions in solutions [2–4]. A configuration $\alpha$ in that case is defined as a set $\{N_1, N_2, \ldots\}$ where $N_i$ is the number of molecules of type $i$ in the solution. There is no specification of where the molecules are, as in our case for surface reactions. In fact, the method is generally used to solve rate equations for chemical reactions. When simulating reactions one talks about Dynamic Monte Carlo (DMC) simulations, a term that we will use as well, or more recently Stochastic Simulation Algorithm (SSA) [5–7]. Many of the algorithms developed in that area can be used for surface reactions as well. However, the efficiency of the various algorithms (i.e.,

A.P.J. Jansen, *An Introduction to Kinetic Monte Carlo Simulations of Surface Reactions*, Lecture Notes in Physics 856,
DOI 10.1007/978-3-642-29488-4_3, © Springer-Verlag Berlin Heidelberg 2012

the computer time and memory) can be different. There are also tricks to increase the efficiency of simulations of reactions in solutions that do not work for surface reactions and vice versa [8–10].

This chapter starts with a discussion of the most widely used algorithm, which we call the Variable Step Size Method (see Sect. 3.2). Several aspects of the algorithm are improved in Sect. 3.3. These improvements are put in a broader context as they can also be used in other algorithms. Sections 3.4 and 3.5 discuss two other algorithms that have been shown to be very useful. Section 3.6 extends the discussion to include situations in which rate constants vary in time. Section 3.7 discusses other approaches for surface reactions and Sect. 3.8 parallel algorithms. Section 3.9 presents a pragmatic approach to the question of which algorithm to use, and discusses other practical aspects of implementing and doing kMC simulations.

## 3.2 The Variable Step Size Method

The Variable Step Size Method (VSSM) method is probably the most widely used algorithm for kMC simulations. In fact, sometimes the name kMC is specifically used to denote this method. We will not do this. The algorithms in this section and those of Sects. 3.4 and 3.5 give the same results, so we prefer to include all algorithms that solve the master equation in kMC. VSSM is also often called the n-fold way [11]. This refers to the algorithm developed by Bortz, Kalos, and Lebowitz which is indeed equivalent to VSSM, but was originally developed for equilibrium Monte Carlo [12]. The name VSSM was coined by Gillespie, and was developed for DMC [2, 3].

### *3.2.1 The Integral Form of the Master Equation*

To start with the derivation of the kMC algorithms for the master equation it is convenient to cast the master equation in an integral form. First we simplify the notation of the master equation. We define a matrix $\mathbf{W}$ by

$$\mathbf{W}_{\alpha\beta} = W_{\alpha\beta}, \tag{3.1}$$

which has vanishing diagonal elements, because $W_{\alpha\alpha} = 0$ by definition, and a diagonal matrix $\mathbf{R}$ by

$$\mathbf{R}_{\alpha\beta} = \begin{cases} 0, & \text{if } \alpha \neq \beta, \\ \sum_\gamma W_{\gamma\beta}, & \text{if } \alpha = \beta. \end{cases} \tag{3.2}$$

If we put the probabilities of the configurations $P_\alpha$ in a vector $\mathbf{P}$, we can write the master equation as

$$\frac{d\mathbf{P}}{dt} = -(\mathbf{R} - \mathbf{W})\mathbf{P}. \tag{3.3}$$

This equation can be interpreted as a time-dependent Schrödinger-equation in imaginary time with Hamiltonian $\mathbf{R}-\mathbf{W}$. This interpretation can be very fruitful [13], and leads, among other things, to the integral form we present here.

We do not want to be distracted by technicalities at this point, so we assume that $\mathbf{R}$ and $\mathbf{W}$ are time independent. We also introduce a matrix $\mathbf{Q}$, which is defined by

$$\mathbf{Q}(t) = \exp[-\mathbf{R}t]. \tag{3.4}$$

We can rewrite the master equation with this definition in the integral form

$$\mathbf{P}(t) = \mathbf{Q}(t)\mathbf{P}(0) + \int_0^t dt'\mathbf{Q}(t-t')\mathbf{W}\mathbf{P}(t'), \tag{3.5}$$

as can be seen by substitution of this expression in Eq. (3.3). The equation is a recurrence relation implicit in $\mathbf{P}$. By substitution of the right-hand-side for $\mathbf{P}(t')$ again and again we get

$$\mathbf{P}(t) = \Big[\mathbf{Q}(t) + \int_0^t dt'\mathbf{Q}(t-t')\mathbf{W}\mathbf{Q}(t') + \int_0^t dt' \int_0^{t'} dt''\mathbf{Q}(t-t')\mathbf{W}\mathbf{Q}(t'-t'')\mathbf{W}\mathbf{Q}(t'') + \ldots\Big]\mathbf{P}(0). \tag{3.6}$$

This equation is valid also for other definitions of $\mathbf{R}$ and $\mathbf{W}$, but the definition we have chosen leads to a useful interpretation. Suppose at $t=0$ the system is in configuration $\alpha$ with probability $P_\alpha(0)$. The probability that at time $t$ the system is still in $\alpha$ (i.e., no process has occurred that would change the configuration) is given by $\mathbf{Q}_{\alpha\alpha}(t)P_\alpha(0) = \exp(-\mathbf{R}_{\alpha\alpha}t)P_\alpha(0)$. This shows that the first term in Eq. (3.6) represents the contribution to the probabilities if no process occurs up to time $t$. The matrix $\mathbf{W}$ determines how the probabilities change when some process occurs. The second term of Eq. (3.6) therefore represents the contribution to the probabilities if no process occurs between times 0 and $t'$, some process occurs at time $t'$, and then no process occurs between times $t'$ and $t$. So the second term stands for the contribution to the probabilities if a single process occurs between times 0 and $t$. Subsequent terms represent contributions if two, three, four, et cetera processes occur. (Note that we use the term process for any change of the configuration. This includes chemical reactions, but also for example diffusional hops of adsorbates from one site to a neighboring one. In Chaps. 5, 6, and 7 we will encounter also other processes.)

### 3.2.2 The Concept of the Variable Step Size Method

The idea of kMC is not to compute probabilities $P_\alpha(t)$ explicitly, but to start with some particular configuration, representative for the initial state of the experiment one wants to simulate, and then generate a sequence of other configurations with the correct probability. The integral form gives us directly a useful algorithm to generate subsequent configurations and the times when the configuration changes. At these times some process (e.g., a chemical reaction) occurs.

Let's call the initial configuration $\alpha$, and let's set the initial time to $t = 0$. Then the probability that the system is still in $\alpha$ at a later time $t$ is given by

$$\mathbf{Q}_{\alpha\alpha}(t) = \exp[-\mathbf{R}_{\alpha\alpha}t]. \tag{3.7}$$

The probability distribution that the first process occurs at time $t$ is minus the derivative with respect to time of this expression: i.e.,

$$\mathbf{R}_{\alpha\alpha}\exp[-\mathbf{R}_{\alpha\alpha}t]. \tag{3.8}$$

This can be seen by taking the integral of this expression from 0 to $t$, which yields the probability that some process has occurred in this interval, which equals $1 - \mathbf{Q}_{\alpha\alpha}(t)$. We now make an important step. We move from a probability distribution for the time that the first process may occur to a time $t'$ when the first process actually occurs. This can be done by equating the probability that the system is still in $\alpha$ at a later time $t$ to a uniform deviate on the unit interval: i.e., by solving

$$\exp\left[-\mathbf{R}_{\alpha\alpha}t'\right] = r, \tag{3.9}$$

for $t'$ where $r$ is the uniform deviate on the unit interval [14].

At time $t'$ some process occurs. According to Eq. (3.6) the different processes that transform configuration $\alpha$ to another configuration $\alpha'$ have rate constants $W_{\alpha'\alpha}$. This means that the probability that the system will be in configuration $\alpha'$ at time $t' + dt$ is $W_{\alpha'\alpha}\,dt$, where $dt$ is some small time interval. We therefore generate a new configuration $\alpha'$ by choosing one of all possible new configurations $\alpha'$ with a probability proportional to $W_{\alpha'\alpha}$. This gives us a new configuration $\alpha'$ at time $t'$. At this point we're in the same situation as when we started the simulation, and we can proceed by repeating the previous steps. So we generate a new time $t''$, using

$$\exp\left[-\mathbf{R}_{\alpha'\alpha'}\left(t'' - t'\right)\right] = r, \tag{3.10}$$

for the time of the new process, and a new configuration $\alpha''$ with a probability proportional to $W_{\alpha''\alpha'}$. Here $r$ is again a uniform deviate on the unit interval, but not the same one of course as in Eq. (3.9). In this manner we continue until some preset condition is met that signals the end of the simulation.

We call this whole procedure the Variable Step Size Method (VSSM). It's a simple method that can be made very efficient. The algorithm is as follows.

Variable Step Size Method: concept (VSSMc)

1. Initialize

   Generate an initial configuration $\alpha$.
   Set the time $t$ to some initial value.
   Choose conditions when to stop the simulation.

2. Time

   Generate a time interval $\Delta t$ when no process occurs

$$\Delta t = -\frac{1}{\sum_\beta W_{\beta\alpha}}\ln r, \tag{3.11}$$

   where $r$ is a uniform deviate on the unit interval.
   Change time to $t \to t + \Delta t$.

3. Process

   Change the configuration to $\alpha'$ with probability $W_{\alpha'\alpha}/\sum_{\beta} W_{\beta\alpha}$: i.e., do the process $\alpha \rightarrow \alpha'$.

4. Continuation

   If the stop conditions are fulfilled then stop. If not repeat at step 2.

Equation (3.11) is obtained from Eqs. (3.9) or (3.10) if we solve them for the time.

We see that the algorithm yields an ordered set of configurations and times when processes occur that can be written as

$$(\alpha_0, t_0) \xrightarrow{t_1} \alpha_1 \xrightarrow{t_2} \alpha_2 \xrightarrow{t_3} \alpha_3 \xrightarrow{t_4} \ldots \tag{3.12}$$

Here $\alpha_0$ is the initial configuration and $t_0$ is the time at the beginning of the simulations. The changes $\alpha_{n-1} \rightarrow \alpha_n$ are caused by processes occurring at times $t_n$. These processes that actually occur are often called events. We will see that all algorithms in this section and in Sects. 3.4 and 3.5 also give such a result (3.12). They are all equivalent because all give at time $t$ a configuration $\alpha$ with probability $P_\alpha(t)$ which is the solution of the master equation with boundary condition $P_\alpha(t_0) = \delta_{\alpha\alpha_0}$ with $t_0$ the time at the beginning of the simulation.

### *3.2.3 Enabled and Disabled Processes*

Although the algorithms in this section and in Sects. 3.4 and 3.5 yield the same result, they often do so at very different computational costs. We are very interested in how computer time and memory scale with system size. It is clear that in general the number of processes in a system is proportional to the size of the system and also to the length of the simulation in real time. The computational costs will therefore scale at least linear with system size and simulation length. We will therefore focus not on costs for a whole simulation, but instead on costs per process.

Looking at the VSSMc algorithm above, we see that it scales in the worse possible way with system size. In step 2, for example, we have to sum over all possible configurations. For a simple lattice with $S$ sites and each lattice point having $N$ possible labels we have a total number of configurations equal to $N^S$. This means that VSSMc scales exponentially with system size. Fortunately, it is easy to improve this. Most of the terms in the summation are zero because there is no process that changes $\alpha$ into $\beta$ and hence $W_{\beta\alpha} = 0$. So we should only use those changes that can actually occur: i.e., we should keep track of the possible processes. Processes that can actually occur at a certain location we call enabled. The total number of (enabled) processes is proportional to the system size, so we can reduce the scaling of computer time per process so that it is not worse than proportional to the system size. Actually, we can reduce the costs even further because we need not determine all enabled processes every time at steps 2 and 3. A process generally has only a local effect and does not affect processes far away. If a process occurs, this causes a

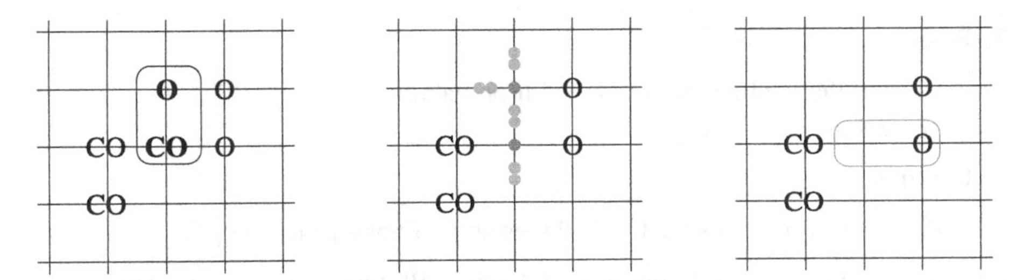


**Fig. 3.1** The *left* drawing shows part of a configuration for a model of CO oxidation. The *fat* CO and oxygen form $CO_2$ which is removed from the surface. The *middle* drawing shows newly enabled processes: CO can adsorb at the sites marked by *dark gray circles*, oxygen can adsorb dissociatively on neighboring sites indicated by two *light gray circles* on the line connecting the sites. The *right* drawing indicates a disabled process: the two *encircled* sites had a CO and an oxygen that could form a $CO_2$

local change in the configuration. This change makes new processes possible only locally, whereas other processes are not possible anymore also locally. We say that such latter processes become disabled. The number of newly enabled and disabled processes only depends on what the configuration looks like at the location where a process has just occurred, but it does not depend on the system size (see Fig. 3.1). So instead of determining all enabled processes again and again we do this only at the initialization and then update a list of all enabled processes. Such a list is sometimes called an event list. The algorithms then becomes as follows.

Variable Step Size Method: improved version (VSSMi)

1. Initialize

   Generate an initial configuration $\alpha$.
   Make a list $L_{\text{proc}}$ of all processes.
   Calculate $k_\alpha = \sum_\beta W_{\beta\alpha}$, with the sum being done only over the processes in $L_{\text{proc}}$.
   Set the time $t$ to some initial value.
   Choose conditions when to stop the simulation.

2. Time

   Generate a time interval $\Delta t$ when no process occurs

$$\Delta t = -\frac{1}{k_\alpha} \ln r, \tag{3.13}$$

   where $r$ is a uniform deviate on the unit interval.
   Change time to $t \to t + \Delta t$.

3. Process

   Choose the process $\alpha \to \alpha'$ from $L_{\text{proc}}$ with probability $W_{\alpha'\alpha}/k_\alpha$: i.e., do the process $\alpha \to \alpha'$.

4. Update

Remove the process $\alpha \to \alpha'$ from $L_{\text{proc}}$.
Add new enabled processes to $L_{\text{proc}}$ and remove disabled processes.
Use these changes to $L_{\text{proc}}$ to calculate $k_{\alpha'}$ from $k_\alpha$.

5. Continuation

If the stop conditions are fulfilled then stop. If not repeat at step 2.

The reasoning leading to VSSMi suggests that the computer time per process of this algorithm does not depend on system size. However, that is still not true. There are two problems. First, choosing the process in step 3 can in general not be done in constant time just with the list of all processes as we will see in Sect. 3.3.1. Second, adding new enabled processes to the list of processes can be done easily in constant time, but removing disabled processes presents a problem. One can scan the list of all processes and remove all disabled processes from the list, but that is an operation proportional to the system size.

## 3.3 Some General Techniques

There are a number of concepts and techniques that are useful to improve VSSMi, but that can also be used to speed up the other algorithms that we will discuss. This is because all algorithms essentially have to do the same thing. They have to determine repeatedly which process occurs, when it occurs, and where it occurs. We will introduce in this section hierarchical selection, using disabled processes, and oversampling. We will also show explicitly how to use these techniques in VSSM. Other algorithms using these techniques will be discussed in subsequent sections.

### *3.3.1 Selection Methods*

Step 3 of VSSMi (Sect. 3.2.3) is a selection. Because different processes have different probabilities of being selected it is called a weighted selection. To make this selection one has to define cumulative rate constants $C_{\alpha'\alpha} = \sum_{\beta \le \alpha'} W_{\beta\alpha}$. The processes in $L_{\text{proc}}$ need to be ordered. In principle, it does not matter what criterion is used for the ordering. A simple and convenient one is to order the processes according to the order in which the processes are generated in steps 1 and 4 of the VSSMi algorithm. The summation is over all processes $\alpha \to \beta$ preceding $\alpha \to \alpha'$ ($\beta < \alpha'$) and $\alpha \to \alpha'$ itself. The process $\alpha \to \alpha'$ can then be picked by choosing $\alpha'$ using $C_{\alpha'-1,\alpha} < rk_\alpha \le C_{\alpha'\alpha}$ where $r$ is a uniform deviate on the unit interval and $\alpha \to \alpha' - 1$ is the process before $\alpha \to \alpha'$ in the ordering of the processes. This weighted selection scales linearly with the number of processes and the system size. The reason for this is that we have to scan all the cumulative rate constants $C_{\alpha'\alpha}$.

It is often possible to choose a process in constant time. To accomplish this we split the list of all processes in groups containing processes of the same type (or

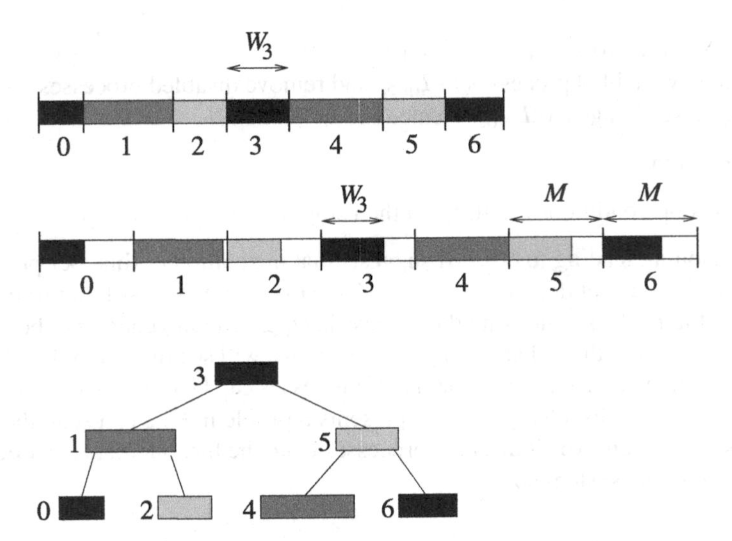


**Fig. 3.2** For a weighted selection the lengths of the bars in the *top* part of the figure have to be added. If all bars have the same length then we can simply choose a bar at random. This is uniform selection. The *bottom* part shows how selection can be done with a binary tree. You store partial sums of subtrees in the nodes, which makes traversing the tree easier. The *middle* part shows a form of selection using oversampling (see Sect. 3.3.4). All bars have a length smaller or equal to $M$. A bar is chosen randomly (uniform selection) and then accepted with probability $W_n/M$

more general with the same rate constant). Two processes are of the same type if they differ only in their position and/or orientation. So CO adsorption, NO dissociation ($NO \to N + O$) and $N_2$ associative desorption ($2N \to N_2$) are examples of process types. If $L^{(i)}_{\text{proc}}$ is the list of $N^{(i)}$ processes with rate constant $W^{(i)}$, then we proceed as follows. First, we pick a type of process $j$ with probability $N^{(j)}W^{(j)}/\sum_i N^{(i)}W^{(i)}$, and then we pick from $L^{(j)}_{\text{proc}}$ a process at random. The first part scales linearly with the number of lists, because it is a weighted selection. This number does not depend on the system size, however, but only on the number of process types or the number of rate constants. The second part is a uniform selection, which means that each process has the same probability of being chosen. This can be done in constant time. So the second part also does not depend on the system size. If the number of process types is small, and it often is, this method is very efficient (see Fig. 3.2). Selecting first the process type and then a specific process of the type is a hierarchical selection.

It is possible to do the weighted selection of the processes also in a time proportional to the logarithm of the system size by using a binary tree or priority queue as shown in the bottom part of Fig. 3.2 [15]. Each node of the tree has a process, its rate constant, and the cumulative rate constants of all processes of the node and both branches below the node. After $rk_\alpha$ has been calculated we look for the node with $C_{\text{left}} < rk_\alpha \leq C_{\text{left}} + W_{\text{node}}$ where $W_{\text{node}}$ is the rate constant of the process of the node and $C_{\text{left}}$ is the cumulative rate constant of the top node of the left branch.

If there is no left branch then we define $C_{\text{left}} = 0$. To find the node we do the following.

1. Start

   Set $X = rk_\alpha$.
   Take the top node of the tree.

2. Process found?

   if $C_{\text{left}} < X \leq C_{\text{left}} + W_{\text{node}}$
   then stop: take the process of the node
   else go to the next step

3. Continue in the left branch?

   if $X \leq C_{\text{left}}$
   then take the top node of the left branch and continue at step 2
   else go to the next step

4. Continue in the right branch.

   Set $X \rightarrow X - (C_{\text{left}} + W_{\text{node}})$.
   Take the top node of the right branch and continue at step 2.

The number of nodes we have to inspect is equal to the depth of the tree. If the tree is well-balanced with all subtrees below a level of approximately equal depth this is proportional to the logarithm of the system size. We can use this method also for a weighted selection of the process type, if the number of process types is large. In fact this occurs in DMC simulations of reactions in solutions and the method we describe here is one method that is used for these simulations [4].

Of the three selection methods described above, uniform selection is clearly the fastest. The drawback is of course that it can only rarely be used, because all probabilities must be the same. If the selection has to be made from a large number of processes, then using a binary tree is more efficient than weighted selection, because it scales better with the number of processes. However, if the number of processes is small, than a weighted selection is faster. There is more overhead in setting up a binary tree and updating the information in the nodes. This overhead determines the efficiency if the number of processes becomes too small. Hierarchical selection is a way to use the advantage of uniform selection as much as possible even when the probabilities are not the same.

Because uniform selection is so efficient, another variation might even be used if the probabilities are not the same but still similar. Suppose $M \geq W_{\beta\alpha}$ for all processes $\alpha \rightarrow \beta$. First use a uniform selection to choose a process randomly, say $\gamma \rightarrow \gamma'$. Next accept that process with probability $W_{\gamma'\gamma}/M$. If the process is not accepted, then repeat the procedure until a process has been accepted. Accepting a process with a probability less than one is a form of oversampling (see Sect. 3.3.4). This variation of uniform selection is efficient provided a process is accepted before too many uniform selections have occurred: i.e., the average of $W_{\beta\alpha}/M$ should not be too small. We will see later that in this form of selection it is not always necessary to know $W_{\beta\alpha}/M$ explicitly (see Sects. 3.3.3 and 3.4).

We will see that grouping all processes with the same rate constant in a process type is not efficient if there are many different rate constants, because each list $L_{\text{proc}}^{(i)}$ will then have one few processes and choosing a process type is little different from choosing a process directly. The following procedure by Schulze may then help [16, 17]. Define a series of ordered rate constants $M^{(i)}$ with $M^{(i)} < M^{(j)}$ if and only if $i < j$. Make lists $L^{(i)}$ with processes. Put process $\alpha \to \beta$ in list $L^{(i)}$ if $M^{(i-1)} < W_{\beta\alpha} \leq M^{(i)}$. To select a process first select a list $L^{(i)}$ with probability $N^{(i)}L^{(i)}/\sum_j N^{(j)}L^{(j)}$ with $N^{(i)}$ the number of processes in $L^{(i)}$. Next select a process $\alpha \to \beta$ from $L^{(i)}$ uniformly but accept it with probability $W_{\beta\alpha}/M^{(i)}$. The procedure is the same as the one for process types except that in the last step uniform selection has been replaced by uniform selection with oversampling. This allows us to restrict the number of list even if there are many rate constants.

### *3.3.2 Using Disabled Processes*

The problem of removing disabled processes mentioned at the end of Sect. 3.2.3 has a surprisingly simple solution, although it is a bit more difficult to see that it is actually correct. Instead of removing the disabled processes, we simply leave them in the list of all processes, but when a process has to occur we check if it is disabled. If it is, we remove it. If it is an enabled process, we treat it as usual. That this is correct can be proven as follows. Suppose that $k_{\text{en}}$ is the sum of the rate constants of all enabled processes, and we have one disabled process with rate constant $k_{\text{dis}}$. Also suppose without loss of generality that the system is at time $t = 0$. The probability distribution for the first process to occur is $k_{\text{en}} \exp(-k_{\text{en}}t)$ (see Eq. (3.8)). If we work with the list that includes the disabled process then the probability distribution for the first process occurring at time $t$ being also an enabled process is $k_{\text{en}} \exp(-(k_{\text{en}} + k_{\text{dis}})t)$, which is the probability that no process occurs until time $t$ $(= (k_{\text{en}} + k_{\text{dis}}) \exp(-(k_{\text{en}} + k_{\text{dis}})t))$ times the probability that the process is enabled $(= k_{\text{en}}/(k_{\text{en}} + k_{\text{dis}}))$. This is the first contribution to the probability distribution for the enabled process if the disabled process is not removed. The probability distribution that the first process is disabled but the second process is enabled and occurs at time $t$ is given according to Eq. (3.6) by

$$\begin{aligned}
&\int_0^t dt' \left[k_{\text{en}} e^{-k_{\text{en}}(t-t')}\right] \frac{k_{\text{dis}}}{k_{\text{en}} + k_{\text{dis}}} \left[(k_{\text{en}} + k_{\text{dis}}) e^{-(k_{\text{en}}+k_{\text{dis}})t'}\right] \\
&= k_{\text{en}} e^{-k_{\text{en}}t} k_{\text{dis}} \int_0^t dt' e^{-k_{\text{dis}}t'} \\
&= k_{\text{en}} e^{-k_{\text{en}}t} \left[1 - e^{-k_{\text{dis}}t}\right].
\end{aligned} \tag{3.14}$$

Adding this to $k_{\text{en}} \exp(-(k_{\text{en}} + k_{\text{dis}})t)$ gives us $k_{\text{en}} \exp(-k_{\text{en}}t)$, which is what we should have.

This shows that adding a single disabled process does not change the probability distribution for the time that the first enabled process occurs. In the same way we can show that adding a second disabled process gives the same probability distribution as having a single disabled process, and that adding a third disabled process is the same as having two disabled processes, et cetera. So by induction we see that disabled processes do not change the probability distribution for the occurrence of the enabled processes. Also step 3 of VSSMi in Sect. 3.2.3 is no problem. The enabled processes are chosen with a probability proportional to their rate constant whether or not disabled processes are present in he list.

The VSSM algorithm now gets the following form.

Variable Step Size Method with an approximate list of processes (VSSMa)

1. Initialize

   Generate an initial configuration $\alpha$.
   Make lists $L^{(i)}_{\text{proc}}$ containing all processes of type $i$.
   Calculate $k^{(i)} = N^{(i)}W^{(i)}$, with $N^{(i)}$ the number of processes of type $i$ and $W^{(i)}$ the rate constant of these processes.
   Set the time $t$ to some initial value.
   Choose conditions when to stop the simulation.

2. Time

   Generate a time interval $\Delta t$ when no process occurs

$$\Delta t = -\frac{1}{\sum_i k^{(i)}} \ln r, \tag{3.15}$$

   where $r$ is a uniform deviate on the unit interval.
   Change time to $t \to t + \Delta t$.

3. Process

   Pick a type of process $j$ with probability $k^{(j)}/\sum_i k^{(i)}$, and then pick the process $\beta \to \alpha'$ from $L^{(j)}_{\text{proc}}$ at random. If the process is enabled go to step 4. If it is disabled go to step 6. (Note that during the simulation $k^{(j)}$ may have obtained contributions from disabled processes starting from configurations different from the current configuration $\alpha$, which means that $\beta$ need not be equal to $\alpha$.)

4. Enabled process

   Change the configuration to $\alpha'$.

5. Enabled update

   Remove the process $\beta \to \alpha'$ from $L^{(j)}_{\text{proc}}$. Change $k^{(j)} \to k^{(j)} - W^{(j)}$. (Note that because the process is enabled $\beta = \alpha$.)
   Add new enabled processes to the lists $L^{(i)}_{\text{proc}}$.
   Use these processes to calculate the new values for $k^{(i)}$ from the old ones.
   Skip to step 8.

6. Disabled process

   Do not change the configuration: $\alpha'$ is the same configuration as $\alpha$.

7. Disabled update

   Remove the disabled process from $L_{\text{proc}}^{(j)}$.
   Change $k^{(j)} \to k^{(j)} - W^{(j)}$.

8. Continuation

   If the stop conditions are fulfilled then stop. If not repeat at step 2.

The computer time per process of algorithm VSSMa does not depend on system size. This is achieved by working with an approximate list of all processes: a list that can contain disabled processes. Note that picking the process type at step 3 can be done by weighted selected or with a binary tree as explained in Sect. 3.3.1.

Although VSSMa does not depend on system size, there still seems a price to pay. The lists $L_{\text{proc}}^{(i)}$ become polluted with disabled processes, and, when there are many of them, the algorithm will spend a lot of time in steps 6 and 7. The time will be incremented with every disabled process that is encountered, but the configuration does not change. Actually the only inefficiency is in the updating of the time in step 2. Disabled processes in VSSMa are only removed when it is unavoidable. That seems more efficient than removing them at the moment they become disabled.

Encountering a disabled process in step 3 of VSSMa is also called a null event. Algorithms that avoid null events are called rejection-free. It might seem that null events are wasteful and that rejection-free algorithms should be preferred. The problem is that a rejection-free algorithm needs to get rid of disabled processes at the moment they become disabled. This can be costly. We have already seen that scanning the list of processes is simply too expensive. Much better is to use so-called inverted lists [17–19]. When the occupation of a site changes, then these lists link that site to the disabled processes in the list of processes, thus allowing them to be removed. This works in constant time. However, such lists are complicated data structures and updating them is costly. It seems better to compute (3.15) a few times extra in the case of a null event.

A nasty problem can occur when an enabled process becomes disabled and then becomes enabled again. An example of this is when there is an adsorbate at a particular site, and the desorption from that site may then be enabled. That adsorbate may however hop to a neighboring site instead of desorbing. The desorption from the original site becomes then disabled. Later the adsorbate may hop back to the original site, after which the desorption becomes enabled again. In VSSMa the desorption is added to the list of processes every time the adsorbate is moved to the site. Consequently, there may be multiple occurrences of the same desorption in the list of processes. This poses no problem provided only one of the occurrences is enabled. The most convenient way to accomplish this is to keep track of the time when a process has been put on the list of processes, and also the time when each site has got its label. By comparing these time stamps it is possible to determine the last time that a process has been put in the list of processes. That process is the only one that should be enabled. Because of round-off errors one should not use floating point representation for the time stamps. Instead on should use integers time stamps: for example, the number of processes that have occurred since the beginning of a simulation.

### 3.3.3 Reducing Memory Requirements

We might also want to change the VSSM algorithm, not because of time, but because of memory considerations. The descriptions of the algorithms so far use a list of all processes. This list is quite large and scales with system size. We can do away with this list at the cost of increasing computer time, although the algorithm will still not depend on the system size.

Instead of keeping track of all individual processes we only keep track of how many processes there are of each different type: i.e., no lists $L_{\mathrm{proc}}^{(i)}$ but only the numbers $N^{(i)}$. Because there are no lists we have to count how many processes become enabled and disabled after a process has occurred. This is similar to adding only enabled processes and can be done in constant time. (Note that this is only because we are using no lists. Searching for disabled processes in lists and then removing them is what costs time, not counting how many processes become disabled.) This means that the number $N^{(i)}$ will be exact. The only problem is, after the type of process is determined, how to determine which particular process will occur. This can be done by randomly searching on the surface. The number of places one has to look does not depend on the system size, but on the probability that the process can occur on a randomly selected site. This random search for the location of the process is a form of uniform selection with oversampling as described in Sect. 3.3.1. Another application of such a uniform selection will be given in Sect. 3.9. If the type of process can occur on many places, then a location for a that process will be found rapidly.

To make the formulation of the new algorithm not too difficult we use here a more restrictive definition of a process type, in the sense that two processes are of the same type if one can be obtained from the other by a translation. Previously we also talked about the same process type if the orientation was different. We don't do this here, because we want to have an unambiguous meaning if we say that a process occurs at a particular site (see step 4 of VSSMs) even if more sites are involved. For example, if we have a process type of an A reacting with a B where the B is at a site to the right of the A, then by the site of this process we mean the site where the A is. The new algorithm now becomes as follows.

Variable Step Size Method with random search for the location of processes (VSSMs)

1. Initialize

   Generate an initial configuration.

Count how many processes $N^{(i)}$ of type $i$ there are.
Calculate $k^{(i)} = N^{(i)} W^{(i)}$, with $W^{(i)}$ the rate constant of the processes of type $i$.
Set the time $t$ to some initial value.
Choose conditions when to stop the simulation.

2. Process time

   Generate a time interval $\Delta t$ when no process occurs

$$\Delta t = -\frac{1}{\sum_i k^{(i)}} \ln r, \tag{3.16}$$

   where $r$ is a uniform deviate on the unit interval.
   Change time to $t \to t + \Delta t$.

3. Process type

   Choose a type of process $j$ with probability $k^{(j)} / \sum_i k^{(i)}$.

4. Process location

   Pick randomly a site for the process, until a site is found where the process can actually occur.

5. Update

   Change the configuration.
   Determine the new enabled processes, and change the $N^{(i)}$'s accordingly.
   Determine the disabled processes, and change the $N^{(i)}$'s accordingly.

6. Continuation

   If the stop conditions are fulfilled then stop. If not repeat at step 2.

### 3.3.4 Supertypes

We have mentioned oversampling already a couple of times. We talk about oversampling when we use probabilities for generating samples that are higher than the actual ones. The examples so far were about using oversampling for selection. Here we show how it can also be used for sampling the probability distribution of the times of the processes with a higher rate constant.

Suppose that the rate constant of a process is $W$. Then the probability that it occurs at time $t$ equals $W \exp(-Wt)$. If we use $W'$ with $W' > W$ instead of $W$, and then accept the process with probability $W/W'$ we get the same probability distribution. To prove this we need to add the contributions that the process has to be generated one, two, three et cetera times before one is found that is accepted.

$$\frac{W}{W'}W'\exp(W't)$$
$$+\int_0^t dt'\frac{W}{W'}W'e^{-W'(t-t')}\left[1-\frac{W}{W'}\right]W'e^{-W't'}$$
$$+\int_0^t dt'\int_0^{t'} dt''\frac{W}{W'}W'e^{-W'(t-t')}$$
$$\times\left[1-\frac{W}{W'}\right]W'e^{-W'(t'-t'')}\left[1-\frac{W}{W'}\right]W'e^{-W't''}+\dots$$
$$=We^{-W't}\left[1+\left[1-\frac{W}{W'}\right]W't+\frac{1}{2}\left[1-\frac{W}{W'}\right]^2W'^2t^2+\dots\right]$$
$$=We^{-W't}e^{[1-W/W']W't}=We^{-Wt}. \qquad (3.17)$$

A useful application of this is VSSM with supertypes [20]. Suppose that the number of process types is very large so that one does no longer want to do a weighted selection of the process type in step 3 of VSSMs (Sect. 3.3.3) or VSSMa (Sect. 3.3.2). One can of course use a binary tree, but one can also partition all process types in a small number of supertypes. We define a rate constant for each supertype as follows. Suppose $(j,k)$ is a process type $k$ of supertype $j$ with rate constant $W_{(j,k)}$. We define $W_j$ as the largest of all rate constants $W_{(j,k)}$. Instead of selecting a process type directly, we first select a supertype using for example weighted selection with a probability proportional to $W_j$, update the time, then select a process uniformly, and finally accept it with probability $W_{(j,k)}/W_j$. Note that this method can be regarded as a variation of the method by Schulze discussed in Sect. 3.3.1. Here the $W_j$'s function as the $M^{(i)}$'s in Schulze's method.

Such supertypes might be very useful when we have a system with interactions between the adsorbates that affect the rate constants. In that case the rate constants do not only depend on the process type, but also on the occupation of the sites around the sites on which the process occurs. The number of these surrounding sites may become several dozen and each may be occupied in several ways. All possible ways the surrounding sites can be occupied, and hence the number of rate constants, can be enormous. In this case we can define a supertype as the process type itself, and the corresponding rate constant as the highest possible rate constant for the process depending on the occupation of the surrounding sites.

How efficient oversampling and the use of supertypes is depends very much on the probability that processes are accepted. In the example with the lateral interactions it turns out that the interactions should not be larger than the thermal energy (see Sect. 4.5.5).

## 3.4 The Random Selection Method

The determination of a process and the time it occurs can be split in three parts: the time of the process, the type of the process, and the site of the process. The last two parts were combined in the previous versions of VSSM. The determination of the process type has to be done before the determination of the location of the processes in VSSMs, as otherwise one doesn't know when to stop searching in step 4, but the time of the process can be determined independently from which processes occurs where. It is also possible to determine all three parts independently. This has the advantage that less bookkeeping is necessary: adding and removing processes to update lists or numbers of types of processes is not necessary. However the drawback is the same as in VSSMs, only worse, as in step 4 of VSSMs processes will be attempted at certain locations where the processes cannot occur. If this does not occur too often however, then this drawback may be small.

The trick is again to use oversampling. Suppose we have just one type of process and that we have $N$ of them. (A process type is defined here in the same way as for VSSMs. Two processes are only of the same type if they can be transformed into each other by a translation.) The time to the next occurrence of a process is then given by the probability distribution $NW\exp(-NWt)$ where $W$ is the rate constant of the process. If we assume, however, that we have $M$ of these processes with $M > N$ then we can also generate the time of the next process from the distribution $MW\exp(-MWt)$, but then accept the process with probability $N/M$. The proof of this is given in Sect. 3.3.4. See Eq. (3.17) and substitute $NW$ for $W$ and $MW$ for $W'$.

The following algorithm is useful if we do not need to determine $N$ explicitly. This can be accomplished if we assume that all process types can occur everywhere on the surface. In terms of lists this means that each list $L^{(i)}_{\text{proc}}$ has the same $S$ processes during an entire simulation. Because the lists do not change and they have a simple definition, we do not need to determine them explicitly. Also the times of the processes are always taken from the same probability distribution, and the probabilities to choose a process type do not change. The algorithm looks as follows.

Random Selection Method (RSM)

1. Initialize

   Generate an initial configuration.
   Set the time $t$ to some initial value.
   Define $k = SW_{\text{sum}}$ where $W_{\text{sum}}$ is the sum of the rate constants $W^{(i)}$'s of type $i$.
   Choose conditions when to stop the simulation.

2. Process time

   Generate a time interval $\Delta t$ when no process occurs

$$\Delta t = -\frac{1}{k}\ln r, \tag{3.18}$$

   where $r$ is a uniform deviate on the unit interval.

3. Process type

   Choose a type of process randomly.

4. Process location

   Choose a site randomly.

5. Update

   Change time to $t \to t + \Delta t$.
   If the process is possible at the site from step 4, then accept the process with probability $W^{(i)}/W_{\text{sum}}$ where $i$ is the type of process from step 3.
   If the process is possible and accepted, change the configuration.

6. Continuation

   If the stop conditions are fulfilled then stop. If not repeat at step 2.

This algorithm is called the Random Selection Method (RSM). Note that the time for a process, the type of the process, and the location of the process can be done in any order. Only the time and the configuration of the system need to be updated. The method is therefore very efficient, provided that processes are accepted often in step 5. Also note that in the proof that this algorithm is correct we used the probability $N/M$, but that this probability is never used explicitly in the algorithm. Step 5 accepts processes with this probability implicitly.

## 3.5 The First Reaction Method

Instead of splitting the time, the type, and the location of a process, it is also possible to combine them. This is done in the First Reaction Method.

The First Reaction Method (FRM)

1. Initialize

   Generate an initial configuration $\alpha$.
   Set the time $t$ to some initial value.
   Make a list $L_{\text{proc}}$ containing all processes.
   Generate for each process $\alpha \to \beta$ in $L_{\text{proc}}$ a time of occurrence

$$t_{\beta\alpha} = t - \frac{1}{W_{\beta\alpha}} \ln r \tag{3.19}$$

   with $W_{\beta\alpha}$ the rate constant for the process and $r$ a uniform deviate on the unit interval.
   Choose conditions when to stop the simulation.

2. Process

   Take the process $\alpha \to \alpha'$ with $t_{\alpha'\alpha} \leq t_{\beta\alpha}$ for all $\beta$.
   If the process is enabled go to step 3. If not go to step 4.

3. Enabled update

   Change the configuration to $\alpha'$.

Change time to $t \to t_{\alpha'\alpha}$.
Remove the process $\alpha \to \alpha'$ from $L_{\text{proc}}$.
Add new enabled processes to $L_{\text{proc}}$ and generate for each new process $\alpha' \to \beta$ a time of occurrence

$$t_{\beta\alpha'} = t - \frac{1}{W_{\beta\alpha'}} \ln r. \tag{3.20}$$

Skip to step 5.

4. Disabled update

   Do not change the configuration: $\alpha'$ is the same configuration as $\alpha$.
   Do not change the time.
   Remove the disabled process from $L_{\text{proc}}$.

5. Continuation

   If the stop conditions are fulfilled then stop. If not set $\alpha$ to $\alpha'$ and repeat at step 2.

This algorithm is also known in computer science where it is called Discrete Event Simulation (DES) [21]. In FRM the determination of the type and the site of a process is replaced by comparing times of occurrences for individual processes. That this is correct can be seen as follows. Suppose we have $N$ processes with rate constants $W_1, W_2, \ldots, W_N$. The probability that no process occurs in the interval $[0, t]$ is then $\exp[-t\sum_{n=1}^{N} W_n]$, whereas the probability that neither process 1 nor process 2 nor any other process occurs in that interval equals $\prod_{n=1}^{N} \exp(-W_n t)$, which is obviously the same as the previous expression. This proves that FRM generates correct times for processes. It's a bit more work to show that the processes are chosen with the correct probability. The probability distributions for the times of the processes are $W_i \exp(-W_i t)$ with $i = 1, 2, \ldots, N$. The probability that process 1 occurs before any of the other processes in the FRM algorithm is given by

$$\int_0^\infty dt\, W_1 e^{-W_1 t} \int_t^\infty dt'\, W_2 e^{-W_2 t'} \ldots \int_t^\infty dt''\, W_N e^{-W_N t''}$$
$$= W_1 \int_0^\infty dt e^{-W_1 t} e^{-W_2 t} \ldots e^{-W_N t} = \frac{W_1}{W_1 + W_2 + \ldots + W_N}. \tag{3.21}$$

For process $i$ we find $W_i/(W_1 + W_2 + \ldots + W_N)$, which shows that FRM also chooses the processes with the correct probability. So we see that one can either generate one time for all processes and then choose one process, or generate times for all processes and then take the first that will occur. We will use this in Sect. 3.9 in another way.

The disadvantage of FRM is the determination of the process with the smallest time of occurrence. Scanning a list of all process for each new process scales linearly with system size. More efficient is to make the list of all processes an ordered one, and keep it ordered during a simulation. This can for example be done with a binary tree. Getting the next process does not depend on system size, but inserting new processes in $L_{proc}$ is proportional to the logarithm of the system size [15]. This is not as good as constant time, but it is not particularly bad either. Still VSSM is often more efficient than FRM, but VSSM cannot always be used as we will show later, whereas FRM can always be used.

Note that disabled processes are not removed from the list of all processes. Note also that we only have to generate times for the new enabled processes. Times for processes already in $L_{proc}$ need not be generated again. Suppose that at time $t = t_1$ a time has been generated for a process with rate constant $W$. The probability distribution for that time is $W\exp[-W(t-t_1)]$. Now assume that at time $t = t_2 > t_1$ the process has not occurred. We might generate a new time using the new probability distribution $W\exp[-W(t-t_2)]$. However, the ratio of the values of these probability distributions for times $t > t_2$ is $W\exp[-W(t-t_2)]/W\exp[-W(t-t_1)] = \exp[W(t_2-t_1)]$ is a constant. Hence relative probabilities for the times $t > t_2$ that the process can occur are the same for both probability distributions, and no new time needs to be generated. Another way to look at this is that the probability that the process has not occurred yet at time $t_2$ is $\exp[-W(t_2-t_1)]$. The conditional probability that the process will take place at time $t$ provided it has not occurred yet at time $t_2$ equals $W\exp[-W(t-t_1)]/\exp[-W(t_2-t_1)] = W\exp[-W(t-t_2)]$.

## 3.6 Time-Dependent Rate Constants

If the rate constants $W_{\alpha\beta}$ are themselves time dependent, then the integral form (3.6) needs to be adapted. This situation arises, for example, when dealing with Temperature-Programmed Desorption or Reactions (TPD/TPR) [22–24], and when dealing with voltammetry [25]. The definitions of the matrices $\mathbf{W}$ and $\mathbf{R}$ remain the same, but instead of a matrix $\mathbf{Q}(t)$ we get

$$\mathbf{Q}(t',t) = \exp\left[-\int_t^{t'} dt''\mathbf{R}(t'')\right]. \tag{3.22}$$

With this new $\mathbf{Q}$ matrix the integral form of the master equation becomes

$$\mathbf{P}(t) = \left[\mathbf{Q}(t,0) + \int_0^t dt'\mathbf{Q}(t,t')\mathbf{W}(t')\mathbf{Q}(t',0) + \int_0^t dt' \int_0^{t'} dt''\mathbf{Q}(t,t')\mathbf{W}(t')\mathbf{Q}(t',t'')\mathbf{W}(t'')\mathbf{Q}(t'',0) + \ldots\right] \times \mathbf{P}(0). \tag{3.23}$$

The interpretation of this equation is the same as that of Eq. (3.6). This means that it is also possible to use VSSM to solve the master equation. The relevant equation to determine the times of the processes becomes

$$\mathbf{Q}(t_n, t_{n-1}) = r, \tag{3.24}$$

where $t_{n-1}$ is the time of the last process that has occurred, and the equation should be solved for $t_n$, which is the time of the next process. If just after $t_{n-1}$ the system is in configuration $\alpha_{n-1}$, then the next process leading to configuration $\alpha_n$ should be chosen from all possible process with probability proportional to $W_{\alpha_n\alpha_{n-1}}(t_n)$. Note the argument of $W_{\alpha_n\alpha_{n-1}}$ here.

The drawback of VSSM for time-dependent rate constants is that the equation for the times of the processes is often very difficult to be solved efficiently. Equation (3.24) can in general not be solved analytically, but a numerical solution may also not be easy. The problem is that $\mathbf{R}$ in Eq. (3.22) can contain many terms or terms from processes that have a very different time dependence. A possible solution is to use VSSM for each process type separately: i.e., we solve Eq. (3.24) for each process type separately. This avoids the problem of the different time dependences. The next process is then of the type with the smallest value for $t_n$, and the first process is chosen from those of that type as in VSSM [20, 26]. This works provided the number of process types is small.

Instead of computing a time for the next process using the sum of the rate constants of all possible processes, we can also compute a time for each process. So if we're currently at time $t$ and in configuration $\alpha$, then we compute for each process $\alpha \rightarrow \beta$ a time $t_{\beta\alpha}$ using

$$\exp\left[-\int_t^{t_{\beta\alpha}} dt' W_{\beta\alpha}(t')\right] = r, \tag{3.25}$$

where $r$ is again a uniform deviate on the unit interval. The first process to occur is then the one with the smallest $t_{\beta\alpha}$. It can be shown that this time has the same probability distribution as that of VSSM just as for time-independent rate constants. This method is FRM for time-dependent reaction rate constants [22].

The equations defining the times for the processes, Eq. (3.25), are often much easier to solve than Eq. (3.24). It may seem that this is offset by the fact that the number of equations (3.25) that have to be solved is very large, but that is not really the case. Once one has computed the time of a certain process, it is never necessary to compute the time of that process again just as for the case of time-independent rate constants. In fact, the only difference concerning the number of times (3.24) and (3.25) has to be solved is due to there possibly still being processes in the list of processes at the end of a simulation. In FRM equation (3.25) has already been solved for these processes, as this is done when a process becomes enabled. This is not necessary, and in VSSM this is not done for these processes. Apart from this difference the number of times Eqs. (3.24) and (3.25) have to be solved is equal to the number of processes (enabled or disabled) that are chosen to actually change the configuration during a simulation: step 3 in VSSMa (Sect. 3.3.2) and step 2 of FRM (Sect. 3.5).

Equation (3.25) can have the interesting property that it may have no solution. The expression

$$P_{\text{not}}(t) = \exp\left[-\int_{t_{\text{now}}}^{t} dt' W_{\beta\alpha}(t')\right] \tag{3.26}$$

is the probability that the process $\alpha \rightarrow \beta$ has not occurred at time $t$ if the current time is $t_{\text{now}}$. As $W_{\beta\alpha}$ is a non-negative function of time, this probability decreases with time. It is bounded from below by zero, but it need not go to zero for $t \to \infty$, because the integral need not diverge. If it does not, then there is no solution when $r$ in Eq. (3.25) is smaller than $\lim_{t\to\infty} P_{\text{not}}(t)$. This means that there is a finite probability that the process will never occur. This is the case with some process in voltammetric experiments [25]. There is always a solution if the integral goes to infinity. This is the case when $W_{\beta\alpha}$ goes slower to zero than $1/t$, or does not go to zero at all.

How to solve Eq. (3.25) depends very much on the time dependence of the rate constants $W$. The dependence for voltammetry leads to an equation that can be solved analytically fairly easily, despite the fact that for some processes there may not be a solution. The rate constant for processes on an electrode can be written as [27, 28]

$$W = W_0 \exp\left[\frac{\alpha\gamma e E}{k_B T}\right] \tag{3.27}$$

with $\alpha$ the transfer coefficient ($0 \le \alpha \le 1$), $\gamma$ the number of electrons transferred to the electrode when the process takes place, $e$ the elementary charge, $E$ the electrode potential, and $W_0$ the rate constant at $E = 0$. In voltammetry we have $E = E_0 + \varphi t$ with $\varphi$ the sweep rate: i.e., the rate with which the electrode potential is changed. This gives us

$$\int_{t_{\text{now}}}^{t} dt' W_0 \exp\left[\frac{\alpha\gamma e(E_0 + \varphi t')}{k_B T}\right] \tag{3.28}$$

$$= \frac{W_0 k_B T}{\alpha\gamma e E_0} \exp\left[\frac{\alpha\gamma e E_0}{k_B T}\right]\left\{\exp\left[\frac{\alpha\gamma e\varphi t}{k_B T}\right] - \exp\left[\frac{\alpha\gamma e\varphi t_{\text{now}}}{k_B T}\right]\right\}. \tag{3.29}$$

We see that this is finite if $\gamma\varphi \le 0$ with a maximum of

$$-\frac{W_0 k_B T}{\alpha\gamma e E_0} \exp\left[\frac{\alpha\gamma e(E_0 + \varphi t_{\text{now}})}{k_B T}\right]. \tag{3.30}$$

We have to equate this to $-\ln r$ to find a solution for Eq. (3.25). If $-\ln r$ is larger than this maximum, there is no solution. Otherwise the solution is

$$t = \frac{k_B T}{\alpha\gamma e\varphi} \ln\left\{\exp\left[\frac{\alpha\gamma e\varphi t_{\text{now}}}{k_B T}\right] - \frac{\alpha\gamma e\varphi}{W_0 k_B T} \exp\left[-\frac{\alpha\gamma e E_0}{k_B T}\right] \ln r\right\}. \tag{3.31}$$

For TPD there always is a solution that needs to be determined numerically, but this can be done very efficiency. The rate constant can be written as

$$W = \nu \exp\left[-\frac{E_{\text{act}}}{k_B(T_0 + Bt)}\right] \tag{3.32}$$

with $E_{\text{act}}$ the activation energy, $\nu$ the prefactor, $T_0$ the temperature at time $t = 0$, and $B$ the heating rate. We now get

$$\int_{t_{\text{now}}}^{t} dt' \nu \exp\left[-\frac{E_{\text{act}}}{k_B(T_0 + Bt')}\right] = \Omega(t) - \Omega(t_{\text{now}}) \tag{3.33}$$

with

$$\Omega(t) = \frac{\nu}{B}(T_0 + Bt)\mathrm{E}_2\left[-\frac{E_{\mathrm{act}}}{k_{\mathrm{B}}(T_0 + Bt)}\right] \tag{3.34}$$

and $\mathrm{E}_2$ and exponential integral defined by [29]

$$\mathrm{E}_2(z) = \int_1^\infty dt \frac{e^{-zt}}{t^2}. \tag{3.35}$$

Because $\Omega$ and also its derivative, which is nothing but $W$, are monotonically increasing functions, we can use the Newton–Raphson method, which is not only very fast, but also guaranteed to converge in this case [29]. Subsequent approximations can be computed from

$$t_{n+1} = t_n - \frac{\Omega(t_n) - \Omega(t_{\mathrm{now}}) + \ln r}{W(t_n)} \tag{3.36}$$

whereas any initial value for this sequence (e.g., $t_0 = t_{\mathrm{now}}$) will do. For other experiments than TPD and voltammetry however an analytical or fast numerical solution may not be found.

We note that time-dependent rate constants are sometimes treated as if they are piecewise constant. This means that the total time interval of a simulation $[t_{\mathrm{begin}}, t_{\mathrm{end}}]$ is split into $n$ subintervals of equal duration, and the rate constants on each subinterval are taken equal to those at the time at the beginning of the subinterval. An obvious drawback is that this is an approximation. This may be especially bothersome, because rate constants are often monotonically increasing or decreasing functions of time, which means that one systematically under- or overestimates the rate constant. An advantage on the other hand is that all algorithms for time-independent rate constants can be used. However, as soon as a simulation leaves one subinterval and enters the next, all information on times for processes needs to be recomputed. This means effectively starting a new simulation. This may bring a substantial overhead with it. Taking larger intervals may reduce this, but only by introducing a cruder approximation.

## 3.7 A Comparison with Other Methods

We briefly discuss here a few other approaches. The fixed time step method discretized time into intervals of equal length. The algorithmic methods are older methods that are still occasionally being used and that express time, if at all, in Monte Carlo steps. The kMC method of Fichthorn and Weinberg in Sect. 3.7.3 is very similar to VSSM and is quite popular. Cellular Automata are mentioned also but they are really outside the scope of this book on kMC methods.

### 3.7.1 The Fixed Time Step Method

If we discretize time then the master equation can be written as

$$P_\alpha(t+\Delta t) = P_\alpha(t) + \Delta t \sum_\beta [W_{\alpha\beta} P_\beta(t) - W_{\beta\alpha} P_\alpha(t)]. \tag{3.37}$$

This means that if at time $t$ we are in configuration $\alpha$, then at time $t+\Delta t$ we are still in configuration $\alpha$ with probability $1-\Delta t \sum_\beta W_{\beta\alpha}$ and in configuration $\beta$ different from $\alpha$ with probability $W_{\beta\alpha}\Delta t$. This leads to the following algorithm. We generate a uniform deviate on the unit interval. If $r \geq \Delta t \sum_\beta W_{\beta\alpha}$ then we only change time to $t+\Delta t$. If $r < \Delta t \sum_\beta W_{\beta\alpha}$ then we also change the configuration to $\beta$ with a probability proportional to $W_{\beta\alpha}$.

The algorithm avoids the evaluation of integrals like (3.25). However, it is obviously an approximation, which might necessitate small time steps. As the probability that no process occurs $\Delta t \sum_\beta W_{\beta\alpha}$ is an approximation to the correct value $1-\exp[-\Delta t \sum_\beta W_{\beta\alpha}]$, we see that $\Delta t \sum_\beta W_{\beta\alpha} \ll 1$ must hold. This means that $\Delta t$ must be so small that at most one, but for most steps no, process occurs.

### 3.7.2 Algorithmic Approach

Almost all older kMC methods are based on an algorithm that defines in what way the configuration changes. (A review with many references to work with these methods is reference [30].) The generic form of that algorithm consists of two steps. The first step is to choose a site. The second step is to try all processes at that site. This may involve choosing additional neighboring sites. If a process is possible at that site, then it is executed with some probability that is characteristic for that process. These two steps are repeated many times. The sites are generally chosen at random. In a variant of this algorithm just one process is tried until on average all sites have been visited once, and then the next process is tried, et cetera. This variant is particular popular in situations with fast diffusion: the "real" reactions are tried first on average once on all sites, and then diffusion is used to equilibrate the system before the next cycle of "real" reactions.

These algorithmic kMC methods have provided very valuable insight in the way the configuration of the adsorbates on a catalyst evolves and the development of concepts that are useful in the description of that, but they have some drawbacks. First of all there is no real time. Instead time is specified in so-called Monte Carlo steps (MCS). One MCS is usually defined as the cycle in which every site has on average been visited once for each process. Real time and time in MCS's can be shown to be proportional, but the proportionality constant depends on the configuration of the adlayer [31]. This means that a conversion from one to the other is relatively easy at steady state and in the absence of fluctuations, but not in other situations like oscillatory reactions and varying reaction conditions.

The second drawback is how to choose the probabilities for processes to occur. It is clear that faster processes should have a higher probability, but it is not clear how to quantify this. This drawback is related to the first. Without a link between these probabilities and microscopic reaction rate constants it is not possible a priori to tell how many real seconds one MCS corresponds to. We have used the similarity between the algorithmic approach and RSM to compute the proportionality between real time and MCS's explicitly [31]. In practice people have used the algorithmic approach to look for qualitative changes in the behavior of the system when the reaction probabilities are varied, or they have fitted the probabilities to reproduce experimental results.

The third drawback is that it is difficult with this algorithmic definition to compare with other kinetic theories. Of course, it is possible to compare results, but an analysis of discrepancies in the results is not possible as a common ground (e.g., the master equation in our approach) is missing. The generic form of the algorithm described above resembles the algorithm of RSM. Indeed one may look upon RSM as a method in which the drawbacks of the algorithmic approach have been removed.

### *3.7.3 The Original Kinetic Monte Carlo*

The problem of real time in the algorithmic formulation of kMC has also been solved by Fichthorn and Weinberg [32]. Their method was the first to be called kinetic Monte Carlo and has become quite popular. They replaced the reaction probabilities by rate constants, and assumed that the probability distribution $P_{\text{proc}}(t)$ of the time that a process occurs is a Poisson process: i.e., it is given by

$$P_{\text{proc}}(t) = k \exp\bigl[-k(t - t_{\text{now}})\bigr], \tag{3.38}$$

where $t_{\text{now}}$ is the current time, and $k$ is the rate constant. Using the properties of this distribution they derived a method that is really identical to our VSSM, except in two aspects. One aspect is that the master equation is absent, which makes it again difficult to make a comparison with other kinetic theories. Instead the method was derived by asking under which conditions an equilibrium Monte Carlo simulation can be interpreted as a temporal evolution of a system. The other aspect is that in the original formulation time is incremented deterministically using the expectation value of the probability distribution of the first process to occur: i.e.,

$$\Delta t = \frac{1}{\sum_i N_i k_i}, \tag{3.39}$$

where $k_i$ is the rate constant of process type $i$ (this is the same as our rate constants $W$ in Eq. (2.4)), and $N_i$ is the number of processes of type $i$. This avoids having to solve Eq. (3.10), and has been used subsequently by many others. However, as solving that equation only involves generating a random number and a logarithm, which is only a small contribution to the total computer time, this is not really much of an advantage. Equation (3.39) does neglect temporal fluctuations, which may be incorrect for systems of low dimensionality [33].

Although the derivation of Fichthorn and Weinberg only holds for Poisson processes, their method has also been used to simulate TPD spectra [34]. In that work it was assumed that, when $\Delta t$ computed with Eq. (3.39) is small, the rate constants are well approximated over the interval $\Delta t$ by their values at the start of that interval. This seems plausible, but, as the rate constants increase with time in TPD, Eq. (3.39) systematically overestimates $\Delta t$, and the peaks in the simulated spectra are shifted to higher temperatures. In general, if the rate constants are time dependent then it may not even be possible to define a proper value for $\Delta t$. We have already mentioned the case of voltammetry where there is a finite probability that a process will not occur at all.

### *3.7.4 Cellular Automata*

There is an extensive literature on Cellular Automata. A discussion of this is outside the scope of this chapter, and we will restrict ourselves to some general remarks. We will also restrict ourselves to Cellular Automata in which each cell corresponds to one site. The interested reader is referred to references [35–40] for an overview of the application of Cellular Automata to surface reactions.

The main characteristic of Cellular Automata is that all cells, each of which corresponds to a lattice point in our model of the surface, are updated simultaneously. This allows for an efficient implementation on massively parallel computers. It also facilitates the simulation of pattern formation, which is much harder to simulate with some asynchronous updating scheme as in kMC [41]. The question is how realistic a simultaneous update is, as processes seem to be stochastic. One has tried to incorporate this randomness by using so-called probabilistic Cellular Automata, in which updates are done with some probability. These Cellular Automata differ little from kMC. In fact, probabilistic Cellular Automata can be made that are equivalent to the RSM algorithm [20].

## 3.8 Parallel Algorithms

There has been surprisingly little work done on parallel algorithms for kMC [20, 42–54]. There are two reasons why one might want to use parallelism. The first is that one may want to speed up simulations by using multiple processors. The second is that one might want to simulate a larger system than can be done with a single processor. We will argue that although parallelism can speed up simulations and deal with larger systems, the algorithms are not efficient.

There is actually a third reason for using parallelism. That is if one wants to do many simulations. Each one can be done on a single processor. This trivial form of parallelism is for example useful if one wants to do many simulations at different reaction conditions. Another useful application would be to do a simulation many

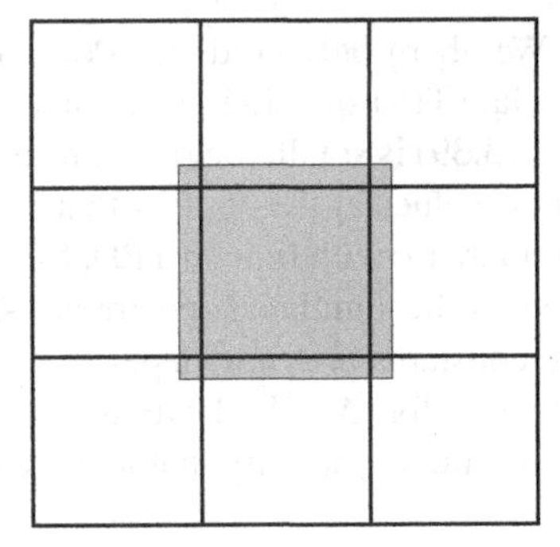
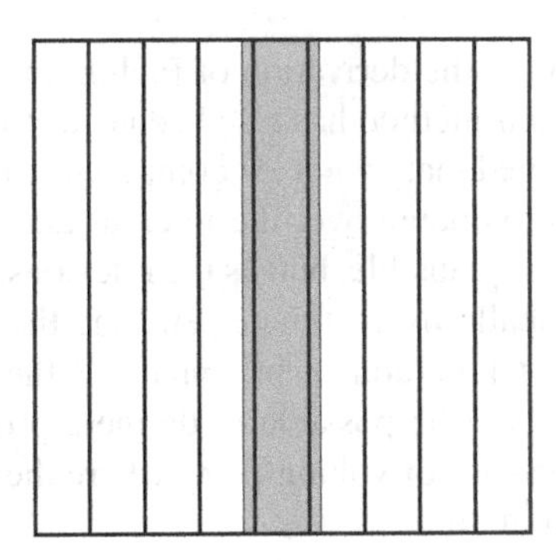

**Fig. 3.3** Two possible ways to partition a system for data parallelism. On the *left* the system is split into (nine) blocks. On the *right* the system is split into (nine) strips. Each processor handles all sites in a single chunk, with a chunk being a block or a strip. Each one has the occupation of the sites in one chunk, and it needs to get information of neighboring processors for processes that involve sites in multiple chunks. It is also possible to store the occupation of the sites at the boundaries of the chunks on different processors. Each processor has then the occupation of the sites in the areas indicated by larger block or wider strip as indicated for one processor by the *gray* areas. Communication between the processors is then needed to keep information on the occupation of the sites at the boundaries consistent

times. To see why one might want to do this one should remember that kMC is a stochastic method. The results will therefore be noisy. To reduce the noise one can either do simulations with large systems, average over long times (only possible for steady states), or average over many simulations (see Sect. 5.2). This kind of parallelism is, of course, optimal and is very easy to use.

Apart from this trivial kind of parallelism one can distinguish between control and data parallelism [55]. The former tries to do the various computational tasks of an algorithm in parallel. For example, calculating a time for the next process, choosing the type of the next process, and choosing the place on the surface where the next process will take place are independent tasks that can be done in parallel in RSM (see Sect. 3.4). Inspection of the algorithms in this chapter however shows that this form of parallelism can only be used with few processors working in parallel. The best opportunity to use parallelism in this way is if there are many types of processes that can be handled independently. One needs to realize however that not only must there be many types of processes, they also have to occur about equally often. If not, then a processor that is assigned a type of process that occurs only rarely will be idle most of the time. Systems that fulfill these requirements are rare. So this type of parallelism seems to be of limited use for kMC.

Data parallelism seems to be more promising. This kind of parallelism splits up the system and assigns different parts to different processors. Figure 3.3 shows two ways of doing this. The surface is divided into blocks or strips and each processor handles the sites and processes in one block or strip. The problem with data parallelism for kMC is that time is a global property of the system and that there may be so-called causality errors affecting the sites at the boundaries of the chunks. The chunks here are blocks or strips into which the system has been partitioned. When the processors handle the processes in each chunk independently each chunk will get a different local time during a simulation even if at the start they will all have the same time. A problem now may occur if there is a process that involves sites at two (or more) neighboring chunks. Suppose that this process has become enabled by processor A and will occur at time $t_A$. When this process then actually occurs the time of the neighboring processor B with site(s) also involved in the process may be $t_B > t_A$. Consequently, processor B has simulated his chunk between times $t_A$ and $t_B$ assuming the wrong occupation for its sites. If the correct occupation might have affected the evolution of the chunk of processor B, we have a causality error. This is an error when an event causes something to happen in the past instead of the future.

The example above shows that the processors need to communicate with each other. First they need to tell neighboring processors the occupation of the sites on the boundary, as otherwise they will not be able to determine which processes should occur involving sites at two or more chunks. There also needs to be some mechanism that synchronizes the local times of the processors. Communication between processors reduces however the efficiency of a parallel algorithm.

Because the communication is needed for the sites at the boundary, it can be minimized by reducing he fraction of all sites that are at the boundary: i.e., by making large chunks. This immediately shows that using parallelism to speed up the simulation of a small system can not be done efficiently. Too much time will be wasted on communication. Using large chunks means also fewer chunks which of course also reduces the parallelism. This means that only really large systems may benefit from parallelization.

There are two approaches to deal with the synchronization problem of the times of the different processors: the conservative and the optimistic [56]. This distinction is based on the use of safe and unsafe processes. A safe process is a process that occurs at a later time than other processes with which it shares sites and for which it can be shown that it also will not lead to processes at earlier times. Such a safe process will not lead to causality errors and a simulation can always let such a process take place. An unsafe process is a process that is not safe and may possibly lead to causality errors.

A conservative algorithm only does safe processes. Such an algorithm will never show causality errors, but it may result in deadlock. This is a situation in which there are no more safe processes and the simulation stops even though there are still enabled processes. There are mechanisms to deal with such a situation, but they require extensive communication between the processors. Even if deadlock can be avoided, conservative algorithms are not very efficient, because processors often have to wait for neighboring processors to catch up so as to avoid large differences in local times that may lead to causality errors.

Although unsafe processes may lead to causality errors, this does not need to be necessarily so. Optimistic algorithms use this. They allow unsafe processes to occur and assume that causality errors will occur only rarely or not at all. These algorithms have mechanisms to detect and correct for causality errors. If we look again at the example above where a process occurs at the boundary of two chunks A and B at time $t_A$ when the processor of chunk B is already at time $t_B > t_A$, then a so-called rollback mechanism will undo all effects of the processes that occurred between times $t_A$ and $t_B$. This means of course that the computation of these processes has been a waste of processor time. The algorithm will be very inefficient if the rollback mechanism is invoked often.

Both conservative and optimistic algorithms have been implemented for kMC simulations. Korniss et al. have used a conservative algorithm by Lubachevky in which processors wait if necessary for neighboring processors to catch up so that differences in local times are minimized [20, 42–46, 57]. They have done mainly model studies on the kinetic Ising model for which it could be shown that deadlock would not occur. Optimistic algorithms were developed by Merrick and Fichthorn and by Nandipati et al. [51, 54].

There are good reasons to think that a parallel algorithm will have trouble simulating those systems efficiently for which one really would like to use a parallel algorithm. If a system is reasonably homogeneous, then a small part of it is already representative for the whole system. A parallel algorithm is then not useful. We have already noted before that only large systems can benefit from parallelization. However, such a system needs to be heterogeneous. As a consequence the kinetics of different chunks will proceed at different rates. This means that a conservative algorithm will have processors that are idle a large part of the time because little is happening in their chunk. Optimistic algorithms will show frequent rollback because the local time changes fast in chunks were there are few processes whereas it changes slowly in chunks with many processes.

Several studies have been done that did allow causality errors to occur. The idea to use approximate algorithms is that these errors may possibly have only a small or negligible effect on the kinetics [20, 47–50, 52, 53]. In the work of Shin et al. the chunks were grouped into sublattices (see Fig. 3.4) [48–50, 52]. (Note that these sublattices are different from the ones we have defined in Chap. 2.) This grouping was done in such a way that all chunks could be done in parallel without the possibility of causality errors if the simulation restricted itself to one sublattice. A simulation then did one sublattice, then another, et cetera either in a random or predetermined order. Nedea showed that the same idea could be used with sublattices as defined in Chap. 2 [47]. The chunks in that work consisted of single sites (see Fig. 3.4).

Another idea by Shin et al. was to simulate all chunks for a preset time $T$ without worrying about causality errors [49]. They synchronized the simulations of the chunks after each period $T$. The effect of the causality error was shown to be negligible provided $T$ was smaller than the reciprocal of the largest rate constant. Nandipati et al. compared various algorithms and also extended the idea by Merrick and Fichthorn to have dynamics boundaries [51, 54]. Sites might be reassigned during a simulation to a different processor.

All of the studies mentioned above were primarily aimed at the development and implementation of the algorithms. So little experience has yet been obtained with realistic models.

| A | B | A | B | A | B |
|---|---|---|---|---|---|
| C | D | C | D | C | D |
| A | B | A | B | A | B |
| C | D | C | D | C | D |
| A | B | A | B | A | B |
| C | D | C | D | C | D |

**Fig. 3.4** Two possible uses of sublattices in parallelization. On the *left* the system is partitioned in blocks (*thick lines*) and each block is split into four regions A, B, C, and D. All sites in regions A form a sublattice, as do all sites in regions B, C, and D. Each processor handles one block. If all processor at a certain time deal with the sites of only one sublattice then there will be no causality errors. On the *right* each block stands for a site. All sites are partitioned in sublattices A, B, C, D, and E. Processes for all sites in a single sublattice can be done in parallel if they involve only one site or one site and a neighboring site. This is because all the sites of a single sublattice have no common neighboring sites as indicated by the *thick lines*

## 3.9 Practical Considerations Concerning Algorithms

There are different aspects to consider if you just want to use the algorithms above to simulate a particular reaction system, or if you want to implement them. We consider first aspects related to the implementation of the algorithms. For the implementation the efficiency of the methods described above depends very much on details of the algorithm that we have not discussed. For example, a binary tree for storing all processes can be implemented in many ways. Each has its advantages and drawbacks. Which implementation is best may depend on the processes that one wants to simulate. However, some general guidelines can and will be given here. The interested reader is referred to references [20] and [31] for a more extensive analysis.

An important point is that memory and computation time depend mainly on the data structures that are used. Except for the time steps there is actually relatively little to calculate. Calculations involve the generation of a random number to compute times for processes and to choose processes or process types and sites. The data structures that contain the processes and/or process types have a larger affect on computer time and memory however. These lists are priority queues [15], and in particular for FRM these may become quite large. A problem are the disabled processes. Removing them depends linearly on the size of the lists and is generally inefficient, and should not be done after each process. It is better to remove them only when they should occur, and it is found that they have become disabled. Alternatively, one can do garbage collection when the size of the list becomes too large [58]. There is a trade-off here. Doing garbage collection often keeps the lists small, which reduces the costs of handling them. On the other hand doing garbage collection can be costly itself, and should therefore be done as few times as possible. There will be an optimum for the frequency with which garbage collection should be done, but that will depend on the processes to be simulated. Handling the lists depends in the worst case only logarithmically on the size of them in FRM. In VSSM and RSM this can even be done in constant time.

There are a few other aspects that are important and that we haven't mentioned yet. A central step in all algorithms is the determination of what are the new processes that have become possible just after a process has occurred. There are dependencies between the processes that may be used to speed up the simulation. A small example may make this clearer. Suppose we have just adsorption of A or B onto vacant sites, and formation of AB from an A next to a B leaving two vacant sites. The formation of an AB will allow new A and B adsorptions, but no new AB formation. So it is not necessary to check if any AB formations have become enabled after an AB formation has just occurred.

Testing if a process is disabled is not trivial. It won't do to see if the occupation of the relevant sites allows the process to occur. It may be that the occupation of the sites has changed a few times but has converted back to a situation so that the process can occur again. What has happened then is that when the process became enabled for the second time it was added to the list of processes for the second time too. If the first instance of the process on the list is not recognized as disabled, then the process will occur at the first time of occurrence. This means that effectively the process has a double rate constant. This is similar to oversampling (Sect. 3.3.4) and accepting each process with probability 1. (This problem does not occur, of course, in VSSMs and in RSM.)

Recognizing that a process is disabled can be done by keeping track of when a process became enabled and when the occupation of a site last changed. If a site involved in the process was modified after the process became enabled, then the process should be regarded as being disabled. Using the times of these changes may however lead to problems because of rounding errors in the representation of real (floating point) numbers. Instead one can use integers that count processes when they become enabled. Each process is assigned then its count number, and each site is assigned the number of the process that last changed it. It a site involved in a process has a number larger than the number of the process, then the process is disabled.

From the point of view of using the algorithms to simulate a system an important aspect seems to be the scaling with system size. The important difference between FRM on the one, and VSSM (i.e., VSSMa and VSSMs) and RSM on the other hand is the dependence on the system size. Computer time per process in VSSM and RSM does not depend on the size of the system. This is because in these methods choosing a process is done using uniform selection, which does not depend on the size of the list of processes. In RSM there is not even such a list. In FRM the computer time per process depends logarithmically on the system size. Here we have to determine which of all processes will occur first. So for large systems VSSM and RSM are generally to be preferred. The data structure of FRM is so time consuming that FRM should only be used if really necessary.

There are however a number of cases that occur quite frequently in which VSSM and RSM are not efficient. This is when there are many process types and when the rate constants depend on time. Time-dependent rate constants have been discussed in Sect. 3.6. Many process types arise, for example, when there are lateral interactions. In this case VSSM becomes inefficient because it will take a lot of time to determine the process type, as with lateral interactions each different rate constant counts as a different process. If many adsorbates affect the rate constant of a process because of the lateral interactions, then the number of process types easily becomes larger than the number of sites (see Sects. 2.2.3 and 6.3). RSM can be used for lateral interactions, provided that the effect of them is small. With RSM one need only include in the process description those sites for which the occupation changes. If one also includes the sites with the adsorbates affecting the rate constants then the probability that one chooses a process type that can occur at the randomly chosen site is too low. The adsorbates affecting the rate constants should, of course, be included when one calculates the rate constant for the determination of the acceptance of a process. If the effect of lateral interactions is large then this acceptance will often be low, and RSM will not be very efficient. This is usually the case. In general, one should realize that simulations of systems with lateral interactions are always costly.

If VSSM and RSM can be used, then the choice between them depends on how many sites in the system the processes can occur. RSM is efficient for processes that occur on many sites. The probability that a process is possible on the randomly chosen location is then high. If this is not the case then VSSM should be used.

The choice between FRM, VSSM, and RSM need not be made for all processes in a system together, but can be made per process type, because it is easy to combine the different methods. Suppose that process type 1 is best treated by VSSM, but process type 2 best by RSM. We then determine the first process of type 1 using VSSM, and the first of type 2 by RSM. The first process to occur actually is then simply found by comparing the times of the processes: the first to occur is the one with the smallest value for its time. The proof that this is correct is identical to the proof of the correctness of FRM (see Eq. (3.21) and the text before that equation). Combining algorithms in this way can be particularly advantageous for models with many process types.

To summarize, VSSM is generally the best method to use unless the number of process types is very large. In that case use FRM. If you have a process that occurs almost everywhere, RSM should be considered. Simply doing the simulation with different methods and comparing is of course best. This can be done easily with a code that implements all these algorithms: e.g., Carlos [59]. One simply does a number of short exploratory simulations just to see how much computer time they take. Table 3.1 gives on impression of the variation in the efficiency of the algorithms for a selection of models. The fastest simulation (i.e., the one with the largest number of processes simulated per second) is the one of the Ziff–Gulari–Barshad model with RSM. It is however very exceptional that RSM does so well. For more complicated models it can perform extremely poorly. There is a general trend that the speed of a simulation goes down if the model becomes more complicated: i.e., when there are more processes and when more sites are involved in the processes. (See Chaps. 5, 6, and 7 to understand how it can be possible that processes can involve dozens of

**Table 3.1** Number of processes simulated per second for different models of surface reactions using Carlos version 5.1 [59]. The numbers have to be multiplied by 1000. All simulations were run on a single processor of a HP Compaq dc7800 Desktop PC running the Debian 4.0 Linux operating system

| | FRM | VSSM | RSM | FRM + RSM | VSSM + RSM | # sites |
|---|---|---|---|---|---|---|
| Ziff–Gulari–Barshad model[1] | 940 | 1900 | 3200 | – | – | 1–2 |
| TPD of CO/Rh(100)[2] | 35 | – | – | – | – | 5–8 |
| | 26[3] | 2.3[3] | – | – | – | 5–8 |
| CO electrooxidation on PtRu[4] | 570 | – | – | – | – | 1–2 |
| | 370[5] | 460[5] | – | – | – | 1–2 |
| Bisulfate/Cu(111)[6] | 100 | – | – | – | – | 2–10 |
| | 140[7] | 150[7] | – | – | – | 2–10 |
| CO + O/Pt(100)[8] | 67 | 250 | 16 | 110 | 290 | 2–10 |
| TPR of NO/Rh(111)[9] | 21 | – | – | – | – | 7–23 |
| $NH_3$/Pt(111)[10] | 16 | 17 | 0.03 | – | 14 | 1–39 |

[1]See reference [60]. The simulations were done on a square lattice of size 128 × 128 with periodic boundary conditions. The system was first brought to steady state and then simulated for 1000 units of time. The parameter of the model was set to $y = 0.5255$. [2]See reference [61] for a description of the system except that only the nearest-neighbor interaction between the CO molecules was included. The simulations were done on a square lattice of size 256 × 256 with periodic boundary conditions. The initial coverage of CO was 0.5 ML. The initial temperature was 250 K and the system was then heated by 5 K/s up to a temperature of 650 K. The interactions between the CO molecules were included by splitting all sites involved in a process in a group of sites for which the occupation changed when the process took place, and a group of sites with adsorbates that only affected the rate constant (see Sect. 6.3). [3]An alternative way to model lateral interactions in which all possible occupations of all sites involved in the processes were specified explicitly was used (see Sect. 6.3). For VSSM the rate constants were assumed to be piecewise constant over intervals of 1 s. The reduced speed of the simulation was caused by the fact that the list of processes had to be recomputed for each interval. FRM showed the same speed as VSSM when piecewise constant rate constants were used. [4]See references [62]. The simulations were done on a square lattice of size 256 × 256 with periodic boundary conditions. The initial coverage with CO was 0.99 ML. Half of the sites were on Pt, half on Ru, with both metals forming islands with thousands of sites. The temperature was 300 K, and the initial electrode potential was 0.05 V and was then increased by 0.1 V/s up to 0.35 V. [5]The rate constants were assumed to be piecewise constant over intervals of 0.02 s. [6]See reference [63]. The simulations were done on a square lattice of size 128 × 128 with periodic boundary conditions. The substrate was initial empty. The temperature was 300 K, and initial electrode potential was −0.250 V and was then increased by 0.05 V/s up to 0.375 V. [7]The rate constants were assumed to be piecewise constant over intervals of 0.05 s. [8]See reference [64]. The simulations were done on a square lattice of size 512 × 512 with periodic boundary conditions and two sites per unit cell. The substrate was initial completely covered by CO except for four sites in a row that were occupied by oxygen atoms. The temperature was 490 K. When RSM was used in combination with FRM or VSSM, oxygen adsorption, $CO_2$ formation, and all diffusion was modeled with FRM or VSSM, and the other processes with RSM. For these simulations Carlos version 3.0 was used. [9]See reference [65]. The simulations were done on a hexagonal lattice of size 66 × 66 with periodic boundary conditions. Labels where used to distinguish between top and the two types of hollow site (see Sect. 5.5.2). The initial coverage with NO was 0.75 ML. The initial temperature was 225 K and the system was then heated by 10 K/s up to a temperature of 625 K. [10]Reduction of $NH_3$ to $N_2$ and $H_2$. See references [66, 67] for a determination of the rate constants. The simulations were done on a hexagonal lattice of size 128 × 128 with periodic boundary conditions and six sites per unit cell: one top, three bridge, and two hollow sites. The substrate was initial empty. The temperature was 1000 K and the $NH_3$ pressure 1 atm. For the VSSM + RSM all processes involving five or fewer sites were simulated with RSM, the others, with at least 16 sites, with VSSM

sites.) Note also that for some systems only a number for FRM is given. In those cases the simulation code Carlos that was used only allowed FRM to be used either because of lateral interactions or because of time-dependent rate constants [59]. There were lateral interactions in the TPD model and the NO/Rh(111) model, but only a few sites were involved. A VSSM simulation could be done by simply specifying all possible occupations of the sites explicitly (also those that did not change occupation in the processes), but only affected the rate constants. This is different from the normal way lateral interactions are specified in the Carlos code (see Sect. 6.3). Note that the simulation did not become faster, even though in that case VSSM could be used. The reason for this is that finding enabled and disabled processes takes more time the more sites are included in the process specification. There were time-dependent rate constant for the TPD model, the sulphate/Cu(111) model, and the NO/Rh(111) model. For the TPD model with VSSM and sulphate/Cu(111) the rate constants were assumed to be piecewise constant. This yielded a substantial speed-up for the sulphate/Cu(111) model, but not for TPD for the reasons mentioned above. For CO + O/Pt(100) it is shown that combining methods can lead to some speed-up, especially for FRM + RSM compared to FRM and RSM separately.

## References


1. K. Binder, *Monte Carlo Methods in Statistical Physics* (Springer, Berlin, 1986)
2. D.T. Gillespie, J. Comput. Phys. **22**, 403 (1976)
3. D.T. Gillespie, J. Phys. Chem. **81**, 2340 (1977)
4. J. Honerkamp, *Stochastische Dynamische Systeme* (VCH, Weinheim, 1990)
5. Y. Cao, H. Li, L. Petzold, J. Chem. Phys. **121**, 4059 (2004)
6. Y. Cao, D.T. Gillespie, L. Petzold, J. Chem. Phys. **122**, 014116 (2005)
7. Y. Cao, D.T. Gillespie, L. Petzold, J. Comput. Phys. **206**, 395 (2005)
8. A.P.J. Jansen, J.J. Lukkien, Catal. Today **53**, 259 (1999)
9. D.T. Gillespie, J. Chem. Phys. **115**, 1716 (2001)
10. H. Resat, H.S. Wiley, D.A. Dixon, J. Phys. Chem. B **105**, 11026 (2001)
11. A. Chatterjee, D.G. Vlachos, J. Comput.-Aided Mater. Des. **14**, 253 (2007)
12. A.B. Bortz, M.H. Kalos, J.L. Lebowitz, J. Comput. Phys. **17**, 10 (1975)
13. F.C. Alcaraz, M. Droz, M. Henkel, V. Rittenberg, J. Phys. **230**, 250 (1994)
14. W. Feller, *An Introduction to Probability Theory and Its Applications* (Wiley, New York, 1970)
15. D.E. Knuth, *The Art of Computer Programming, Volume III: Sorting and Searching* (Addison-Wesley, Reading, 1973)
16. T.P. Schulze, Phys. Rev. E **65**, 036704 (2002)
17. T.P. Schulze, J. Comput. Phys. **227**, 2455 (2008)
18. E. Hansen, M. Neurock, Chem. Eng. Sci. **54**, 3411 (1999)
19. M. Stamatakis, D.G. Vlachos, J. Chem. Phys. **134**, 214115 (2011)
20. J.P.L. Segers, *Algorithms for the Simulation of Surface Processes* (Eindhoven University of Technology, Eindhoven, 1999)
21. I. Mitrani, *Simulation Techniques for Discrete Event Systems* (Cambridge University Press, Cambridge, 1982)
22. A.P.J. Jansen, Comput. Phys. Commun. **86**, 1 (1995)
23. A.P.J. Jansen, Phys. Rev. B **52**, 5400 (1995)
24. R.M. Nieminen, A.P.J. Jansen, Appl. Catal. A, Gen. **160**, 99 (1997)

25. M.T.M. Koper, J.J. Lukkien, A.P.J. Jansen, P.A.J. Hilbers, R.A. van Santen, J. Chem. Phys. **109**, 6051 (1998)
26. V. Rai, H. Pitsch, A. Novikov, Phys. Rev. E **74**, 046707 (2006)
27. C.G.M. Hermse, A.P. van Bavel, M.T.M. Koper, J.J. Lukkien, R.A. van Santen, A.P.J. Jansen, Surf. Sci. **572**, 247 (2004)
28. C.G.M. Hermse, A.P. van Bavel, M.T.M. Koper, J.J. Lukkien, R.A. van Santen, A.P.J. Jansen, Phys. Rev. B **73**, 195422 (2006)
29. W.H. Press, B.P. Flannery, S.A. Teukolsky, W.T. Vetterling, *Numerical Recipes. The Art of Scientific Computing* (Cambridge University Press, Cambridge, 1989)
30. S.J. Lombardo, A.T. Bell, Surf. Sci. Rep. **13**, 1 (1991)
31. J.J. Lukkien, J.P.L. Segers, P.A.J. Hilbers, R.J. Gelten, A.P.J. Jansen, Phys. Rev. E **58**, 2598 (1998)
32. K.A. Fichthorn, W.H. Weinberg, J. Chem. Phys. **95**, 1090 (1991)
33. V. Privman, *Nonequilibrium Statistical Mechanics in One Dimension* (Cambridge University Press, Cambridge, 1997)
34. B. Meng, W.H. Weinberg, J. Chem. Phys. **100**, 5280 (1994)
35. J. Mai, W. von Niessen, Phys. Rev. A **44**, R6165 (1991)
36. J. Mai, W. von Niessen, J. Chem. Phys. **98**, 2032 (1993)
37. R. Danielak, A. Perera, M. Moreau, M. Frankowicz, R. Kapral, Physica A **229**, 428 (1996)
38. J.P. Boon, B. Dab, R. Kapral, A. Lawniczak, Phys. Rep. **273**, 55 (1996)
39. J.R. Weimar, *Simulation with Cellular Automata* (Logos Verlag, Berlin, 1997)
40. S. Wolfram, *A New Kind of Science* (Wolfram Media, Champaign, 2002)
41. B. Drossel, Phys. Rev. Lett. **76**, 936 (1996)
42. G. Korniss, M.A. Novotny, P.A. Rikvold, J. Comput. Phys. **153**, 488 (1999)
43. G. Korniss, C.J. White, P.A. Rikvold, M.A. Novotny, Phys. Rev. E **63**, 016120 (2000)
44. G. Korniss, Z. Toroczkai, M.A. Novotny, P.A. Rikvold, Phys. Rev. Lett. **84**, 1351 (2000)
45. G. Korniss, M.A. Novotny, Z. Toroczkai, P.A. Rikvold, in *Computer Simulations in Condensed Matter Physics XIII*, ed. by D.P. Landau, S. Lewis, H.B. Schütler (Springer, Berlin, 2001), pp. 183–188
46. A. Kolakowska, M.A. Novotny, G. Korniss, Phys. Rev. E **67**, 046703 (2003)
47. S.V. Nedea, Analysis and simulations of catalytic reactions. Ph.D. thesis, Eindhoven (2003)
48. Y. Shim, J.G. Amar, Phys. Rev. B **71**, 115436 (2005)
49. Y. Shim, J.G. Amar, Phys. Rev. B **71**, 125432 (2005)
50. Y. Shim, J.G. Amar, J. Comput. Phys. **212**, 305 (2006)
51. M. Merrick, K.A. Fichthorn, Phys. Rev. E **75**, 011606 (2007)
52. F. Shi, Y. Shim, J.G. Amar, Phys. Rev. E **76**, 031607 (2007)
53. E. Martínez, J. Marian, M.H. Kalos, J.M. Perlado, J. Comput. Phys. **227**, 3804 (2008)
54. G. Nandipati, Y. Shim, J.G. Amar, A. Karim, A. Kara, T.S. Rahman, O. Trushin, J. Phys., Condens. Matter **21**, 084214 (2009)
55. M.J. Quinn, *Parallel Computing: Theory and Practice* (McGraw-Hill, New York, 1994)
56. R.M. Fujimoto, Commun. ACM **33**, 31 (1990)
57. B.D. Lubachevsky, J. Comput. Phys. **75**, 103 (1988)
58. D.E. Knuth, *The Art of Computer Programming, Volume I: Fundamental Algorithms* (Addison-Wesley, Reading, 1973)
59. Carlos is a general-purpose program, written in C by J.J. Lukkien, for simulating reactions on surfaces that can be represented by regular lattices: an implementation of the First Reaction Method, the Variable Step Size Method, and the Random Selection Method. http://www.win.tue.nl/~johanl/projects/Carlos/
60. R.M. Ziff, E. Gulari, Y. Barshad, Phys. Rev. Lett. **56**, 2553 (1986)
61. A.P.J. Jansen, Phys. Rev. B **69**, 035414 (2004)
62. M.T.M. Koper, J.J. Lukkien, A.P.J. Jansen, R.A. van Santen, J. Phys. Chem. B **103**, 5522 (1999)
63. C.G.M. Hermse, A.P. van Bavel, A.P.J. Jansen, L.A.M.M. Barbosa, P. Sautet, R.A. van Santen, J. Phys. Chem. B **108**, 11035 (2004)

64. R.J. Gelten, A.P.J. Jansen, R.A. van Santen, J.J. Lukkien, J.P.L. Segers, P.A.J. Hilbers, J. Chem. Phys. **108**, 5921 (1998)
65. C.G.M. Hermse, F. Frechard, A.P. van Bavel, J.J. Lukkien, J.W. Niemantsverdriet, R.A. van Santen, A.P.J. Jansen, J. Chem. Phys. **118**, 7081 (2003)
66. W.K. Offermans, A.P.J. Jansen, R.A. van Santen, Surf. Sci. **600**, 1714 (2006)
67. W.K. Offermans, A.P.J. Jansen, R.A. van Santen, G. Novell-Leruth, J.M. Ricart, J. Pérez-Ramirez, J. Phys. Chem. C **111**, 17551 (2007)

# Chapter 4
# How to Get Kinetic Parameters

**Abstract** This chapter shows how rate constants can either be calculated or be derived from experimental results. Calculating rate constants involves determining the initial and the transition state of a process, the energies of these states, and their partition functions. We show that the general expression for the partition functions can often be simplified when a degree of freedom is a vibration, a rotation, or a free translation. Recipes can be given for how to combine partition functions to get rate constants for processes like Langmuir–Hinshelwood and Eley–Rideal reactions, adsorption and desorption, and diffusion. The phenomenological or macroscopic equation is the essential equation to get rate constants from experiments. It is shown how to use it for simple desorption, simple and dissociative adsorption, uni- and bimolecular reactions, and diffusion. Lateral interactions can affect rate constants substantially, but because they are relatively weak, special attention needs to be given to the reliability of calculations of these interactions. Cross validation and Bayesian model selection are discussed in relation to the cluster expansion for these interactions.

## 4.1 Introductory Remarks on Kinetic Parameters

Kinetic parameters are part of the input of kinetic Monte Carlo (kMC) simulations. The determination of them is therefore strictly speaking not part of kMC, but, as they are essential for kMC simulations, it is important to have some idea of how to obtain them. It is possible to simply guess values in model studies and in studies of "what-if" scenarios. In such studies one is usually interested in how the kinetics changes as a function of these values. There is however an increasing tendency to make models more realistic, and one tries to model processes as they actually take place in the system of interest. For example, one does not just have a site for adsorbates, but one wants to use the specific site for each adsorbate that it actually prefers. One also wants to use the precise rate constants of the processes. It then won't do to just guess values. This chapter is about how to obtain good values for kinetic parameters in realistic models.

There are basically two ways to get kinetic parameters. One way is to calculate them. The other is to derive them from experimental data. One should be aware that it is often better to get kinetic parameters from experiments. It may be easier to do

A.P.J. Jansen, *An Introduction to Kinetic Monte Carlo Simulations of Surface Reactions*, Lecture Notes in Physics 856,
DOI 10.1007/978-3-642-29488-4_4, 

an experiment than a calculation, and the result may also be more reliable. One the other hand, some kinetic parameters are simply not possible to obtain from an experiment. For example, in the reduction of NO on Rh(111) one possible process is NO desorption (see Sect. 7.2), but this process only takes place when all neighboring sites of the NO are blocked either by actual occupation by another adsorbate or by strong repulsive interactions. There is an enormous number of different ways in which such blocking can occur. Each of these ways can have a different rate constant for NO desorption. In a kMC simulation these possibilities are treated as modifications of desorption of an isolated NO [1]. One therefore needs the rate constant for such isolated molecule. It is not possible however to get this from experiments, because such molecule will never desorb because it dissociates. One must therefore calculate the rate constant.

In principle there is only one kinetic parameter for each process which is its rate constant. Because one very often wants to know how this rate constant depends on temperature, one uses a so-called Arrhenius form (see Eq. (4.2)), which defines an activation energy and a prefactor. These can be given a physical interpretation and have become standard kinetic parameters as well. One can also look at the dependence of the rate constant on other reaction conditions (e.g., pressure). This can normally be treated as a dependence of the prefactor or the activation energy on these reaction conditions, which is generally quite straightforward. We will therefore not deal with this here in general, but only in relation to certain types of processes: e.g., in Sect. 4.4.3 on adsorption.

A set of kinetic parameters that does need a separate discussion is formed by lateral interactions. These interactions between adsorbates affect the rate constants, but in a very complicated way. We assume here that the lateral interactions only affect the activation energy. The main reason for this is that changes in the activation energy have a much larger effect on the rate constant. Effects of the lateral interactions on the prefactor are much smaller and can often by ignored.

Rate constants are part of a particular description of the kinetics. The rate constants of kMC are not the same as those in macroscopic rate equations, although sometimes they can be related to each other. The rate constants of kMC are the most fundamental in the sense that they refer to microscopic processes and are based on the smallest number of assumptions. Section 4.6 discusses the relation between the rate constants in kMC and those of other approaches.

## 4.2 Two Expressions for Rate Constants

This section gives the general expression for the rate constants that are used in kMC, and we discuss the relation with the Arrhenius form.

### 4.2.1 The General Expression

The rate constants in kMC simulations are those we find in the master equation derived in Chap. 2.

$$\frac{dP_\alpha}{dt} = \sum_\beta [W_{\alpha\beta} P_\beta - W_{\beta\alpha} P_\alpha]. \tag{4.1}$$

Here $t$ is time, $\alpha$ and $\beta$ are configurations of the adlayer, $P_\alpha$ and $P_\beta$ are their probabilities, and $W_{\alpha\beta}$ and $W_{\beta\alpha}$ are so-called transition probabilities per unit time that specify the rate with which the adlayer changes due to processes that can take place. We will call these $W$'s rate constants or rate coefficients, but one should be aware that they are not necessarily the same constants that one has in macroscopic rate equations. This will be discussed more fully in Sect. 4.6.

The derivation in Chap. 2 not only has given us the master equation, but also an expression for the rate constants.

$$W = \frac{k_B T}{h} \frac{Q^\ddagger}{Q} \exp\left[ -\frac{E_{\text{bar}}}{k_B T} \right] \tag{4.2}$$

with $T$ the temperature, $k_B$ the Boltzmann constant, and $h$ Planck's constant. The $Q$'s are partition functions, and $E_{\text{bar}}$ is the height of the activation barrier. This expression is formally identical to the Transition-State Theory (TST) expression for rate constants [2]. There are differences in the definition of the partition functions $Q$ and $Q^\ddagger$, but they can often be neglected.

Note that Eq. (4.2) has an extra factor compared to Eq. (2.32). This is because two exponential factors have been take out off the partition functions. The exact expressions for the partition functions in expression (4.2) are

$$Q^\ddagger = \int_{S_{\beta\alpha}} dS \int_{-\infty}^{\infty} \frac{dp_1 \dots dp_{i-1} dp_{i+1} \dots dp_D}{h^{D-1}} \exp\left[ -\frac{H - E_{\text{TS}}}{k_B T} \right], \tag{4.3}$$

$$Q = \int_{R_\alpha} d\mathbf{q} \int_{-\infty}^{\infty} \frac{d\mathbf{p}}{h^D} \exp\left[ -\frac{H - E_{\text{IS}}}{k_B T} \right], \tag{4.4}$$

where $\mathbf{q}$ stands for all coordinates, and $\mathbf{p}$ stands for all momenta, $D$ is the number of degrees of freedom, and the integration is over the region $R_\alpha$ in configuration space that corresponds to configuration $\alpha$ (see Fig. 2.9) or surface $S_{\beta\alpha}$ that separates $R_\alpha$ from $R_\beta$. $H$ is the Hamiltonian of the system, $E_{\text{IS}}$ is the minimum of the potential energy in $R_\alpha$, and $E_{\text{TS}}$ is the minimum of the potential energy on $S_{\beta\alpha}$. The subscripts of the energies stand for initial and transition state, respectively. We tacitly assume that the transition state is on $S_{\beta\alpha}$ and corresponds to the minimum of the potential energy on that surface. The partition functions are then also those of the initial (Eq. (4.4)) and transition state (Eq. (4.3)). The energies $E_{\text{IS}}$ and $E_{\text{TS}}$ are new and are absent in Eqs. (2.33) and (2.34). We have

$$E_{\text{bar}} = E_{\text{TS}} - E_{\text{IS}} \tag{4.5}$$

with $E_{\text{bar}}$ the activation barrier in Eq. (4.2). The advantage of writing the rate constants in this way is that the potential energy has a minimum equal to zero in $R_\alpha$ and

$S_{\beta\alpha}$ in the expressions for the partition functions. This simplifies the calculation of the partition functions. It also is convenient for the introduction of the activation energy.

### 4.2.2 The Arrhenius Form

Often an activation energy plus prefactor (or pre-exponential factor) is given instead of the rate constants. The relation with the rate constants is given by

$$W = \nu \exp\left[-\frac{E_{\text{act}}}{k_B T}\right] \tag{4.6}$$

with $W$ the rate constant, $E_{\text{act}}$ the activation energy, and $\nu$ the prefactor.

A superficial comparison of Eqs. (4.2) and (4.6) may suggest that $E_{\text{act}} = E_{\text{bar}}$ and that $\nu$ is given by the factors before the exponential in (4.2), but that is generally not correct. The partition functions in (4.2) often hide exponential factors that contribute to $E_{\text{act}}$. Moreover, the Arrhenius form is really only useful if we can assume that the prefactor and activation energy do not depend on temperature, because then it yields a simple dependence of the rate constant on temperature.

The Arrhenius form (4.6) with temperature-independent activation energy and prefactor is an approximation, but almost always a very good one. If we set the partition functions equal to 1, which is often reasonable as we will see in Sect. 4.3.4, then we get Fig. 4.1. If the Arrhenius form were exact, the figure should show a straight line. There is some deviation, but it is very small. Moreover, the temperature range in the figure is much larger than the range that one normally has to deal with, and the temperature is also generally lower than the temperature where there is a clear deviation.

There are two ways to get values for the activation energy. The first, and best, is to use Eq. (4.2). Calculate the rate constant for the temperature range that one is interested in, and do linear regression to fit of $\ln W$ as a function of $1/T$ [3]. We have

$$\ln W = \ln \nu - \frac{E_{\text{act}}}{k_B} \frac{1}{T}, \tag{4.7}$$

so the constant of this fit equals $\ln \nu$, and the coefficient of the linear term equals $-E_{\text{act}}/k_B$.

The second way is less accurate, but also simpler. Start by determining the factors in the partition functions that have the same exponential dependence as the activation barrier. (See Sect. 4.3.2 for details.) Suppose $Q^{\ddagger}$ has a factor $\exp(-\varepsilon_{\text{TS}}/k_B T)$, and $Q$ has a factor $\exp(-\varepsilon_{\text{IS}}/k_B T)$. Set

$$E_{\text{act}} = E_{\text{bar}} + \varepsilon_{\text{TS}} - \varepsilon_{\text{IS}}. \tag{4.8}$$

The contributions $\varepsilon_{\text{TS}}$ and $\varepsilon_{\text{IS}}$ are usually so-called zero-point energies. Next calculate the rate constant at the temperature that you think is most important for the

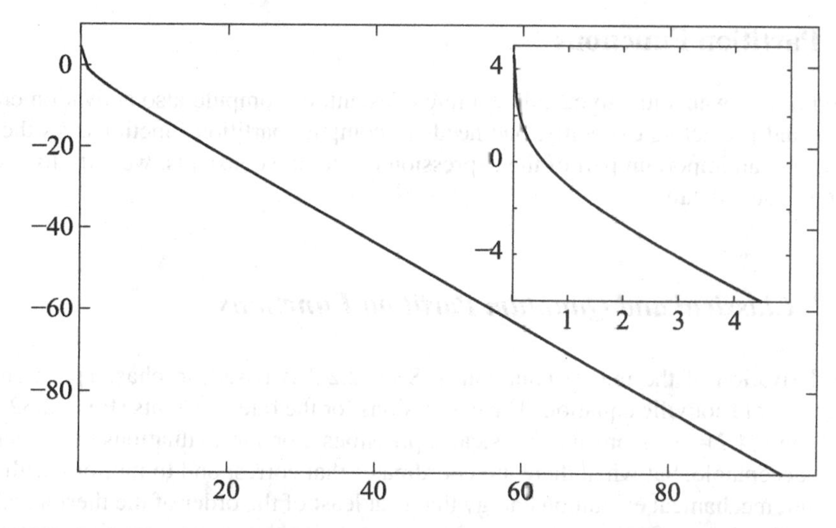


**Fig. 4.1** The logarithm of the rate constant $\ln(Wh/E_{\mathrm{bar}})$ according to Eq. (4.2) with $Q^{\ddagger} = Q = 1$ plotted versus reciprocal temperature $E_{\mathrm{bar}}/k_{\mathrm{B}}T$. Although the expression is not linear, the deviation of linearity is only very small, and only visible at high temperatures (i.e., small values of $E_{\mathrm{bar}}/k_{\mathrm{B}}T$). A very similar plot is obtained even if the partition functions can not be approximated by $Q^{\ddagger} = Q = 1$

processes that you are interested in. If we call the calculated rate constant $W_{\mathrm{calc}}$, then take $\nu = W_{\mathrm{calc}} \exp(E_{\mathrm{act}}/k_{\mathrm{B}}T)$ as prefactor.

Sometimes the activation energy is defined by

$$E_{\mathrm{act}} = -\frac{\partial \ln W}{\partial(1/k_{\mathrm{B}}T)} \tag{4.9}$$

with $W$ given by Eq. (4.2) or a similar expression. This gives Eq. (4.8) for the activation energy plus some additional term. One should realize that this expression and the procedures described above are all just definitions of the activation energy. Expression (4.8) has a simple physical interpretation as will be shown in Sect. 4.3, but it leads to an Arrhenius form that is valid over only a limited temperature range. The activation energy derived from the linear regression is valid over a much wider temperature range, but it has no good physical interpretation. Expression (4.9) seems to be the least useful. It has no good physical interpretation, and it does not yield a simple temperature dependence compared to the Arrhenius form. One may argue that it is more accurate, but if that is important then one can just is well use the TST expression (4.2). Note also that working with activation energy and prefactor is only useful when one wants to do simulations at different temperatures. Otherwise, it is easier to specify the rate constant directly.

## 4.3 Partition Functions

Whether one wants just to calculate a rate constant, or compute also activation energies and prefactors explicitly, one needs to compute partition functions. As they form such an important part of the expression of the rate constants, we will discuss them in some detail.

### *4.3.1 Classical and Quantum Partition Functions*

The derivation of the master equation in Sect. 2.2.2 is based on phase space and the classical Liouville equation. The expressions for the rate constants (Eqs. (2.32)–(2.34) and (4.2)–(4.4)) are also classical expressions. For some situations this is perfectly acceptable, but when there are coordinates that correspond to motions with a quantum mechanical excitation energy that is at least of the order of the thermal energy $k_B T$, it is not. Of course, one might try to rederive the master equation starting from quantum mechanics, but that is very difficult and seldom necessary. In many cases it suffices to replace expressions (4.3) and (4.4) by their quantum mechanical analogue. Indeed, in the following we will discuss first the quantum mechanical form of a partition function and only then the classical one. One aspect that can be important and that is still absent even if one uses quantum mechanical partition functions is the effect of tunneling. To deal with this one can use a tunneling correction for the rate constants [4–6].

### *4.3.2 Zero-Point Energy*

The main reason for the difference between the activation barrier $E_{bar}$ and the activation energy $E_{act}$ is that the partition functions in Eq. (4.2) may contribute to the exponential factor. This is especially true if there are motions that, when treated quantum mechanically, have an excitation energy that is of the same order of magnitude or larger then $k_B T$. Such motions should not be treated classically, and the partition functions (4.3) and (4.4) should be replaced by quantum mechanical ones. A quantum mechanical partition function can be written as

$$Q = \sum_{n=0}^{\infty} \exp\left[-\frac{E_n}{k_B T}\right] \tag{4.10}$$

with the summation over all eigenstates of the Hamiltonian of the system (i.e., the solutions of the time-independent Schrödinger equation) and $E_n$ the energy of state $n$ [7]. If the we define state $n = 0$ as the ground state, then we can rewrite this also as

$$Q = \exp\left[-\frac{E_0}{k_B T}\right] \sum_{n=0}^{\infty} \exp\left[-\frac{E_n - E_0}{k_B T}\right]. \tag{4.11}$$

We can combine the first factor on the right-hand-side with the exponential term $\exp(-E_{\text{bar}}/k_{\text{B}}T)$ in Eq. (4.2). The summation in Eq. (4.11) will not have such an exponential dependence. The first term in the summation is always equal to 1. If the excitation energies $E_n - E_0$ for $n \geq 1$ are large compared to $k_{\text{B}}T$, then the other terms are very small and it may be appropriate to neglect them. The summation then is simply equal to 1. This often happens for vibrations (see Sect. 4.3.4).

The definitions of the partition functions (Eqs. (4.3) and (4.4)) show that the energies of the states should be taken with respect to a minimum of the potential energy. The difference between $E_0$ and this minimum is called the zero-point energy [8]. Equation (4.2) is often written as

$$W = \frac{k_{\text{B}}T}{h}\frac{\tilde{Q}^{\ddagger}}{\tilde{Q}}\exp\left[-\frac{E_{\text{bar}} + E_{\text{zp}}}{k_{\text{B}}T}\right]. \tag{4.12}$$

$E_{\text{zp}}$ contains all the zero-point energy contributions from the partition functions. We have

$$E_{\text{zp}} = \varepsilon_{\text{TS}} - \varepsilon_{\text{IS}} \tag{4.13}$$

with $\varepsilon_{\text{TS}}$ the zero-point energy extracted from $Q^{\ddagger}$ and $\varepsilon_{\text{IS}}$ that from $Q$ (see also Eq. (4.8)). The quantity $E_{\text{bar}} + E_{\text{zp}}$ is called the zero-point energy corrected activation barrier. The partition function $\tilde{Q}^{\ddagger}$ and $\tilde{Q}$ are defined by the summation in Eq. (4.11): i.e., the reference level of the energy is chosen so that the ground state has by definition an energy equal to zero. It is very important for the calculation of the partition function to be aware if one is using Eq. (4.2) or Eq. (4.12).[1]

### 4.3.3 Types of Partition Function

General expressions for partition functions are not very useful. They have to be simplified, and for this we need to make some approximations. It depends on the type of motion which approximation is appropriate. The type of motion also determines if we need to work with the quantum mechanical expression given by Eq. (4.11), or if we may use the classical expression Eqs. (4.3) and (4.4). In principle, this depends on the ratio between the excitation energy for a certain motion and the thermal energy $k_{\text{B}}T$. At high temperatures the difference between the quantum mechanical and the classical expressions becomes negligible.[2] This will turn out to be convenient,

[1] If one determines the activation energy $E_{\text{act}}$ by a linear regression of a set of rate constants at different temperatures, then this energy is not exactly equal to $E_{\text{bar}} + E_{\text{zp}}$. This is because the factors $(k_{\text{B}}T/h)(\tilde{Q}^{\ddagger}/\tilde{Q})$ also give a small contribution to $E_{\text{act}}$. See the discussion at the end of Sect. 4.2.2.

[2] Strictly speaking this is not always true. There may be a factor in the quantum version that is absent in the classical one, and that is related to the statistics of identical particles. See for example the geometry factor in Eqs. (4.22) and (4.23).

because for some motions the quantum mechanical expression is much harder to evaluate than the classical one.

We will split the degrees of freedom of a system in several groups. These groups allow us to write a partition function as a product.

$$Q = Q_{\text{trans}} Q_{\text{rot}} Q_{\text{vib}} \tag{4.14}$$

with $Q_{\text{trans}}$ the partition function of the translations, $Q_{\text{rot}}$ the partition function of the rotations, and $Q_{\text{vib}}$ the partition function of the vibrations.[3] These partition functions can often be split further. For example, for the vibrations one can often use the harmonic approximation (see Eq. (4.16)). The vibrations can then be decoupled into so-called normal modes, [9], and we can write

$$Q_{\text{vib}} = \prod_n Q_{\text{vib}}^{(n)} \tag{4.15}$$

where the product is over all normal modes with $Q_{\text{vib}}^{(n)}$ the partition function of normal mode $n$ (see Eq. (4.18) Sect. 4.3.4). Also $Q_{\text{trans}}$ and $Q_{\text{rot}}$ can often by decoupled at least partially.

### 4.3.4 Vibrations

Most degrees of freedom of the molecules involved in a reaction are vibrations. The potential energy of such a degree of freedom can be approximated by

$$V(x) = \frac{1}{2} m\omega^2 x^2 \tag{4.16}$$

with $x$ the coordinate of the degree of freedom, $m$ the corresponding mass, and $\omega$ some constant, which turns out to be equal to the (angular) frequency of the vibration. The minimum energy of the potential here has been set equal to 0 and $x$ is defined in such a way that $x = 0$ corresponds to the position of the minimum of the potential. The energies of the states are then given by [8]

$$E_n = \left(n + \frac{1}{2}\right)\hbar\omega \tag{4.17}$$

with $n$ a non-negative integer.

---

[3]There are actually two more factors [7]. There is the partition function of the electronic states $Q_{\text{el}}$ and the partition function of the spins of the nuclei $Q_{\text{nucl}}$. We will ignore both. The electronic ground state defines the potential energy of the system, so the partition function is defined only by the summation in Eq. (4.11). As the electronic excitation energies are, except for rare case, much larger than the thermal energy, we get $Q_{\text{el}} = 1$. The spin state of the nuclei generally does not change during a reaction. So its partition functions for the transition and initial state cancel in the expression for the rate constant, and can therefore be ignored.

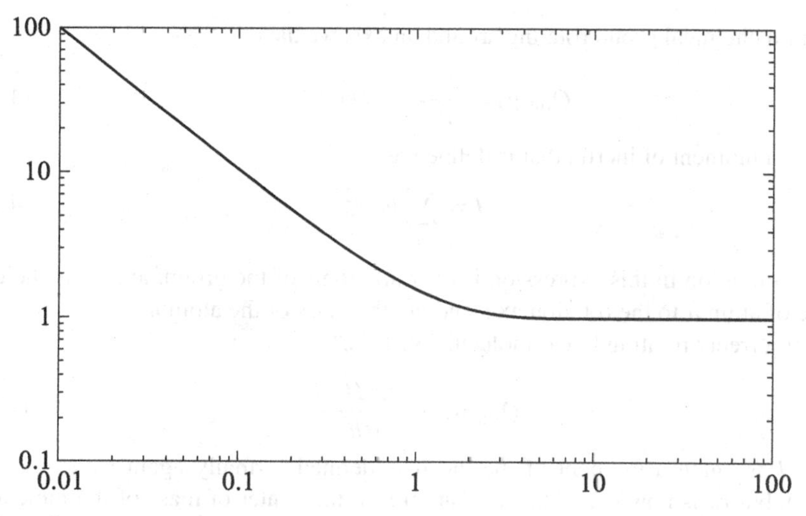


**Fig. 4.2** $\tilde{Q}_{\text{vib}}$ of Eq. (4.19) as a function of $\hbar\omega/k_B T$

The minimum of the potential energy is 0, so the zero-point energy is $\hbar\omega/2$. The summation of Eq. (4.11) can be done analytically and the result for the partition function is

$$Q_{\text{vib}} = \tilde{Q}_{\text{vib}} e^{-\hbar\omega/2k_B T} \tag{4.18}$$

with

$$\tilde{Q}_{\text{vib}} = \frac{1}{1 - e^{-\hbar\omega/k_B T}}. \tag{4.19}$$

The reason to split off $\tilde{Q}_{\text{vib}}$ is that $\exp[-\hbar\omega/2k_B T]$ is a zero-point energy factor that can be combined with $\exp(-E_{\text{bar}}/k_B T)$. Moreover, $\tilde{Q}_{\text{vib}}$ can be approximated by 1 if $\hbar\omega \gg k_B T$, which is quite common. If on the other hand $\hbar\omega \ll k_B T$ we have $\tilde{Q}_{\text{vib}} = k_B T/\hbar\omega$ (see Fig. 4.2).

### *4.3.5 Rotations*

There are three different partition functions for rotation that are relevant. If we have a chemical group that rotates around some axis (e.g., $CH_3$), then we have a one-dimensional rotation. If we have a linear molecule in the gas phase (relevant for adsorption and desorption), then we have two rotational degrees of freedom. If we have a non-linear molecule in the gas phase, then we have even three rotational degrees of freedom. There are no closed-form expressions for the quantum mechanical partition functions, but the rotational excitations are small compared to the thermal energy for almost all molecules and temperatures. So we will give the high-temperature limit of the quantum mechanical expressions [7]. These are equal to the classical expressions apart from geometric factors.

For a chemical group rotating around an axis we have

$$Q_{\mathrm{rot,1D}} = \frac{2\pi}{h}\sqrt{2\pi I k_{\mathrm{B}} T} \tag{4.20}$$

with $I$ a moment of inertia that is defined as

$$I = \sum_n m_n r_n^2. \tag{4.21}$$

The summation in this expression is over all atoms of the group, and $r_n$ is the distance of atom $n$ to the rotation axis and $m_n$ the mass of the atom $n$.

For a freely rotating linear molecule we have[4]

$$Q_{\mathrm{rot,2D}} = \frac{8\pi^2 I k_{\mathrm{B}} T}{\sigma h^2}. \tag{4.22}$$

Here $I$ is again a moment of inertia. It is defined formally again by expression (4.21), but $r_n$ is now the distance of atom $n$ to the center of mass of the molecule. The symmetry number $\sigma$ is 2 for molecules with an inversion center and 1 otherwise.

For a freely rotating non-linear molecule we have

$$Q_{\mathrm{rot,3D}} = \frac{\sqrt{\pi}}{\sigma} \prod_n \sqrt{\frac{8\pi^2 I_n k_{\mathrm{B}} T}{h^2}}. \tag{4.23}$$

The summation is over the principle moments of inertia. These are the eigenvalues of the matrix with the moments of inertia. This matrix has diagonal components

$$I_{\alpha\alpha} = \sum_n m_n \left[ |\mathbf{r}_n|^2 - r_{n\alpha}^2 \right] \tag{4.24}$$

and non-diagonal components

$$I_{\alpha\beta} = \sum_n m_n r_{n\alpha} r_{n\beta}. \tag{4.25}$$

The summation is over all atoms in the molecule, $\mathbf{r}_n$ is the vector from the center of mass of the molecule to atom $n$, and $\alpha, \beta = x, y, z$. The symmetry number $\sigma$ is the number of proper rotations in the point group of the molecule.

One can define so-called rotational temperatures $\theta_{\mathrm{rot}}$ for a molecule via

$$\theta_{\mathrm{rot}} = \frac{h^2}{8\pi^2 I k_{\mathrm{B}}} \tag{4.26}$$

with $I$ a principle moment of inertia. With this definition we can write

$$Q_{\mathrm{rot,2D}} = \frac{1}{\sigma}\frac{T}{\theta_{\mathrm{rot}}} \tag{4.27}$$

[4]Note that below room temperature this expression should not be used for molecular hydrogen. The rotational excitations for this molecule are so high, that the high-temperature approximation only becomes valid at higher temperatures.

**Table 4.1** Rotational temperatures and geometry factors for some small molecules. There is only one temperature for linear molecules. (Note that these temperature are not necessarily very accurate. They were calculated assuming a rigid structure of the molecule based on DFT/B3LYP calculations with a 6-31G+(d) basis set [10])

| Molecule | $\theta_{\text{rot,A}}/\text{K}$ | $\theta_{\text{rot,B}}/\text{K}$ | $\theta_{\text{rot,C}}/\text{K}$ | $\theta_{\text{rot,A}}\theta_{\text{rot,B}}\theta_{\text{rot,C}}/\text{K}^3$ | $\sigma$ |
|---|---|---|---|---|---|
| $H_2$ | 87.9 | | | | 2 |
| $CH_4$ | 7.60 | 7.60 | 7.60 | 438.2 | 12 |
| $NH_3$ | 14.4 | 14.4 | 8.97 | 1856 | 3 |
| $H_2O$ | 39.7 | 20.4 | 13.5 | 1089 | 2 |
| CO | 2.73 | | | | 1 |
| $CO_2$ | 0.554 | | | | 2 |
| $N_2$ | 2.84 | | | | 2 |
| NO | 2.42 | | | | 1 |
| $NO_2$ | 11.3 | 0.619 | 0.587 | 4.112 | 1 |

and

$$Q_{\text{rot,3D}} = \frac{\sqrt{\pi}}{\sigma}\left(\frac{T^3}{\theta_{\text{rot,A}}\theta_{\text{rot,B}}\theta_{\text{rot,C}}}\right)^{1/2} \tag{4.28}$$

with three different rotational temperatures for a non-linear molecule. Table 4.1 shows values for some small molecules. The advantage of these temperatures is that they simplify the calculations of rotational partition functions.

### 4.3.6 Hindered Rotations

The rotation of a chemical group around an axis is often not completely free, because of steric repulsion or other interactions. For such hindered rotation it is not possible to give a closed-form expression of the partition function. On the other hand, because it is only a one-dimensional problem, it is relatively easy to compute the partition function numerically.

The Schrödinger equation for the rotation is given by

$$H\Psi_n(\varphi) = \left[-\frac{\hbar^2}{2I}\frac{d^2}{d\varphi^2} + V(\varphi)\right]\Psi_n(\varphi) = E_n\Psi_n(\varphi) \tag{4.29}$$

with $H$ the Hamiltonian, $\varphi$ the rotational angle, $I$ the moment of inertia given by Eq. (4.21), and $V(\varphi)$ the potential energy that describes the barriers that prevent the free rotation. The subscript $n$ is a non-negative integer that labels the various solutions. The function $V$ might be determined by doing quantum chemical calculations of the system for various values of $\varphi$, and then fitting the results to some functional form: e.g., a Fourier series [11].

To solve this Schrödinger equation we expand the wave functions in a basis and then use the variation principle [12]. It seems obvious to use the geometric functions sin and cos as a basis, or complex exponential function. Suppose we take a basis $e^{-iM\varphi}/\sqrt{2\pi}, e^{-i(M-1)\varphi}/\sqrt{2\pi}, \ldots, e^{-i\varphi}/\sqrt{2\pi}, 1/\sqrt{2\pi}, e^{i\varphi}/\sqrt{2\pi}, \ldots, e^{i(M-1)\varphi}/\sqrt{2\pi}, e^{iM\varphi}/\sqrt{2\pi}$, write

$$\Psi_n(\varphi) = \sum_{m=-M}^{M} \frac{e^{im\varphi}}{\sqrt{2\pi}} c_{mn} \tag{4.30}$$

and use the variation principle. We then get

$$\mathbf{HC} = \mathbf{CE} \tag{4.31}$$

with the matrices given by the following definitions of the matrix elements.

$$\mathbf{H}_{mn} = \frac{1}{2\pi} \int_0^{2\pi} d\varphi e^{-im\varphi} H e^{in\varphi}, \tag{4.32}$$

$C_{mn} = c_{mn}$, and $E_{mn} = E_n \delta_{mn}$. So we get an eigenvalue equation with the energies in the diagonal matrix $\mathbf{E}$ and the coefficients of the wave functions in the columns of the matrix $\mathbf{C}$.

Once the energies have been determined, the partition function then follows immediately from Eq. (4.11). Note that the accuracy depends on the size of the basis: i.e., on the value of $M$. One should increase $M$ until the value of $Q$ does not change anymore.

### 4.3.7 Translations

Free translations are rare for surface processes. They are only found when the corrugation of the surface is very small, and an adsorbate can move almost unhindered in some direction. This can occur either for the initial state of a process or the transition state.

The classical partition function is appropriate and quite easy to compute. Using Eq. (4.3) for a one-dimensional translation gives

$$Q_{\text{trans}} = \int_0^L dx \int_{-\infty}^{\infty} \frac{dp}{h} \exp\left[-\frac{p^2}{2mk_BT}\right] = \frac{L\sqrt{2\pi mk_BT}}{h}. \tag{4.33}$$

Note that $L$ is the length of the interval over which the translation is free. In simulations it is generally related to the size of an adsorption site (see Sect. 4.4.2). The exception is adsorption (see Sect. 4.4.3).

### 4.3.8 Floppy Molecules

A floppy molecule is a molecule with a potential-energy surface (PES) that has many minima that are separated by barriers that can easily overcome by thermal excitation or tunneling. Partition functions for floppy molecules can be quite hard to compute, and we know of no simple expressions or methods that one can use. If the floppy part of a molecule is not involved in the reaction, then one may assume that the motions of that part are the same in the initial and transition state so that their contribution to the partition functions in Eq. (4.2) cancel. If one really needs to compute the partition function of a floppy molecule, then free-energy techniques from Molecular Dynamics (MD) are probably the most appropriate [13], as the partition function and the free energy $F$ are related via [7]

$$F = -k_B T \ln Q. \tag{4.34}$$

There are also so-called rare event techniques in MD to compute rate constants directly for floppy molecules [13].

## 4.4 The Practice of Calculating Rate Constants

The first thing to do before one starts doing any actual calculations, certainly if they are electronic structure calculations, is to determine which type of process one want to calculate the rate constant of. For some processes it may not even be necessary to do any calculations.

The general approach is as follows. First, determine the initial state of a process. This is a stable configuration corresponding to a minimum of the PES. Compute the energy and the partition function (Eq. (4.3)) for that minimum. Next determine the transition state for the process. Again compute the energy and the partition function (Eq. (4.4) in this case). The difference of the energies is $E_{bar}$, and the rate constant can be computed using Eq. (4.2). The details on how to actually do these calculations is outside the scope of this monograph. Most of the calculations that have actually been done have been electronic structure calculations about which various excellent textbooks have already been written [14–17]. There have been also a few calculations based on PES's [18–20].

### *4.4.1 Langmuir–Hinshelwood Reactions*

The general approach just described is exactly what one has to do for reactions for which reactants and products remain on the surface. The hardest part is in general the determination of the transition state. Some methods (e.g., the dimer method [21]) only need the initial state as a starting point. Others (e.g., the nudged-elastic band method [22, 23]) also need the final state. In the latter case one only needs to calculate extra the partition function of the final state and then one can calculate the rate constant of the reverse reaction too. The transition state of the reverse reaction and its partition function is of course the same as that of the forward reaction.

The energies of the initial state $E_{IS}$ and the transition state $E_{TS}$ give the activation barrier $E_{bar} = E_{TS} - E_{IS}$. To compute the partition functions one should do calculations of the vibrational frequencies of the initial and the transition state. Before doing such calculations one should determine which degrees of freedom to include in the calculation. Vibrations that do not change in a reaction give partition functions that are the same for the initial and the transition state. They cancel in the expression of the rate constant, and need not be included. It is not always clear which degrees of freedom can be left out however. Including too many never gives incorrect results, but is less efficient. Leaving out important degrees of freedom may give less accurate results, although this is generally much less serious than an error in the determination of the activation barrier.

After calculating the vibrational frequencies, one should inspect the corresponding vibrations. In particular, one should check if all motions actually correspond to vibrations. If that is the case, then the partition functions of the initial and transition states are simply products of vibrational partition functions. To be precise, add the zero-point energies of the vibrations of the transition state to the activation barrier $E_{\text{bar}}$ and subtract those of the initial state. The partition functions are products of $\tilde{Q}_{\text{vib}}$'s given by Eq. (4.19). Note that the transition state should have one and only one vibration with an imaginary frequency ($\omega^2 < 0$). Do not include the zero-point energy and the partition function of this vibration in the calculation of the rate constant.

Very often $(k_B T/h)(\tilde{Q}^\ddagger/\tilde{Q}) \approx 10^{13}\ \text{s}^{-1}$. This is because there are two limiting cases that accidentally give results of the same order of magnitude. The vibrational excitation energies are either of the same order of magnitude as the thermal energy, or they are much larger. In the former case partition functions become approximately $k_B T/\hbar\omega$ and $(k_B T/h)(\tilde{Q}^\ddagger/\tilde{Q})$ becomes the product of vibrational frequencies of the initial state divided by the product of the vibrational frequencies of the transition state. The result is generally in the order of $10^{13}\ \text{s}^{-1}$. In the latter case partition function become approximately equal to 1. Now $\tilde{Q}^\ddagger/\tilde{Q} \approx 1$ and $k_B T/h$ is also about $10^{13}\ \text{s}^{-1}$. If one does not want to calculate partition functions, then $10^{13}\ \text{s}^{-1}$ is a good value to take for the prefactor.

Really different prefactors are obtained if there is a degree of freedom that is not a vibration. This can be a rotation (e.g., of a methyl group or a small molecule as a whole rotating around an axis perpendicular to the surface). If the barrier for rotation is large (i.e., much larger than the thermal energy $k_B T$), then such a degree of freedom can be treated as a vibration. If not the vibrational partition function should be replaced by expression (4.20) or be treated as a hindered rotation.

### *4.4.2 Desorption*

The reason why desorption needs to be treated differently from Langmuir–Hinshelwood reactions is that often the transition state is equal to the final state: i.e., it is a molecule in the gas phase. If that is not the case, and the transition state is a structure with the molecule still adsorbed on the surface, then the calculation of the rate constant for desorption can be done in exactly the same way as for a

Langmuir–Hinshelwood reactions.[5] We therefore deal here only with the situation in which the transition state is a molecule in the gas phase.

The activation barrier $E_{\mathrm{bar}}$ in that case equals the adsorption energy. The partition functions and the zero-point energy of the initial state can be calculated in exactly the same way as for Langmuir–Hinshelwood reactions. For the molecule in the gas phase this also holds for the internal vibrations and possible internal degrees of freedom that correspond to rotations of chemical groups. The difference with Langmuir–Hinshelwood reactions is in the motions of the molecule as a whole. A calculation of the vibrations of the molecule will yield five (for a linear molecule) or six vibrations (for a nonlinear molecule) with a frequency that should be exactly zero. These correspond to overall translations and overall rotations. The zero-point motions of both are zero. The partition function for the overall rotation is either given by Eq. (4.22) for a linear molecule or Eq. (4.23) for a non-linear molecule. The overall rotation can yield prefactors for desorption that are substantially larger than the customary $10^{13}\ \mathrm{s}^{-1}$.

A little bit of care needs to be taken when calculating the partition function of the overall translations. Here the limits on the integration over the coordinates in Eq. (4.3) become important. The coordinates in that expression are the two center-of-mass coordinates parallel to the substrate, because we take the surface $S_{\beta\alpha}$ to distinguish the adsorbed from the desorbed state of course parallel to the substrate. The integration region is the region of an adsorption site. This is because we want to have the rate constant for desorption from a particular site. Because we choose the surface $S_{\beta\alpha}$ far from the substrate so that the corrugation of the potential is negligible, the integration of the coordinates is equal to the area of a site $A_{\mathrm{site}}$. This area is defined as the total area of the substrate divided by the number of sites from which the adsorbate can desorb. (Note that there may be other sites from which no desorption is possible. These should not be counted here, because it is really the dividing surface $S_{\beta\alpha}$ that should be partitioned into areas corresponding to the sites from which desorption takes place.) The partition function for the center-of-mass motion contributing to $Q^{\ddagger}$ is then a factor $2\pi m k_{\mathrm{B}} T A_{\mathrm{site}}/h^2$.

In the discussion of adsorption rate constants (Sect. 4.4.3) we will see an effect of recrossing the surface $S_{\beta\alpha}$. This same effect also influences desorption rate constants as will be shown there.

### 4.4.3 Adsorption

Because the initial state of adsorption is a molecule in the gas phase, we need to include the gas phase if we want to calculate the rate constant for adsorption. Suppose our system consists of a gas of just one molecule in a rectangular box, and we

[5] It may be that there is a very low barrier for diffusion of the adsorbed molecule. In that case the partition function of the initial or transition state will have a 2D translational partition function for the center of mass motion as a factor. See the example of CO desorption in Sect. 4.4.6.

are interested in the adsorption of that gas onto one of the interior sides of the box. Suppose this side has an area $A$ and the volume of the box is $V$.

We first look at adsorption that is not activated: i.e., the surface $S_{\beta\alpha}$ is a plane far from the side where the adsorption takes place and on which the potential energy is constant and equal to its value in the gas phase. The partition function (4.3) of the center-of-mass is then a product of two translational partition functions of translations parallel to the side. (We can ignore the other degrees of freedom, because they cancel in the two partition functions in Eq. (4.2).) The partition function (4.4) has a third factor which is the translational partition function of a translation perpendicular to the side. The ratio $Q^{\ddagger}/Q$ of the partition functions is then one over that third translational partition function. The activation barrier is zero, so Eq. (4.2) for adsorption somewhere on the side becomes

$$W_{\text{ads}} = \frac{Ak_{\text{B}}T}{V\sqrt{2\pi mk_{\text{B}}T}} \tag{4.35}$$

where we have used Eq. (4.33) for the translational partition function and $L = V/A$.

It looks as if this depends on the actual size of the box, but that is not the case. First we have $PV = k_{\text{B}}T$ with $P$ the pressure exerted by the single molecule. Second for a kMC simulation we do not want to know the rate constant for adsorption somewhere on the side, but at a particular site. We have to divide therefore by the number of sites. This then yields

$$W_{\text{ads}} = \frac{PA_{\text{site}}}{\sqrt{2\pi mk_{\text{B}}T}} \tag{4.36}$$

with $A_{\text{site}}$ the area of a single site. ($A_{\text{site}}$ is really the area $A$ divided by the number of ways a molecule may adsorb. This number is not always equal to the number of sites. For example, when we have dissociative adsorption it is two or four times the number of sites depending on whether the fragments of the dissociation are the same or not.) Although we have derived this expression for a gas with just one molecule, it also holds if there are more molecules. The only thing that changes is the numerical value of the pressure $P$. The expression can also be derived by determining the number of molecules that hit the side per unit time. The expression is very simple, and none of the quantities need to be determined via an extensive computation.

If the adsorption is activated (i.e., there is a transition state), then the results change substantially. We start again with just one molecule in the gas phase. The rate constant for adsorption on a specific site becomes

$$W_{\text{ads}} = \frac{k_{\text{B}}T}{h}\frac{Q^{\ddagger}}{Q_{\text{trans}}Q_{\text{rot,3D}}Q_{\text{int}}}\exp\left[-\frac{E_{\text{bar}}}{k_{\text{B}}T}\right]. \tag{4.37}$$

Here $Q_{\text{trans}}$ is the translational partition function for the center-of-mass of molecule in the gas phase, $Q_{\text{rot,3D}}$ its partition function for its overall rotation, and $Q_{\text{int}}$ for its internal motions (probably lots of vibrations, but maybe also rotations of chemical groups, etc.). For $Q_{\text{rot,3D}}$ Eq. (4.23) holds, and $Q_{\text{int}}$ can be written as a product of function of the type (4.19) and possibly (4.20). (Note that the vibrations yield zero-point energies.) For the $Q_{\text{trans}}$ we have

$$Q_{\text{trans}} = V\left[\frac{2\pi m k_{\text{B}} T}{h^2}\right]^{3/2}. \tag{4.38}$$

If we substitute this in the expression for the rate constant and use $PV = k_{\text{B}}T$ again, we get

$$W_{\text{ads}} = \frac{h^2 P}{(2\pi m k_{\text{B}} T)^{3/2}} \frac{Q^{\ddagger}}{Q_{\text{rot,3D}} Q_{\text{int}}} \exp\left[-\frac{E_{\text{bar}}}{k_{\text{B}} T}\right]. \tag{4.39}$$

As before this expression also holds for a gas consisting of many molecule. Only the numerical value of the pressure $P$ is different.

If we compare a molecule in the gas phase with a molecule adsorbed on the surface, we note that the molecule can freely translate and rotate in the gas phase, but not when adsorbed. This means that the molecule in the gas phase has a higher entropy then when it is adsorbed. This should of course affect the adsorption-desorption equilibrium, and also the rate constants for adsorption and desorption. Remarkably, only one of these rate constants is affected, but which one depends on the transition state. If there is no transition state (or what is the same thing, the final and transition state corresponding to the molecule in the gas phase), then the desorption rate constant is large. This is because of the increase of rotational entropy in the transition/final state upon desorption. The adsorption rate constant on the other hand has a normal value given by the number of molecules in the gas hitting the substrate per unit time. If there is a transition state in which the molecule is adsorbed on the substrate, then the desorption rate constant has a normal value, but the adsorption rate constant is small. This is because the molecule loses rotational entropy when it adsorbs. In both cases the molecule loses translation entropy when it adsorbs.

Note that the expression for the adsorption rate constant does not contain a sticking coefficient. This coefficient reflects the influence of trajectories in phase space that cross and then recross the dividing surface as explained at the end of Sect. 2.2.2. So far we have neglected this phenomenon, assuming that its effect on the rate constant is negligible. For adsorption this is often not the case, and an extra factor $\sigma$ should be added to the expressions for the adsorption rate constant. This $\sigma$ is the sticking coefficient.

Figure 4.3 shows characteristic trajectories and how they give rise to the sticking coefficient. Trajectory 1 is a regular adsorption. Trajectory 2 shows a molecule being scattered from the substrate leading to a recrossing of the dividing surface. This trajectory should not be included in the calculation of the adsorption rate constant, but Eqs. (4.36) and (4.39) however do not exclude it. These equations include the effect of crossings at $1_+$, $2_+$, and $3_+$. The effect of the last is probably negligible, but the effect of $2_+$ is not. By introducing a sticking coefficient $\sigma$ we have the effects of the crossings at $2_-$ and $3_-$ cancel those of the crossings at $2_+$ and $3_+$, respectively.

Note that the crossings at $2_-$, $3_-$, $4_-$ all affect the rate constant of desorption, whereas only the one at $4_-$ constitute a real desorption. This means that the expressions for desorption should also be corrected for recrossings. It turns out that this

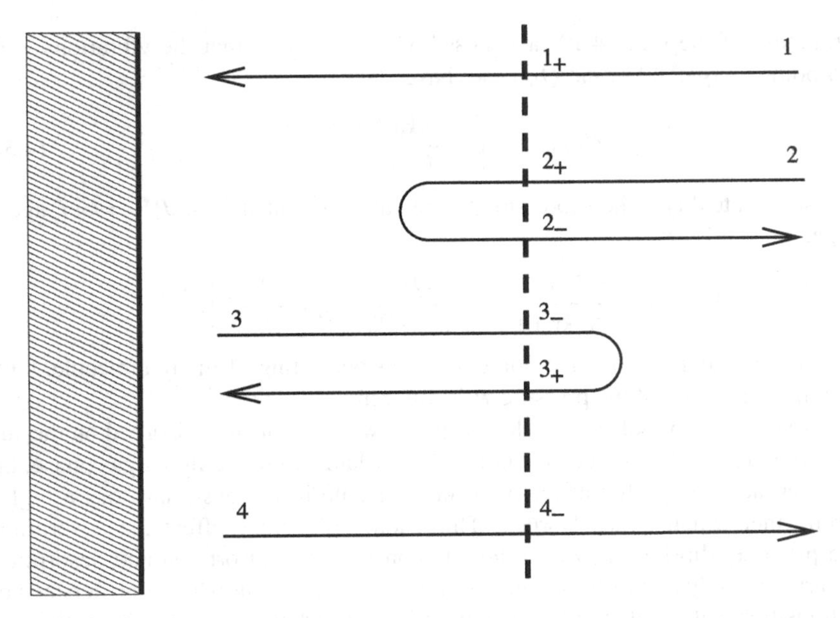


**Fig. 4.3** Schematic drawing of four characteristic trajectories in phase space of adsorption and desorption. The *hatched part* on the *left* is the part of space that is unaccessible because of the presence of the substrate. The *dashed line* is the dividing surface for adsorption and desorption separating the molecule being in the gas phase from it being adsorbed on the substrate. The meaning of *1*$_+$, *2*$_+$ et cetera is explained in the text

correction factor is also the sticking coefficient. To see this let's look at an arbitrary reaction and its reverse. The forward reaction has a rate constant

$$W_f = \frac{k_B T}{h} \frac{Q_{TS}}{Q_{IS}} e^{-(E_{TS}-E_{IS})/k_B T} \tag{4.40}$$

with TS referring to the transition state and IS to the initial state of the forward reaction. Zero-point energies are included in the partition functions. For the reverse reaction we have

$$W_r = \frac{k_B T}{h} \frac{Q_{TS}}{Q_{FS}} e^{-(E_{TS}-E_{FS})/k_B T} \tag{4.41}$$

with FS referring to the final state of the forward reaction. These expressions neglect the effect of recrossing. Incorporating this would give rate constants $\tilde{W}_f$ and $\tilde{W}_r$ for which $\tilde{W}_f = f_f W_f$ and $\tilde{W}_r = f_r W_r$ holds. In these expressions $f_f$ and $f_r$ are so-called dynamic correction factors. The sticking coefficient is the dynamic correction factor for adsorption.

For equilibrium we can define an equilibrium constant

$$K = \frac{W_f}{W_r} = \frac{Q_{FS}}{Q_{IS}} e^{(E_{IS}-E_{FS})/k_B T}. \tag{4.42}$$

We note that this expression is completely independent of the transition state. It is obvious that this should be so, as the equilibrium depends only on the thermodynamic properties of the initial and final states. This has an important consequence, because also the shape of the trajectories at the dividing surface (i.e., recrossings) do also not affect the equilibrium, and we must also have

$$K = \frac{\tilde{W}_f}{\tilde{W}_r}. \tag{4.43}$$

Consequently, $f_f = f_r$ must hold as well: i.e., the dynamic correction factors for a reaction and its reverse are the same. For an adsorption-desorption equilibrium this means that including a sticking coefficient in the adsorption rate constant forces us to include it also in the desorption rate constant, as otherwise we get an incorrect equilibrium.

### 4.4.4 Eley–Rideal Reactions

Eley–Rideal reactions have expressions for the rate constant that are very similar to those of activated adsorption. The difference is that in expression (4.39) for the Eley–Rideal reaction $P$, $m$, $Q_{\mathrm{rot,3D}}$, and $Q_{\mathrm{int}}$ refer only to the reactant in the gas phase, that there is another partition function in the denominator for the degrees of freedom of the adsorbed reactant, and that $Q^{\ddagger}$ contains the degrees of freedom of the atoms of both reactants.

### 4.4.5 Diffusion

In the lattice-gas model diffusion is represented as a hopping of an adsorbate from one site to a neighboring one. The calculation of rate constants for diffusion can then be done in exactly the same way as for Langmuir–Hinshelwood reactions.

### 4.4.6 Examples

To illustrate how rate constants are actually calculated we will discuss a few examples. Note that not all numbers have really been computed. In those instances where data is missing we will guesstimated the numbers. This does not affect the manner in which the rate constant are computed in any way.

Let's start with what is the most common case by looking at the reaction $NH_2 \rightarrow NH + H$ on a Rh(111) surface [24]. Density-Functional Theory (DFT) calculations have shown that $NH_2$ prefers to adsorb on a bridge site. NH on the other hand prefers an fcc hollow site and the H atom that is formed in this reaction also ends up at an fcc hollow site. A nudged-elastic band calculation was used to determine the transition state. This gave an activation barrier of $E_{\mathrm{bar}} = 1.07$ eV. The vibrational frequencies were calculated for both $NH_2$ at a bridge site, and for the transition state. Only the atoms of the adsorbates were allowed to move. These calculations gave a zero-point energy correction of $E_{zp} = -0.23$ eV. Inspection of the vibrations showed that all motions were indeed vibrations, so that the partition functions of Eq. (4.2) could be written as products of vibrational partition functions. It was found that $\tilde{Q}^{\ddagger}/\tilde{Q} \approx 1$. With these results rate constants were calculated between 100 K and 1000 K. Linear regression of the logarithm of the rate constant as a function of $1/T$ then gave an activation energy $E_{act} = 0.86$ eV and a prefactor $\nu = 1.68 \cdot 10^{13}$ s$^{-1}$. There is a small difference of 0.02 eV between $E_{act}$ and the zero-point energy corrected activation barrier $E_{bar} + E_{zp}$.

In the previous example the partition functions $\tilde{Q}$ and $\tilde{Q}^{\ddagger}$ are very close to 1. We have typically found that the rate constant varies not more than a factor of about 3, as long as the motions of the adsorbates are all vibrations. For the reaction $NH_2 + H \rightarrow NH_3$ on Rh(111) the effect is much larger however. All calculations for this reaction were done in the same way as for the previous example. (Both reactions were part of an extensive study on the dehydrogenation and oxidation of $NH_3$ on rhodium [24, 25].) The essential difference in the new reaction is that in the transition state $NH_3$ has almost its equilibrium geometry. In particular, there is a free rotation of $NH_3$ around its three-fold axis. This rotation was treated as a 1D rotation with a partition function give by Eq. (4.20). All other motions were vibrations. Because of this 1D rotation, the partition function of the transition state $\tilde{Q}^{\ddagger}$ became much larger than when there would be only vibrations. Proceeding as for $NH_2 \rightarrow NH + H$, except for the replacement of one vibrational partition function by $Q_{rot,1D}$, it was found that $E_{act} = 1.10$ eV and $\nu = 1.90 \cdot 10^{15}$ s$^{-1}$. Note the substantially larger prefactor due to the gain in rotation entropy in the transition state. The difference between $E_{act}$ and $E_{bar} + E_{zp}$ was not affected by this rotation: it was only 0.04 eV.

There have been extensive studies of CO on numerous transition metal surfaces. We look at desorption of CO from Rh(100) assuming that the transition state is equal to the final state: i.e., CO in the gas phase. The calculated adsorption energy, and therefore also the activation barrier, is 1.85 eV [26]. In the initial state all the motions are vibrations. The experimental C-O stretch vibration is 1990 cm$^{-1}$ and the metal-CO stretch is 470 cm$^{-1}$ [27]. Let's assume that the other four vibrations of CO are 250 cm$^{-1}$ based on a comparison with NO/Rh(100) [28]. As a consequence the zero-point energy is 0.43 eV, and $\tilde{Q} = 19.5$ at $T = 500$ K, which is the peak maximum temperature of the Temperature-Programmed Desorption (TPD) spectrum at low coverage [27]. For the transition state we have a product of three different partition functions. There is a vibrational one $\tilde{Q}_{vib}$ for the C-O stretch. The second is the rotational $\tilde{Q}_{rot,2D}$. The third is a 2D translational one $\tilde{Q}_{trans}$. The stretch vibration gives a zero-point energy of 0.27 eV, so that the zero-point energy corrected activation barrier is 1.69 eV. We have $\tilde{Q}_{vib} = 1.003$ at $T = 500$ K, $\tilde{Q}_{rot,2D} = 191.5$ (see Eq. (4.22) with $I = 1.542 \cdot 10^{-46}$ kg.m$^2$), and $\tilde{Q}_{trans} = 2\pi m k_B T A_{site}/h^2 = 341.0$ (with $A_{site} = 7.43 \cdot 10^{-20}$ m$^2$). Consequently, $\tilde{Q}^{\ddagger} = 6.55 \cdot 10^4$. With $\nu = (k_B T/h)(\tilde{Q}^{\ddagger}/\tilde{Q}) = 3.5 \cdot 10^{16}$ s$^{-1}$. For the adsorption of CO onto the same surface we can use Eq. (4.36). Again at $T = 500$ K and a pressure of $P = 1$ atm we get $W_{ads} = 1.68 \cdot 10^8$ s$^{-1}$.

It is interesting to compare the result for the desorption rate constant to experimental results and look at the consequences for the adsorption rate constant. TPD experiments give $E_{\mathrm{act}} = 1.44$ eV and $\nu = 3.98 \cdot 10^{13}$ s$^{-1}$ [27]. It looks at first glance that there is a substantial difference between the calculated and the experimental values. However, if we would use the calculated parameters to simulate the TPD spectra we would find peaks that are almost at the same temperature as in the experiment, but only somewhat narrower. The experimental activation energy and prefactor give a rate constant for desorption at $T = 500$ K of 0.122 s$^{-1}$. The calculated rate constant is 0.515 s$^{-1}$. This differs only by a factor of 4.2. (Remarkable there is a molecular beam experiment that does give a prefactor that is close to the calculated one [29].) There is clearly a compensating effect in the derivations of the activation energy and the prefactor. The difference between the activation energy and the adsorption energy above may be due to errors in the DFT calculations. A possible explanation for the large discrepancy in the prefactor is that our assumptions of a transition state of CO in the gas phase is incorrect. An alternative explanation is that the motion of the center of mass of the adsorbed CO parallel to the surface consists not of two vibrations, but of a free 2D translation. Instead of two vibrational partition functions we then get a 2D translational partition function. This causes an increase of $\tilde{Q}$ by a factor of 340, and makes the prefactor $1.03 \cdot 10^{14}$ s$^{-1}$. Note that then also part of the zero-point correction disappears, which unfortunately increases the discrepancy between calculated and experimental activation energy.

The calculated prefactor for the desorption is a factor 880 too large. Yet another explanation is related to the adsorption. The reason is that the adsorption-desorption equilibrium does not depend on the transition state, and hence neither does the ratio between the rate constants for adsorption and desorption. This means that if the desorption prefactor is too large then so is the adsorption rate constant. The reason why the latter may be too large is that the sticking coefficient may be small. So the calculated prefactor for desorption may be too large, because trajectories of scattered CO molecules may be been included in that calculation. See Sect. 4.4.3 for a more extensive discussion of this.

### *4.4.7 Summary*

The previous sections describe the most common way to calculate the rate constants of different types of processes. There are however exceptional cases as the discussion of the CO desorption from Rh(100) in Sect. 4.4.6 shows. We will attempt here to describe a systematic approach.

Do the following for the initial and transition state. First, look at the overall translations: i.e., the motions of the center of mass of a molecule. Are some of these degrees of freedom free: i.e., is the PES (nearly) flat for those degrees of freedom? If yes, use a translational partition function for those degrees of freedom. If not, these degrees of freedom become part of the group of vibrational degrees of freedom.

Second, look at the overall rotations: i.e., rotations in which the structure of the molecule remains the same. Are some of these degrees of freedom free or possible but hindered? If yes, use a rotational partition function for those degrees of freedom. If not, these degrees of freedom too become part of the group of vibrational degrees of freedom. The particular rotational partition function that one needs to use depends on whether the molecule is linear or nonlinear, on the number of axes around which the molecule can rotate freely, and on whether the rotation is hindered or not.

Third, take the vibrational degrees of freedom obtained from inspecting the overall translations and rotations and add the internal degrees of freedom. All these degrees of freedom may be coupled. Do a vibrational analysis of them. If the vibrational analysis gives only normal modes that are indeed vibrations, then use vibrational partition functions for them. If some of them are rotations of chemical groups, use 1D rotational partition functions or the partition functions of a hindered rotation. If some of them correspond to a floppy motion, then consider using the MD techniques mentioned in Sect. 4.3.8.

Fourth, for vibrational degrees of freedom one should use the partition function defined by the summation in Eq. (4.11). Add the zero-point energy of the vibration of the transition state to the activation barrier, and subtract those of the initial state.

Fifth, combine all partition functions and energies according to Eq. (4.12) to compute the rate constant.

## 4.5 Lateral Interactions

Interactions between adsorbates, or lateral interactions have at least been known for as long as diffraction techniques have revealed that adlayers can form very well-defined structures at low temperatures. The importance of these interactions for kinetics at higher temperatures has only more recently been acknowledged, but forms now an active area of research. This is understandable if one realizes that even small interactions between adsorbates can be of the same magnitude or larger than the thermal energy, and can therefore change rate constants by an order of magnitude or more, especially at low temperatures.

It is now possible to do quantum chemical calculations on quite realistic models of adsorbates on transition metal surfaces especially using DFT. However, many lateral interactions are small, and although it may be possible to get a value for a particular lateral interaction using DFT, it does not mean that the result is accurate. One should therefore always determine if a particular lateral interaction can be calculated reliably at all.

### 4.5.1 The Cluster Expansion

Lateral interactions are most commonly described with a cluster expansion. The adsorption energy $E_{ads}$ per adsorbate in a particular adlayer structure is then written as

$$E_{\text{ads}} = \sum_m c_m V_m \tag{4.44}$$

with $V_m$ the value of the interaction of type $m$, and $c_m$ the number of interactions of type $m$ per adsorbate. The interactions $V_m$ stand for the interaction of the adsorbate with the substrate (adsorption energy of an isolated adsorbate), pair interactions between adsorbates at various distances, all possible three-particle interactions, four-particle interactions, et cetera. This expansion can be made to reproduce the calculated adsorption energy as accurately as one wants [30]. This generally takes however a very large number of terms. Moreover, it may lead to overfitting: i.e., the cluster expansion will not only describe the interactions, but also the errors one makes in the calculation of the adsorption energies. This occurs because errors in the calculated $E_{\text{ads}}$ can easily be as large or even larger than some $V_m$'s. To avoid this one needs to truncate the cluster expansion. We will call a particular choice of $V_m$'s that one includes in the summation of Eq. (4.44) an interaction model.

The truncation of the cluster expansion for lateral interactions between adsorbates has so far mainly been done based on the desired accuracy with which the truncated expansion should reproduce the calculated results, the number of acceptable terms in the expansion, the type of terms in the expansion (pair, three-particle, et cetera), the estimated accuracy of the calculated results, and possibly other factors. Often these factors involve a trade-off: e.g., one prefers a short expansion, but it should also have a certain accuracy. Researchers have usually dealt with this using their personal experience and insight, and few objective criteria have been used. We will here describe two statistical techniques that do yield such criteria for truncating the cluster expansion.

There are a few useful remarks we can make independently of the method that we use to determine the lateral interactions. Let's define

$$E_{\text{ads}}^{\text{fit}}(n) = \sum_m c_m(n) V_m \tag{4.45}$$

with $n$ indicating the various adlayer structures that we use to determine the lateral interactions, and $c_m(n)$ the number of interactions of type $m$ per adsorbate in structure $n$. If $E_{\text{ads}}^{\text{calc}}(n)$ is the calculated adsorption energy per adsorbate in structure $n$, then we want the difference between the $E_{\text{ads}}^{\text{fit}}(n)$'s and $E_{\text{ads}}^{\text{calc}}(n)$'s to be as small as possible. Let's also define the errors

$$\varepsilon_n = E_{\text{ads}}^{\text{fit}}(n) - E_{\text{ads}}^{\text{calc}}(n). \tag{4.46}$$

These errors need not be independent. There may be a systematic component $\sigma$ that is the same for all adlayer structures, and a rest $\rho_n = \varepsilon_n - \sigma$ that is independent. The systematic error may be caused by a specific shortcoming of the calculations of the adsorption energy (e.g., incorrect description of van-der-Waals interactions in DFT), but that need not be the case. Here it is just a numerical error common to all calculations.

If there is only one adsorbate adsorbing at one type of site, then we can write for each adlayer structure

$$E_{\text{ads}}^{\text{calc}}(n) = (V_0 - \sigma) + \left(\sum_{m>0} c_m(n) V_m - \rho_n\right) \tag{4.47}$$

with $V_0$ the adsorption energy of an isolated adsorbate, because $c_0(n) = 1$ for all structures $n$. We see that the terms that differ for various adlayer structures are not affected by the systematic error $\sigma$. These terms are also the only ones that contain the lateral interactions. This means that the systematic error does not affect the lateral interactions. They depend only on the smaller random errors $\rho_n$. Consequently lateral interactions can be determined better than one might suppose having some idea of the accuracy with which one can compute adsorption energies. The systematic error affects only $V_0$.

The expression above does not hold when there are different adsorbates or when there is adsorption on different types of site. The conclusion drawn above still holds however. This can be seen as follows. Suppose $V_0, V_1, \ldots, V_A$ are the adsorption energies originating from various isolated adsorbates and adsorption sites. Because $E_{\text{ads}}$ is the adsorption energy per adsorbate, we have $c_0(n)+c_1(n)+\ldots+c_A(n)=1$ for all structures $n$. We can therefore write

$$E_{\text{ads}}^{\text{calc}}(n) = \left(\sum_{m=0}^{A} c_m(n)(V_m - \sigma)\right) + \left(\sum_{m>A} c_m(n) V_m - \rho_n\right). \tag{4.48}$$

This shows again that a change in the calculated adsorption energies $E_{\text{ads}}^{\text{calc}}(n)$ that can be represented by a change in the systematic error $\sigma$ gives the same errors $\varepsilon_n$ with a corresponding change in the adsorption energies $V_0, V_1, \ldots, V_A$, but with the same lateral interactions.

Looking at errors as above also shows that singling out one adlayer structure, say $n = 1$, to determine the adsorption energy of an isolated adsorbate $V_0$ is not a good idea. Because then we would have $E_{\text{fit}}^{(1)} = E_{\text{calc}}^{(1)} = V_0$ and work with $E_{\text{calc}}^{(n)} - E_{\text{calc}}^{(1)}$ to determine the lateral interactions. However

$$E_{\text{ads}}^{\text{calc}}(n) - E_{\text{calc}}^{(1)} = \left(E_{\text{ads}}^{\text{fit}}(n) - E_{\text{ads}}^{\text{fit}}(1)\right) - (\rho_n - \rho_1). \tag{4.49}$$

We see that $E_{\text{ads}}^{\text{calc}}(n) - E_{\text{calc}}^{(1)}$ has a different error from $E_{\text{ads}}^{\text{calc}}(n)$. In fact, as the $\rho$'s are independent $\rho_n - \rho_1$ has a standard deviation that is $\sqrt{2}$ times the one of $\rho_n$. Consequently the errors we will make in the determination of the lateral interactions will also be a factor $\sqrt{2}$ larger. It is better to treat the adsorption energy $V_0$ in the same way as the lateral interaction parameters.

### *4.5.2 Linear Regression*

Suppose one has determined which $V_m$'s in Eq. (4.44) should be included, then there remains the problem of computing their values. Because the adsorption energy $E_{\text{ads}}$

depends linearly on the $V_m$'s, they can be determined using linear regression [3]. This means we use a least-squares approach by minimizing

$$\chi^2 = \frac{1}{N_{\text{str}}} \sum_n \left[ \sum_m c_m(n) V_m - E_{\text{ads}}^{\text{calc}}(n) \right]^2 \tag{4.50}$$

as a function of the $V_m$'s.

Differentiating the expression with respect to the different $V_m$'s and equating the results to zero gives the so-called normal equations

$$\sum_m \alpha_{km} V_m = \beta_k \tag{4.51}$$

with

$$\alpha_{km} = \sum_n c_k(n) c_m(n) \tag{4.52}$$

and

$$\beta_k = \sum_n c_k(n) E_{\text{ads}}^{\text{calc}}(n) \tag{4.53}$$

from which we can determine the $V_m$'s. We refer to the literature for discussions on problems with these equations because they may be (approximately) linear dependent, and how to deal with this [3]. It should however be obvious that if one has $N_{\text{str}}$ different adlayer structures, that one certainly will not be able to compute more than $N_{\text{str}}$ different $V_m$'s. In general $N_{\text{str}}$ should be much larger than the number of $V_m$'s.

If the matrix $\alpha$ of the normal equations is not (approximately) singular, then it can be used to get an idea of how well the $V_m$'s are determined. The inverse matrix of $\alpha$ has matrix elements that are proportional to the covariances of the $V_m$'s. In particular, the diagonal elements of $\alpha^{-1}$ are proportional to the standard errors for the $V_m$'s. The proportionality constant is equal to $\chi^2$.

### 4.5.3 Cross Validation

The truncation problem of the cluster expansion has been earlier encountered for the calculation of interactions between atoms forming an alloy. This has led to the development of the leave-one-out cross-validation (LOO-CV) method [31, 32]. This is a statistical technique that uses part of the results of a set of calculations to determine values for the interactions between the atoms and the rest of the calculations to test these values. Because determination and testing of the interactions is done on independent results of calculations, one obtains an estimate of how well the values for the interactions one calculates will predict energies of unknown structures. This method has recently also been applied to the determination of lateral interactions [33–35].

The LOO-CV method works as follows [31, 32, 36]. We take all structures except one structure $n$. We then do a linear regression for a particular interaction model (i.e., a set of $V_m$'s) using these structures. This gives us a set of interaction parameters that we can use to predict the adsorption energy for structure $n$. We then compare this energy $E_{\mathrm{ads}}^{\mathrm{pred}}(n)$ with the calculated energy $E_{\mathrm{ads}}^{\mathrm{calc}}(n)$. We do this not just for one structure $n$ but for all structures, and define the cross-validation score or leave-one-out error

$$R_{\mathrm{CV}}^2 = \frac{1}{N_{\mathrm{str}}} \sum_{n=1}^{N_{\mathrm{str}}} \left[E_{\mathrm{ads}}^{\mathrm{pred}}(n) - E_{\mathrm{ads}}^{\mathrm{calc}}(n)\right]^2. \tag{4.54}$$

This error indicates how well an interaction model predicts the energy. One starts with a model with few interaction parameters, and determines the CV score. Adding parameters will initially lower the CV score, which means that the model becomes better. Adding too many parameter should however increase the CV score because of overfitting.

If the number of structures $n$ is large, one may not want to leave out every structure $n$, determine the lateral interactions, and compare $E_{\mathrm{ads}}^{\mathrm{pred}}(n)$ to $E_{\mathrm{ads}}^{\mathrm{calc}}(n)$. Instead one can partition all structures into groups. Each structure should be in one and only one group. One does the same as for LOO, but now one leaves out all structures in one group instead of just one structure. This is called the leave-many-out (LMO) method. The groups are generally made of similar size, but the structures are partitioned randomly [36].

The minimum of the CV score indicates a best set of interaction parameters. The problem with the method is that often the CV score becomes almost constant when more parameters are added, and it is very hard to determine the minimum of the CV score [37–40].

### 4.5.4 Bayesian Model Selection

An alternative solution to the truncation problem can be obtained using Bayesian statistics [37, 41].[6] Instead of the CV score of the LOO-CV method, one assigns a probability to each interaction model. The best model is the one with the highest probability. This approach does not have the drawback of the LOO-CV method. Moreover, it seems that the approach leads to models with fewer interaction parameters. The method also lends itself well to an analysis of the importance of parameters: it is easy to compute probabilities for individual parameters, and correlation between parameters.

---

[6]Parts of Sect. 4.5.4 have been reprinted with permission from A.P.J. Jansen, C. Popa, Bayesian approach to the calculation of lateral interactions: NO/Rh(111), Phys. Rev. B **78**, 085404 (2008). Copyright 2008, American Physical Society.

We assume again that we have done calculations resulting in the adsorption energy per adsorbate $E^{\text{calc}}_{\text{ads}}(n)$ with $n$ an index to distinguish the adlayer structures, and we describe the energy $E^{\text{calc}}_{\text{ads}}(n)$ using a cluster expansion for the lateral interactions. The expression (4.45) for $E^{\text{fit}}_{\text{ads}}(n)$ should approximate $E^{\text{calc}}_{\text{ads}}(n)$ as well as possible. Let's use $S$ to indicate a subset of all interaction parameters $V_m$: i.e., it indicates the interaction model. Let's use $V$ as a shorthand for the set of all values of the interaction parameters in $S$. For all calculated adsorption energies we use $E$. We will determine which interaction model is best by calculating $P(S|E)$, which stands for the probability that the calculated adsorption energies $E$ can be described by interaction parameters in $S$. The best interaction model has the highest $P(S|E)$.

We can use Bayes's theorem to relate the probability of $S$ given $E$ (i.e., $P(S|E)$) to the probability of $E$ given $S$ (i.e., $P(E|S)$) [42–44].

$$P(S|E) \propto P(E|S)P(S). \tag{4.55}$$

The proportionality constant that is missing in this expression can be determined by normalizing $P(S|E)$ as a function of $S$. The probability $P(E|S)$ is often called the likelihood of $S$ given $E$, the probability $P(S)$ is called the prior (probability) of $S$, and the probability $P(S|E)$ is called the posterior (probability). The likelihood $P(E|S)$ is the probability that we should find certain values $E^{\text{calc}}_{\text{ads}}(n)$ if the interaction model $S$ is the right one. We will determine this from $P(E|S,V)$: i.e., we not only know which parameters are in the interaction model, but also their values (see also Eq. (4.58)). The prior $P(S)$ represents what we think are plausible models before we look at the results of our calculations.

Bayes's theorem is used as follows in the selection of models for the lateral interactions. We want to calculate $P(S|E)$. How good an interaction model is will also depend on whether we can find good values for the lateral interactions. It is important to distinguish between $S$ (the parameters in the model) and $V$ (the values of these parameters). We can introduce the values by regarding $P(E|S)$ as a marginal distribution of $P(E,V|S)$ via [44]

$$P(E|S) = \int dV\, P(E,V|S) \tag{4.56}$$

where the integration is over all possible values for the interaction parameters. The integrand can be written as

$$P(E,V|S) = P(E|S,V)P(V|S). \tag{4.57}$$

Substitution in the Bayes's expression for $P(S|E)$ then gives

$$P(S|E) \propto P(S) \int dV\, P(E|S,V)P(V|S). \tag{4.58}$$

This allows us to compute $P(S|E)$, because we can make a good guess of what the calculated adsorption energies should be given the lateral interactions (i.e., $P(E|S,V)$), and it should be possible to think of reasonable priors $P(V|S)$ and $P(S)$.

Suppose we have a set of interaction parameters $S$ with values $V$. A normal way to obtain such a set is via a least-squares procedure to fit $E_{\text{ads}}^{\text{fit}}(n)$ to $E_{\text{ads}}^{\text{calc}}(n)$. Suppose that the set has all interaction parameters to describe the system and that they have the correct values. The most likely values for $E_{\text{ads}}^{\text{calc}}(n)$ should then be equal to $E_{\text{ads}}^{\text{fit}}(n)$. Due to errors in the calculations $E_{\text{ads}}^{\text{fit}}(n)$ and $E_{\text{ads}}^{\text{calc}}(n)$ will not be exactly equal. It can be shown using the Maximum Entropy Principle (MEP) that if we have no information on the difference, it can best be described by a Gaussian probability distribution [42]: i.e.,

$$P(E|S,V) = \exp\left[-\frac{1}{2}\chi_1^2\right]\prod_{n=1}^{N_{\text{str}}}\frac{1}{\sqrt{2\pi\sigma_n^2}} \tag{4.59}$$

with

$$\chi_1^2 = \sum_{n=1}^{N_{\text{str}}}\left[\frac{E_{\text{ads}}^{\text{fit}}(n) - E_{\text{ads}}^{\text{calc}}(n)}{\sigma_n}\right]^2, \tag{4.60}$$

$N_{\text{str}}$ the number of adlayer structures for which we have calculated adsorption energies, and $\sigma_n$ an error estimate of the calculated adsorption energies $E_{\text{ads}}^{\text{calc}}(n)$. Note that $P(E|S,V)$ is a function of the adsorption energies $E_{\text{ads}}^{\text{calc}}(n)$, but the integration in Eq. (4.58) is over $V$. This means that we should regard $\chi_1^2$ as a function of the interaction parameters $V$.

Usually one does not know much about the interaction parameters before one starts with the calculations of the adsorption energies, nor about which interaction parameters to include in the model. Ideas on which model $S$ is appropriate are split in Eq. (4.58) from ideas on which values $V$ seem reasonable. The former have a prior $P(S)$, the latter a prior $P(V|S)$. The MEP shows again that the best expression for $P(V|S)$, which is also computationally convenient, is a Gaussian distribution [42]

$$P(V|S) = \exp\left[-\frac{1}{2}\chi_2^2\right]\prod_{m=1}^{N_{\text{par}}}\frac{1}{\sqrt{2\pi s_m^2}} \tag{4.61}$$

with

$$\chi_2^2 = \sum_{m=1}^{N_{\text{par}}}\left[-\frac{(V_m - V_m^{(0)})^2}{s_m^2}\right], \tag{4.62}$$

$N_{\text{par}}$ the number of parameters in the interaction model, $V_m^{(0)}$ the most likely prior value for parameter $m$, and $s_m$ the standard deviation. The lack of prior knowledge of the values of the interactions can be implemented by choosing large values for these deviations. We find it harder to give a general expression for $P(S)$. If a model contains a pair interaction for adsorbates at a certain distance, then pair interactions at shorter distances should be included as well. Also, if there is a three-particle interaction in a model, then all pair interactions between these three particles having such three-particle interaction should be included too. One also will want to cut off the summation in Eq. (4.45). Apart from these considerations it seems natural to take all interaction models equally likely.

The integration in Eq. (4.58) can be done easily because the integrand is a Gaussian expression in the integration variables. We define a column vector $\mathbf{c}$ via $\mathbf{c}_n = E^{\mathrm{calc}}_{\mathrm{ads}}(n)$, a column vector $\mathbf{e}$ via $\mathbf{e}_n = E^{\mathrm{fit}}_{\mathrm{ads}}(n)$, and a matrix $\mathbf{A}$ via $\mathbf{A}_{nm} = \sigma_n^{-2}\delta_{nm}$. With these we can write $\chi_1^2 = (\mathbf{e}-\mathbf{c})^T\mathbf{A}(\mathbf{e}-\mathbf{c})$. The fitted adsorption energies $\mathbf{e}$ can be written as $\mathbf{e} = \mathbf{M}\mathbf{v}$ with $\mathbf{v}$ a column vector with the interaction parameters ($\mathbf{v}_m = V_m$), and the matrix $\mathbf{M}$ contains the coefficients of the interaction parameters in Eq. (4.45). Similarly we write $\chi_2^2 = (\mathbf{v}-\bar{\mathbf{v}})^T\mathbf{B}(\mathbf{v}-\bar{\mathbf{v}})$, with $\bar{\mathbf{v}}_n = V_n^{(0)}$ and $\mathbf{B}_{nm} = s_m^{-2}\delta_{nm}$. We can combine $\chi_1^2$ and $\chi_2^2$, and write the result as a quadratic function of the interaction parameters. This gives us

$$\chi_1^2 + \chi_2^2 = (\mathbf{v}-\tilde{\mathbf{v}})^T\big[\mathbf{M}^T\mathbf{A}\mathbf{M}+\mathbf{B}\big](\mathbf{v}-\tilde{\mathbf{v}}) + \mu, \tag{4.63}$$

with

$$\tilde{\mathbf{v}} = \big[\mathbf{M}^T\mathbf{A}\mathbf{M}+\mathbf{B}\big]^{-1}\big(\mathbf{M}^T\mathbf{A}\mathbf{c}+\mathbf{B}\bar{\mathbf{v}}\big), \tag{4.64}$$

and

$$\mu = \mathbf{c}^T\mathbf{A}\mathbf{c} + \bar{\mathbf{v}}^T\mathbf{B}\bar{\mathbf{v}} - \tilde{\mathbf{v}}^T\big[\mathbf{M}^T\mathbf{A}\mathbf{M}+\mathbf{B}\big]\tilde{\mathbf{v}}. \tag{4.65}$$

With this expression the integration in Eq. (4.58) then becomes

$$P(S|E) = \frac{P(S)}{P(E)}\frac{1}{\sqrt{\prod_n(2\pi\sigma_n^2)}}\prod_{m=1}^{N_{\mathrm{par}}}\left[\frac{\tilde{s}_m}{s_m}\right]e^{-\mu/2}, \tag{4.66}$$

with $1/\tilde{s}_m^2$ being an eigenvalue of the matrix $\mathbf{M}^T\mathbf{A}\mathbf{M}+\mathbf{B}$.

Equation (4.66) can be interpreted as follows. The probability of an interaction model is higher when $\mu$ is smaller. If we assume that the prior distribution $P(V|S)$ is very broad, so that we can set $\mathbf{B} = \mathbf{0}$, then $\chi_2^2 = 0$, $\tilde{\mathbf{v}}$ minimizes $\chi_1^2$, and $\mu$ is the least-squares sum of the difference between the calculated and fitted adsorption energies. This means that the probability $P(S|E)$ become higher if the fit becomes better. This is of course as it should be.

The probability $P(S|E)$ also becomes higher when the prior errors $s_m$ become smaller. This too is to be expected, because it more-or-less means that we know a parameter already in advance. Less easy to interpret is the dependence on the $\tilde{s}_m^2$'s. It seems that a high value for $\tilde{s}_m$ would improve the model. This however would be incorrect, and is also counter-intuitive. As will be shown below the $\tilde{s}_m$'s are error estimates for a set of statistically independent interaction parameters. So a high value for $\tilde{s}_m$ would mean a parameter that is ill-defined, which one would not expect to improve an interaction model. The solution of this paradox is hidden in $\mu$ which also depends on the $\tilde{s}_m$'s. To see this let's compare two interaction models $S^{(1)}$ and $S^{(2)}$ with the difference that $S^{(2)}$ has one additional parameter compared to $S^{(1)}$. Let's also assume, for simplicity, they have the same prior ($P(S^{(2)}) = P(S^{(1)})$), that $\mathbf{B} = \mathbf{0}$, and that this additional parameter is independent from the others. This means that each $\tilde{s}_m$ of $S^{(1)}$ is also found for $S^{(2)}$, but $S^{(2)}$ has an additional factor in the product of Eq. (4.66) which we call $\tilde{s}_N$. We then get

$$\frac{P(S^{(2)}|E)}{P(S^{(1)}|E)} = \frac{\tilde{s}_N}{s_N}e^{-(\mu^{(2)}-\mu^{(1)})/2} \tag{4.67}$$

with $\mu^{(n)}$ the least-squares sum for $S^{(n)}$. When the additional parameter in $S^{(2)}$ is independent from the others, the matrix $\mathbf{M}^T\mathbf{AM}$ blocks and $\mu^{(2)} - \mu^{(1)} = -(\mathbf{M}^T\mathbf{AM})_{NN}(\tilde{\mathbf{v}}_N)^2$ holds. This leads to

$$\frac{P(S^{(2)}|E)}{P(S^{(1)}|E)} = \frac{\tilde{s}_N}{s_N} e^{(\tilde{\mathbf{v}}_N/\tilde{s}_N)^2/2}. \tag{4.68}$$

This function is for small values of $\tilde{s}_N$ a decreasing function. Consequently, if the interaction parameters are defined better (i.e., smaller $\tilde{s}_m$), then the probability $P(S|E)$ becomes higher, again as expected. Equation (4.68) shows also how adding a parameter can reduce the probability of an interaction model, because the ratio $\tilde{s}_N/s_N$ will usually be smaller than one, because the error estimates in the prior $P(V|S)$ will be large.

The results above depend on the error estimates $\sigma_n$. This means that the posterior really has these estimates as a parameter and should be written $P(S|E,\sigma)$. Because we need to know the $\sigma_n$'s but have no good information on them, they are often called nuisance parameters. There are various ways to deal with this [43, 44]. We find it the most convenient to put $\sigma_n = \sigma$ for all $n$ and to get a value for $\sigma$ by determining the maximum of its probability distribution $P(\sigma|E)$. This probability can be related to probability distributions that we have dealt with before.

$$P(\sigma|E) \propto P(\sigma)\sum_S \int dV\, P(E|S,V,\sigma)P(S,V|\sigma). \tag{4.69}$$

$P(E|S,V,\sigma)$ is given by Eq. (4.59) and $P(S,V|\sigma)$ by Eq. (4.61). The proportionality constant can be determined by normalization. Only $P(\sigma)$ is new. However, if we assume that it is a uniform distribution then we only need to do the integral and sum over all models. The integral has already been done before, so we only need to add all results for the different models.

The parameters $V_m^{(0)}$ and $s_m$ can also be regarded as nuisance parameters, but it is more convenient to handle them differently. We do not want to use any prior information for the interaction parameters, so we take large values for the error estimates $s_m$. As a consequence the values that we take for the $V_m^{(0)}$'s have then only a negligible effect on the final results,

Once we have determined $S$ we also need the parameters $V$. For this we can used $P(S,V|E)$, which equals $P(E|S,V)P(S,V)$. We have already seen the two factors in this product: the only difference with what we have done before is that we do not need to integrate out the interaction parameters. The interaction parameters themselves can be obtained from $P(S,V|E)$ by computing the expectation values of $V$. The derivation above shows that the parameters should be chosen equal to components of the vector $\tilde{\mathbf{v}}$. $P(S,V|E)$ can give us also error estimates for the parameters: the covariance matrix is given by $(\mathbf{M}^T\mathbf{AM}+\mathbf{B})^{-1}$.

### 4.5.5 The Effect of Lateral Interactions on Transition States

The previous sections have dealt with equilibrium situations for the lateral interactions: i.e., minima of the PES. For the kinetics we also need to know how lateral interactions affect transition states. There has hardly been any work done on this [45]. From a theoretical point of view one can in principle use quantum chemical calculations just as one would for the stable states.

One can also use the following pragmatic approach. For a reaction we have a model of the lateral interactions that tells us how the energies of the initial and the final states (both minima of the PES) depend on the lateral interactions. We then use a Brønsted–Polanyi relation to relate the shifts in the initial and final state to a change in the activation energy [46, 47]

$$E_{\text{act}} = E_{\text{act}}^{(0)} + \alpha\big(\Delta E - \Delta E^{(0)}\big). \tag{4.70}$$

Here $E_{\text{act}}$ ($E_{\text{act}}^{(0)}$) is the activation energy with (without) lateral interactions, and $\Delta E$ ($\Delta E^{(0)}$) is the reaction energy with (without) lateral interactions. $\Delta E < 0$ for exothermic and $\Delta E > 0$ for endothermic reactions. The Brønsted–Polanyi coefficient $\alpha$ varies between 0 and 1, and is a measure of how much the effect of the lateral interactions of the initial and final states influences the activation energy. The idea of the Brønsted–Polanyi relation is that if we have a transition state that resembles the initial state (a so-called early barrier), then lateral interactions will affect the transition state as much as the initial state, and the activation energy will not depend on the lateral interactions. So we choose $\alpha = 0$ in that situation. If the transition state resembles the final state (a so-called late barrier), then changes in the lateral interactions of the initial and final states will fully end up in the transition state, and we choose $\alpha = 1$ (see Fig. 4.4). In general, the Brønsted–Polanyi relation interpolates the lateral interactions between the initial and final state.

### 4.5.6 Other Models for Lateral Interactions

There are many other models for the lateral interactions [34]. These are all based on a particular physical model describing the mechanism of the interactions. This is different from the cluster expansion, which is a purely mathematical model. For example, the bond-order conservation model is based on the observation that adsorption of an atom or molecule changes the electronic structure of a metal and therefore changes the energy of subsequent adsorption of other atoms and molecules. Elastic deformation models are based on a similar effect when adsorption changes the structure of a substrate.

The advantage of models based on the mechanism of the interactions is that they are often more compact: i.e., there are fewer parameters to determine. Disadvantages of these models are that they are often restricted to a certain type of system, and it is hard to improve them when they are not accurate enough: e.g., when there

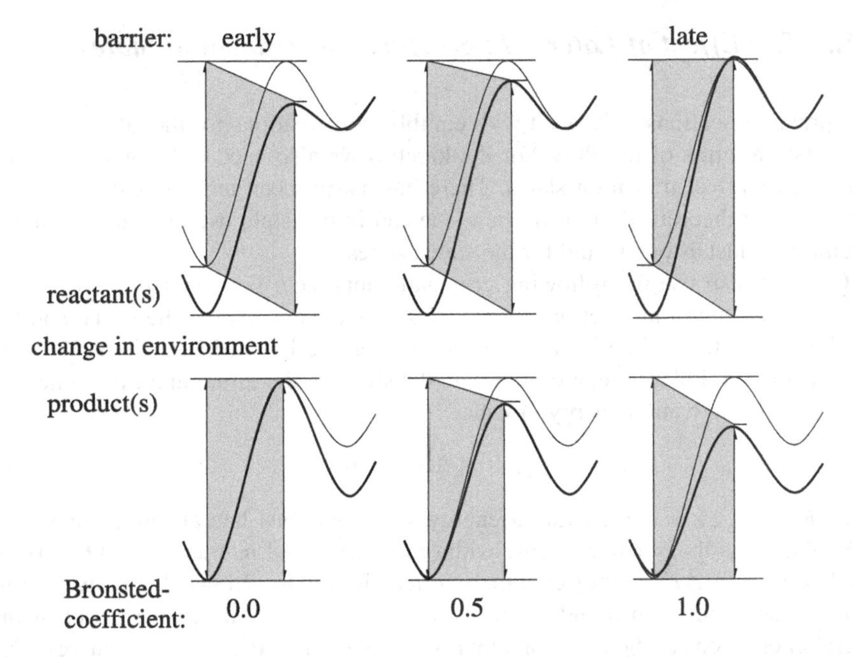


**Fig. 4.4** Sketches showing how changes in the lateral interactions affecting the reactants or the products change the activation energy (*vertical arrows*) depending on whether the barrier is early, late, or in between. The *thin line* indicates the reaction profile without lateral interactions, the *fat line* with interactions

are several mechanisms contributing to the lateral interactions. Moreover, there has hardly any work been done on the reliability of the determination of the parameters in these models.

## 4.6 Rate Constants from Experiments

One of the problems of calculating rate constants is the accuracy. The method that is mostly used to calculate the energetics of adsorbates on a transition metal surface is DFT [48–50]. Estimates of the error made using DFT for such systems are at least about 10 kJ/mol. An error of this size in the activation energy means that at room temperature the rate constant is off by about two orders of magnitude. How well a pre-exponential factor can be calculated is not really known at all. This does not mean that calculating rate constants is useless. The errors in the energetics have less effect, if the temperature is higher, but even more important is that one can calculate rate constants for processes that are experimentally hardly or not accessible at all. If, on the other hand, one can obtain rate constants from an experiment, then the value that is obtained is generally more reliable than one that has been calculated.

In general, one has to deal with a system in which several reactions can take place at the same time. The crude approach to obtain rate constants from experiments is then to try to fit all rate constants to the experiments at the same time. This is seldom a good idea. First of all such a procedure can be quite complicated. The data that one gets from an experiment are hardly ever a linear function of the rate constants. Consequently the fitting procedure consists of minimizing a nonlinear function that stands for the difference between experimental and the calculated or simulated data. Such a function normally has many local minima, and it is very hard to find the global minimum. But this isn't even the most important drawback. Although one may be able to do a very good fit of the experimental data, this need not mean that the rate constants are good: they may be physically unrealistic, and given enough fit parameters, one can fit anything.

Deriving kinetic parameters from experiments does work well, when one has an experiment of a single simple process that can be described by just one or two parameters. The process should be simple in the sense that one has an analytical expression with which one can derive relatively easily the kinetic parameters given the experimental data. The analytical expression should be exact or at least a very good approximation. If one has to deal with a reaction system that is complicated and consists of many reactions, then one should try to get experiments that measure just one of the reactions. For example, in CO oxidation one has at least adsorption of CO, dissociative adsorption of oxygen, and the formation of $CO_2$. Instead of trying to fit rate constants of these three reactions simultaneously, one should look at experiments that show only one of these reactions. An experiment that only measures adsorption as a function of CO pressure can be used to get the CO adsorption rate constant. The following sections show a number of processes which can be used to get kinetic parameters, and we show how to get the parameters for these processes.

### 4.6.1 Relating Macroscopic Properties to Microscopic Processes

The analytical expressions mentioned above should relate some property that is measured to the rate constants. We will address first the general relation. This relation is exact, but rarely very useful. In the next sections we will show situations were the general relation can be simplified either exactly or with the use of some approximation.

If a system is in a well-defined configuration then a macroscopic property can in general be computed easily. For example, the number of molecules of a particular type in the adlayer can be obtained simply be counting. If the property that we are interested in is denoted by $X$, then its value when the system is in configuration $\alpha$ is given by $X_\alpha$. We look at the expectation value of $X$, which is given by

$$\langle X\rangle = \sum_\alpha P_\alpha X_\alpha. \tag{4.71}$$

Kinetic experiment measure changes, so we look at $d\langle X\rangle/dt$. This is given by

$$\frac{d\langle X\rangle}{dt} = \sum_\alpha \frac{dP_\alpha}{dt} X_\alpha, \tag{4.72}$$

because $X_\alpha$ is a property of a fixed configuration. We can remove the derivative of the probability using the master equation. This gives us

$$\frac{d\langle X\rangle}{dt} = \sum_{\alpha\beta}[W_{\alpha\beta}P_\beta - W_{\beta\alpha}P_\alpha]X_\alpha,$$
$$= \sum_{\alpha\beta} W_{\alpha\beta}P_\beta[X_\alpha - X_\beta]. \tag{4.73}$$

The second step is obtained by swapping the summation indices. The final result can be regarded as the expectation value of the change of $X$ in the process $\beta \to \alpha$ times the rate constant of that process. Equation (4.73) is called the phenomenological or macroscopic equation [51]. It forms the basis for deriving relations between macroscopic properties and rate constants.

### 4.6.2 Simple Desorption

Suppose we have an atom or a molecule that adsorbs onto one particular type of site. We assume that we have a surface of area $A$ with $S$ adsorption sites. If $N_\alpha$ is the number of atoms/molecules in configuration $\alpha$ then

$$\frac{d\langle N\rangle}{dt} = \sum_{\alpha\beta} W_{\alpha\beta}P_\beta[N_\alpha - N_\beta]. \tag{4.74}$$

Diffusion does not change the number of atoms/molecules, and it does not matter in this case whether we include it or not. The only relevant process that we have to look at is desorption. We mentally fix configuration $\beta$ and look only at the summation over $\alpha$. We have to distinguish between two types of terms: the ones where $\alpha$ can originate from $\beta$ by a desorption, and the ones where it cannot. The latter terms have $W_{\alpha\beta} = 0$, because there is no process that changes $\beta$ into $\alpha$, and so they do not contribute to the sum. The former do contribute and we have $W_{\alpha\beta} = W_{\text{des}}$, with $W_{\text{des}}$ the rate constant for desorption. We also have in this case $N_\alpha - N_\beta = -1$, because desorption removes one atom/molecule from the surface. So all these non-zero terms contribute the same value $-W_{\text{des}}$ to the sum for a given configuration $\beta$. Moreover, the number of these terms is equally to the number of atoms/molecules in $\beta$ that can desorb, because each desorbing atom/molecule yields a different $\alpha$. This number of atoms/molecules is $N_\beta$, so

$$\frac{d\langle N\rangle}{dt} = -W_{\text{des}}\sum_{\beta} P_\beta N_\beta = -W_{\text{des}}\langle N\rangle. \tag{4.75}$$

This is an exact expression. Dividing by the number of sites $S$ gives the rate equation for the coverage $\theta = \langle N\rangle/S$.

$$\frac{d\theta}{dt} = -W_{\text{des}}\theta. \tag{4.76}$$

If we compare this to the macroscopic rate equation $d\theta/dt = -k_{des}\theta$ with $k_{des}$ the macroscopic rate constant, we see that $k_{des} = W_{des}$.

For isothermal desorption $k_{des}$ does not depend on time and the solution to the rate equation is

$$\theta(t) = \theta(0)\exp[-k_{des}t], \tag{4.77}$$

where $\theta(0)$ is the coverage at time $t = 0$. Kinetic experiments often measure rates, and for the desorption rate we have

$$\frac{d\theta}{dt}(t) = -k_{des}\theta(0)\exp[-k_{des}t]. \tag{4.78}$$

We can now obtain the rate constant by measuring, for example, the rate of desorption as a function of time and plotting minus the logarithm of the rate as a function of time. Because

$$\ln\left[-\frac{d\theta}{dt}(t)\right] = \ln\left[k_{des}\theta(0)\right] - k_{des}t, \tag{4.79}$$

we can obtain the rate constant which equals minus the slope of the straight line. The same would hold if we would plot the logarithm of the coverage as a function of time. Because of the equality this immediately also yields the rate constant $W_{des}$ to be used in a simulation.

If the rate constant depends on time then solving the rate equation is often much more difficult. We can always rewrite the rate equation as

$$\frac{1}{\theta}\frac{d\theta}{dt} = \frac{d\ln\theta}{dt} = -k_{des}. \tag{4.80}$$

Integrating this equation yields

$$\ln\theta(t) - \ln\theta(0) = -\int_0^t dt' k_{des}\left(t'\right), \tag{4.81}$$

or

$$\theta(t) = \theta(0)\exp\left[-\int_0^t dt' k_{des}\left(t'\right)\right]. \tag{4.82}$$

Whether of not we can get an analytical solution depends on whether we can determine the integral. In TPD experiments we have

$$k_{des}(t) = \nu\exp\left[-\frac{E_{act}}{k_B(T_0 + Bt)}\right] \tag{4.83}$$

with $E_{act}$ an activation energy, $\nu$ a pre-exponential factor, $k_B$ the Boltzmann-factor, $T_0$ the temperature at time $t = 0$, and $B$ the heating rate. The integral can be calculated analytically. The result is

$$\int_0^t dt'\nu\exp\left[-\frac{E_{act}}{k_B(T_0 + Bt')}\right] = \Omega(t) - \Omega(0) \tag{4.84}$$

with

$$\Omega(t) = \frac{\nu}{B}(T_0 + Bt)\mathrm{E}_2\left[\frac{E_{act}}{k_B(T_0 + Bt)}\right], \tag{4.85}$$

where $E_2$ is an exponential integral [11]. Although this solution has been derived some time ago [52], it has not yet been used in the analysis of experimental spectra, but there are several numerical techniques that work well for such simple desorption [53]. Note that we have not made any approximations here and the rate constant $W_{des}$ that we obtain will be exact except for experimental errors.

### 4.6.3 Simple Adsorption

We start with the simplest case in which the adsorption rate is proportional to the number of vacant sites, which is called Langmuir adsorption. We will only indicate in this section in what way the common situation in which the adsorption is higher than expected based on the number of vacant sites differs [2, 54, 55]. This so-called precursor-mediated adsorption is really a composite process.

Again suppose we have atoms or molecules that adsorb onto one particular type of site. We assume that we have a surface of area $A$ with $S$ adsorption sites. If $N_\alpha$ is the number of atoms/molecules in configuration $\alpha$ then again

$$\frac{d\langle N\rangle}{dt} = \sum_{\alpha\beta} W_{\alpha\beta} P_\beta [N_\alpha - N_\beta]. \tag{4.86}$$

Diffusion can again be ignored as for desorption (see Sect. 4.6.2). For the summation over $\alpha$ we have to distinguish between two types of terms: the ones in which $\alpha$ can originate from $\beta$ by a adsorption, and the ones it cannot. The latter terms have $W_{\alpha\beta} = 0$ and so they do not contribute to the sum. The former do contribute and we have $W_{\alpha\beta} = W_{ads}$, with $W_{ads}$ the rate constant for adsorption, and $N_\alpha - N_\beta = 1$, because adsorption adds one atom or molecule to the surface. So all these non-zero terms contribute equally to the sum for a given configuration $\beta$. Moreover, the number of these terms is equal to the number of vacant sites in $\beta$ onto which the molecules can adsorb, because each adsorption yields a different $\alpha$. The number of vacant sites in configuration $\beta$ equals $S - N_\beta$, so

$$\frac{d\langle N\rangle}{dt} = W_{ads} \sum_\beta P_\beta (S - N_\beta) = W_{ads}\big(S - \langle N\rangle\big). \tag{4.87}$$

Dividing by the number of sites $S$ gives the rate equation for the coverage $\theta = \langle N\rangle/S$.

$$\frac{d\theta}{dt} = -W_{ads}(1-\theta). \tag{4.88}$$

If we compare this to the macroscopic rate equation $d\theta/dt = k_{ads}(1-\theta)$ with $k_{ads}$ the macroscopic rate constant, we see that $k_{ads} = W_{ads}$.

So far adsorption is almost the same as desorption. The only difference is where we had $\theta$ for desorption we have $1-\theta$ for adsorption on the right-hand-side of the rate equation. An important difference now arises however. Whereas the macroscopic rate constant for desorption $k_{des}$ is an basic quantity in kinetics of surface reactions, $k_{ads}$ is generally related to other properties. This is because the adsorption process consists of atoms or molecules impinging on the surface, and that is something that can be described very well with kinetic gas theory.

Suppose that the pressure of the gas is $P$ and its temperature $T$, then the number of molecules $F$ hitting a surface of unit area per unit time is given by [7, 56]

$$F = \frac{P}{\sqrt{2\pi m k_B T}} \tag{4.89}$$

with $m$ the mass of the atom or molecule. Not every atom or molecule that hits a surface will stick to it. The sticking coefficient $\sigma$ is defined as the ratio of the number of molecules that stick to the total number hitting the surface. It can also be looked upon as the probability that an atom or molecule hitting the surface sticks. The change in the number of molecules in an area $A$ due to adsorption can then be written as the vacant area times the flux $F$ times the sticking coefficient $\sigma$. The vacant area equals the area $A$ times the fraction of sites in that area that is not occupied. This all leads to

$$\frac{d\langle N\rangle}{dt} = A(1-\theta)F\sigma. \tag{4.90}$$

If we compare this to the equations above we find

$$W_{ads} = \frac{AF\sigma}{S} = \frac{PA_{site}\sigma}{\sqrt{2\pi m k_B T}}, \tag{4.91}$$

where $A_{site}$ is the area of a single site.

Section 4.4.3 contains a discussion of the role of the sticking coefficient in the calculation of adsorption rate constants. The remarks there are just as appropriate here. In particular, whether or not the sticking coefficient is included in the expressions for the kMC rate constant depends on how adsorption, but also desorption (see Sect. 4.4.3), is defined.

Adsorption described so far is proportional to the number of vacant sites. Experiments measure the rate of adsorption and with the expressions derived above one can calculate the microscopic rate constant $W_{ads}$. However, it is often found that the rate of adsorption starts at a certain value for a bare surface and then hardly changes when particles adsorb until the surface is almost completely covered when it suddenly drops to zero. This behavior is generally explained by describing the adsorption as a composite process [2, 54, 55]. A molecule impinging unto the surface adsorbs with the probability $\sigma$ when the site it hits is vacant just as before. However, a molecule that hits a site that is already occupied need not be scattered. It can adsorb indirectly. It first adsorbs, with a certain probability, in a second adsorption layer. Then it starts to diffuse over the surface in this second layer. It can desorb at a later stage, or, and that's the important part, it can encounter a vacant site and adsorb there permanently. This last part can increase the adsorption rate substantially when there are already many sites occupied. The precise dependence of the adsorption rate on the coverage $\theta$ is determined by the rate of diffusion, by the rate of adsorption onto the second layer, and by the rate of desorption from the second layer. If there are factors that affect the structure of the first adsorption layer, e.g. lateral interaction, then these too influence the adsorption rate. If the adsorption is not direct, one talks about a precursor mechanism. A precursor on top of an adsorbed particle is an extrinsic precursor. An intrinsic precursor can be found on top of a vacant site [57]. The precursor mechanism will not always be operative for a bare surface: i.e., there is not always an intrinsic precursor. This means that we can use Eq. (4.91) if we take for $\sigma$ the sticking coefficient for adsorption on a bare surface.

### 4.6.4 Unimolecular Reactions

With the knowledge of simple desorption and adsorption given above it is now easy to derive an expression for the rate constant $W_{\text{uni}}$ for a unimolecular reaction in term of a macroscopic rate constant. In fact the derivation is exactly the same as for the desorption. Desorption changes a site from A to $*$, whereas a unimolecular reaction changes it to B. Replace $*$ by B in the expression for the desorption (and $W_{\text{des}}$ by $W_{\text{uni}}$) and you have the correct expression. As the expression for desorption do not contain a $*$, the procedure is trivial and we find $W_{\text{uni}} = k_{\text{uni}}$ where $k_{\text{uni}}$ is the rate constant from the macroscopic rate equation.

### 4.6.5 Diffusion

We treat diffusion in kMC as any other process, but experimentally one doesn't look at changes in coverages but at displacements of atoms and molecules. We will therefore also look here at how the position of a particle changes.

We assume that we have only one particle on the surface, so that the particle's movement is not hindered by any other particle and the relation between the macroscopic properties and the kMC rate constant is simple. In practice this means that the coverage should be low. We also assume that we have a square lattice with axis parallel to the $x$- and the $y$-axis and that the distance between lattice points is given by $a$. We will later look at other lattices. If $x_\alpha$ is the $x$-coordinate of the particle in configuration $\alpha$, then

$$\frac{d\langle x\rangle}{dt} = \sum_{\alpha\beta} W_{\alpha\beta} P_\beta [x_\alpha - x_\beta]. \tag{4.92}$$

The $x$-coordinate changes because the particle hops from one site to another. When it hops we have $x_\alpha - x_\beta = a$, $-a$, and 0 for a hop along the $x$-axis toward larger $x$, a hop along the $x$-axis toward smaller $x$, or a hop perpendicular to the $x$-axis, respectively. All these hops have a rate constant $W_{\text{hop}}$ and are equally likely. This means $d\langle x\rangle/dt = 0$. The same holds for the $y$-coordinate.

More useful is to look at the square of the coordinates. We then find

$$\frac{d\langle x^2\rangle}{dt} = \sum_{\alpha\beta} W_{\alpha\beta} P_\beta \left[x_\alpha^2 - x_\beta^2\right]. \tag{4.93}$$

Now we have $x_\alpha^2 - x_\beta^2 = 2ax_\beta + a^2$, $-2ax_\beta + a^2$, and 0, respectively. Because the hops are still equally likely, we have

$$\frac{d\langle x^2\rangle}{dt} = 2W_{\text{hop}}a^2. \tag{4.94}$$

We find the same for the $y$-coordinate. The macroscopic equation for diffusion is

$$\frac{d\langle x^2 + y^2\rangle}{dt} = 4D, \tag{4.95}$$

with $D$ the diffusion coefficient. From this we see that we have $W_{\text{hop}} = D/a^2$.

On a hexagonal lattice a particle can hop in six different directions for which $x_\alpha - x_\beta = a, a/2, -a/2, -a, -a/2$, and $a/2$ and $y_\alpha - y_\beta = 0, a\sqrt{3}/2, a\sqrt{3}/2, 0, -a\sqrt{3}/2$, and $-a\sqrt{3}/2$. From this we get again $d\langle x\rangle/dt = 0$. For the squared displacement we find $x_\alpha^2 - x_\beta^2 = 2ax_\beta + a^2, ax_\beta + a^2/4, -ax_\beta + a^2/4, -2ax_\beta + a^2, -ax_\beta + a^2/4, ax_\beta + a^2/4$. This yields again $d\langle x^2\rangle/dt = 2W_{\text{hop}}a^2$. We find the same expression for the $y$-coordinate, so that also for a hexagonal lattice $W_{\text{hop}} = D/a^2$. The same expression holds for a trigonal lattice. The derivation is identical to the ones for the square and hexagonal lattices.

### 4.6.6 Bimolecular Reactions

For all of the processes we have looked at so far it was possible to derive exact macroscopic equations from the master equation. This is not the case for bimolecular reactions. Bimolecular reactions will give rise to an infinite hierarchy of exact macroscopic rate equations. There are two bimolecular reactions we will consider: A + B and A + A. The problem we have mentioned above is the same for both reactions, but there is a small difference in the derivation of a numerical factor in the macroscopic rate equation. We will start with the A + B reaction.

We look at the number of A's. The expressions for the number of B's can be obtained by replacing A's by B's and B's by A's in the following expressions. We have

$$\frac{d\langle N^{(\text{A})}\rangle}{dt} = \sum_{\alpha\beta} W_{\alpha\beta} P_\beta \left[N_\alpha^{(\text{A})} - N_\beta^{(\text{A})}\right], \tag{4.96}$$

where $N_\alpha^{(\text{A})}$ stands for the number of A's. We look at the summation over $\alpha$ and mentally fix $\beta$. If $\alpha$ can originate from $\beta$ by a A + B reaction, then $W_{\alpha\beta} = W_{\text{rx}}$, otherwise $W_{\alpha\beta} = 0$. If such a reaction is possible, then $N_\alpha^{(\text{A})} - N_\beta^{(\text{A})} = -1$. The problem now is with the number of configurations $\alpha$ that can be obtained from $\beta$ by a reaction. This number is equal to the number of AB pairs $N_\beta^{(\mathrm{AB})}$. This leads then to

$$\frac{d\langle N^{(\mathrm{A})}\rangle}{dt} = -W_{\mathrm{rx}} \sum_\beta P_\beta N_\beta^{(\mathrm{AB})} = -W_{\mathrm{rx}} \langle N^{(\mathrm{AB})}\rangle. \tag{4.97}$$

We get the same right-hand-side for the change in the number of B's. We see that on the right-hand-side we have obtained a quantity that we didn't have before. This means that the rate equations are not closed.

We can now proceed in two ways. The first is to write down rate equations for the new quantity $\langle N^{(\mathrm{AB})}\rangle$ and hope that this will lead to equations that are closed. If we do this, we find that this will not happen. Instead we will get a right-hand-side that depends on the number of certain combinations of three particles. We can write down rate equations for these as well, and hope that this will lead finally to a closed set of equations. But that too won't happen. Proceeding by writing rate equations for the new quantities that we obtain will lead to an infinite hierarchy of equations [58–63].

The second way to proceed is to introduce an approximation that will make a finite set of these equations into a closed set. We can do this at different levels. The crudest approximation, and the one that will lead to the common macroscopic rate equations, is to approximate $\langle N^{(\mathrm{AB})}\rangle$ in terms of $\langle N^{(\mathrm{A})}\rangle$ and $\langle N^{(\mathrm{B})}\rangle$. This actually turns out to involve two approximations. The first one is that we assume that the number of adsorbates are randomly distributed over the surface. In this case we have $N_\beta^{(\mathrm{AB})} = Z N_\beta^{(\mathrm{A})}[N_\beta^{(\mathrm{B})}/(S-1)]$, with $Z$ the coordination number of the lattice: i.e., the number of nearest neighbors of a site. ($Z = 4$ for a square lattice, $Z = 6$ for a hexagonal lattice, and $Z = 3$ for a trigonal lattice.) The quantity between square brackets is the probability that a neighboring site of an A is occupied by a B. This approximation is called the Mean Field Approximation (MFA) and leads to

$$\frac{d\langle N^{(\mathrm{A})}\rangle}{dt} = -\frac{Z}{S-1} W_{\mathrm{rx}} \sum_\beta P_\beta N_\beta^{(\mathrm{A})} N_\beta^{(\mathrm{B})} = -\frac{Z}{S-1} W_{\mathrm{rx}} \langle N^{(\mathrm{A})} N^{(\mathrm{B})}\rangle. \tag{4.98}$$

This is still not a closed expression. We have

$$\langle N^{(\mathrm{A})} N^{(\mathrm{B})}\rangle = \langle N^{(\mathrm{A})}\rangle\langle N^{(\mathrm{B})}\rangle + \langle [N^{(\mathrm{A})} - \langle N^{(\mathrm{A})}\rangle][N^{(\mathrm{B})} - \langle N^{(\mathrm{B})}\rangle]\rangle. \tag{4.99}$$

The second term on the right stands for the correlation between fluctuations in the number of A's and the number of B's. In general this is not zero. Because the number of A's and B's decrease simultaneously because of the reaction, this term is expected to be positive. Fluctuations however decrease when the system size is increased. In the thermodynamic limit $S \to \infty$ we can set it to zero. (Neglecting the fluctuations is often included in MFA.) We then finally get

$$\frac{d\langle N^{(\mathrm{A})}\rangle}{dt} = -\frac{Z}{S} W_{\mathrm{rx}} \langle N^{(\mathrm{A})}\rangle\langle N^{(\mathrm{B})}\rangle \tag{4.100}$$

with $S-1$ replaced by $S$ because $S \gg 1$. Dividing by the number of sites $S$ leads then to

$$\frac{d\theta_{\mathrm{A}}}{dt} = -ZW_{\mathrm{rx}}\theta_{\mathrm{A}}\theta_{\mathrm{B}}. \tag{4.101}$$

This should be compared to the macroscopic rate equation

$$\frac{d\theta_{\mathrm{A}}}{dt} = -k_{\mathrm{rx}}\theta_{\mathrm{A}}\theta_{\mathrm{B}}. \tag{4.102}$$

We see from this that we have $W_{\mathrm{rx}} = k_{\mathrm{rx}}/Z$, but only if the two approximations are valid. This may not be the case when the adsorbates form some kind of structure (e.g. islands or a superstructure) or when the system is small (e.g. a small cluster of metal atoms) and the fluctuations are large.

The derivation for the A + A reaction is almost the same. We have

$$\frac{d\langle N^{(\mathrm{A})}\rangle}{dt} = \sum_{\alpha\beta} W_{\alpha\beta}P_\beta\big[N_\alpha^{(\mathrm{A})} - N_\beta^{(\mathrm{A})}\big]. \tag{4.103}$$

If $\alpha$ can originate from $\beta$ by a A+A reaction, then $W_{\alpha\beta} = W_{\mathrm{rx}}$, otherwise $W_{\alpha\beta} = 0$. If such a reaction is possible, then $N_\alpha^{(\mathrm{A})} - N_\beta^{(\mathrm{A})} = -2$, because now two A's react. The number of configurations $\alpha$ that can be obtained from $\beta$ by a reaction is equal to the number of AA pairs $N_\beta^{(\mathrm{AA})}$. This leads then to

$$\frac{d\langle N^{(\mathrm{A})}\rangle}{dt} = -2W_{\mathrm{rx}}\sum_\beta P_\beta N_\beta^{(\mathrm{AA})} = -2W_{\mathrm{rx}}\big\langle N^{(\mathrm{AA})}\big\rangle. \tag{4.104}$$

If we do not want to get an infinite hierarchy of equations with rate equations for quantities of more and more A's, we have to make an approximation again. We approximate $\langle N^{(\mathrm{AA})}\rangle$ in terms of $\langle N^{(\mathrm{A})}\rangle$. We first assume that the number of adsorbates are randomly distributed over the surface. In this case we have $N_\beta^{(\mathrm{AA})} = (1/2)ZN_\beta^{(\mathrm{A})}[N_\beta^{(\mathrm{A})}/S]$. Note the factor 1/2 that avoids double counting of the number of AA pairs. The quantity between square brackets is the probability that a neighboring site of an A is occupied by a A. We have immediately assumed here that $S \gg 1$. This approximation leads to

$$\frac{d\langle N^{(\mathrm{A})}\rangle}{dt} = -\frac{Z}{S}W_{\mathrm{rx}}\sum_\beta P_\beta\big(N_\beta^{(\mathrm{A})}\big)^2 = -\frac{Z}{S}W_{\mathrm{rx}}\big\langle\big(N^{(\mathrm{A})}\big)^2\big\rangle. \tag{4.105}$$

The factor 2 that we had previously has canceled against the factor 1/2 in the expression for the number of AA pairs. To proceed we note that

$$\big\langle\big(N^{(\mathrm{A})}\big)^2\big\rangle = \big\langle N^{(\mathrm{A})}\big\rangle^2 + \big\langle\big(N^{(\mathrm{A})} - \big\langle N^{(\mathrm{A})}\big\rangle\big)^2\big\rangle. \tag{4.106}$$

The second term on the right stands for the fluctuations in the number of A's. This is clearly not zero, but positive. Setting it to zero is again the thermodynamic limit. We finally get

$$\frac{d\langle N^{(\mathrm{A})}\rangle}{dt} = -\frac{Z}{S}W_{\mathrm{rx}}\big\langle N^{(\mathrm{A})}\big\rangle^2. \tag{4.107}$$

Dividing by the number of sites $S$ leads then to

$$\frac{d\theta_{\mathrm{A}}}{dt} = -ZW_{\mathrm{rx}}\theta_{\mathrm{A}}^2. \tag{4.108}$$

This should be compared to the macroscopic rate equation

$$\frac{d\theta_{\mathrm{A}}}{dt} = -2k_{\mathrm{rx}}\theta_{\mathrm{A}}^2. \tag{4.109}$$

Note that there is a factor 2 on the right-hand-side, which is used because a reaction removes two A's. We see from this that we have $W_{\mathrm{rx}} = 2k_{\mathrm{rx}}/Z$.

### 4.6.7 Dissociative Adsorption

We deal here with the quite common case of a molecule of the type $B_2$ that adsorbs dissociatively on two neighboring sites. An example of such adsorption is oxygen adsorption on many transition metal surfaces. We will see this adsorption for example when we will discuss the Ziff–Gulari–Barshad model in Chap. 7. We will see here that it is often convenient to look at limiting cases to derive an expression of the rate constant of adsorption.

We look at the number of B's. We have again

$$\frac{d\langle N^{(\mathrm{B})}\rangle}{dt} = \sum_{\alpha\beta} W_{\alpha\beta} P_\beta \bigl[N_\alpha^{(\mathrm{B})} - N_\beta^{(\mathrm{B})}\bigr], \tag{4.110}$$

where $N_\alpha^{(\mathrm{B})}$ stands for the number of B's. If $\alpha$ can originate from $\beta$ by an adsorption reaction, then $W_{\alpha\beta} = W_{\mathrm{ads}}$, otherwise $W_{\alpha\beta} = 0$. If such a reaction is possible, then $N_\alpha^{(\mathrm{B})} - N_\beta^{(\mathrm{B})} = 2$. The problem now is with the number of configurations $\alpha$ that can be obtained from $\beta$ by a reaction. This number is equal to the number of pairs of neighboring vacant sites $N_\beta^{(**)}$. This leads then to

$$\frac{d\langle N^{(\mathrm{B})}\rangle}{dt} = 2W_{\mathrm{ads}} \sum_\beta P_\beta N_\beta^{(**)} = 2W_{\mathrm{ads}}\bigl\langle N^{(**)}\bigr\rangle. \tag{4.111}$$

The right-hand-side can in general only be approximated, but such an approximation is not needed for the case of a bare surface. In that case we have $N^{(**)} = ZS/2$, where $Z$ is the coordination number of the lattice and $S$ the number of sites in the system. This leads to

$$\frac{d\langle N^{(\mathrm{B})}\rangle}{dt} = ZSW_{\mathrm{ads}}. \tag{4.112}$$

The change in the number of adsorbates for a bare surface is also equal to

$$\frac{d\langle N^{(\mathrm{B})}\rangle}{dt} = 2AF\sigma, \tag{4.113}$$

where $A$ is the area of the surface, $F$ is the number of particles hitting a unit area of the surface per unit time, and $\sigma$ is the sticking coefficient. The factor 2 is due to the fact that a molecule that adsorbs dissociatively yields two adsorbates. The flux $F$ we've seen before and is given by

$$F = \frac{P}{\sqrt{2\pi m k_B T}} \tag{4.114}$$

with $P$ the pressure, $T$ the temperature and $m$ the mass of a molecule. This means that

$$\frac{d\langle N^{(B)}\rangle}{dt} = \frac{2AP\sigma}{\sqrt{2\pi m k_B T}}. \tag{4.115}$$

If we compare this with expression (4.112), we get

$$W_{ads} = \frac{2A_{site}P\sigma}{Z\sqrt{2\pi m k_B T}} \tag{4.116}$$

with $A_{site}$ the area of a single site. (See Sects. 4.4.3 and 4.6.3 for a discussion of the role of the sticking coefficient.)

### 4.6.8 A Brute-Force Approach

Because kMC can be derived from first-principles it is exact for a given model or set of processes. So if we know which processes take place in a system, we know which lateral interactions are important, and we know the atomic structure of the system and which sites are relevant, then we can in principle determine all kinetic parameters by fitting kMC simulations to the experimental results. In fact, it might even be possible to use this to compare models and decide which one is the better one. Fitting kinetic parameters of different models and comparing the quality of the fits might allow us to choose between various candidate models.

There are however a number of pitfalls to this approach. Most of them are typical for fitting procedures but there is also one that has to do with the nature of kMC simulations. General problems with the approach are the following. Fitting will entail minimizing a function that indicates how much the kMC and the experimental results differ. Such a function may have multiple minima. Often one wants to have the global minimum, but it may be that such a minimum corresponds to kinetic parameters that are unphysical (e.g., ridiculous values for interaction parameters or negative activation energies). The fitting procedure should therefore be restricted to those areas of the parameter space that correspond the acceptable values for the kinetic parameters.

Although a minimum of the fitting function may be mathematically well defined, it is not uncommon that the fitting function changes very little in certain directions. This means that the experimental results do not depend very much or not at all on one or more kinetic parameters or certain combinations of kinetic parameters. As a consequence such parameters can not be determined well and the value that

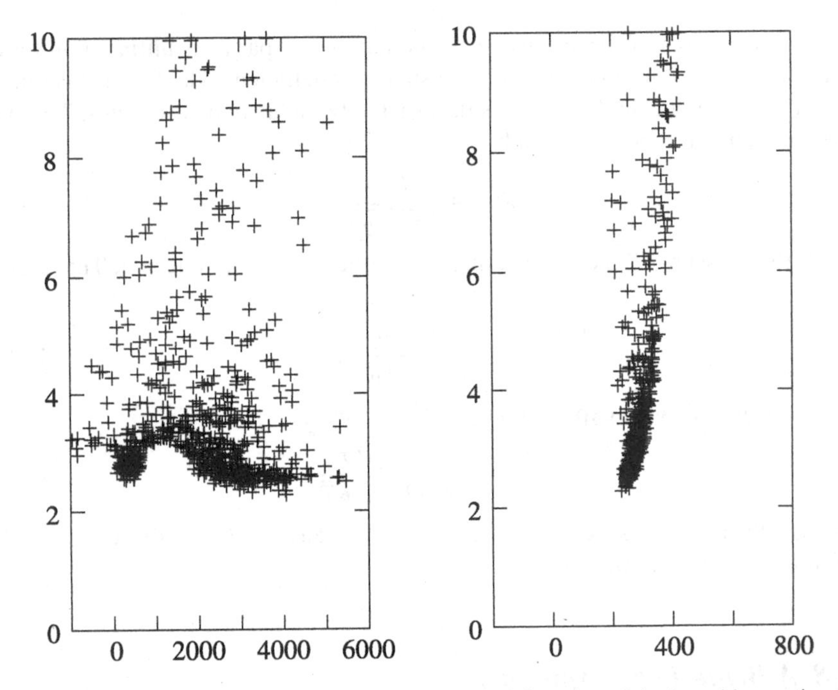


**Fig. 4.5** The value of the fitting function as a function of the value of the lateral interaction (divided by $k_B$ in K) between CO molecules at top sites on Rh(100) at nearest neighbors-distances (*left*) and at next-nearest-neighbors distances (*right*). Each *cross* corresponds to a single set of kinetic parameters for which kMC results have been compared to experimental TPD-spectra during the fitting procedure

the fitting procedure will give for them will be extremely sensitive to errors in the experimental results.

kMC has on top of this the problem that it is a stochastic method. This means that many numerical methods can not be used, for example because they use derivatives or because they become unstable. So we need a method that does a global minimization in a restricted parameter space and that can handle noisy data. It should also indicate when a parameter does not influence the quality of the fit so that it can not be determined. If possible it should also be efficient so that the number of simulations that need to be done is small. There are a number of methods that can be considered. Simulating annealing and computational evolution methods are prime candidates [64–69], but particle swarm optimization, memetic algorithms, tabu search, and possibly others might be useful as well [69, 70].

Differential evolution was used in a study to obtain kinetic parameters for desorption of CO on Rh(100) at low coverages [71]. The prefactor, the activation energy, and lateral interactions at three different distances between the CO molecules were determined. Figure 4.5 shows how the quality of the fit (i.e., the fitting function that needs to be minimized) depends on the interaction between nearest- and next-nearest neighbors. Note that variations in the nearest-neighbor interaction affects the fit hardly at all. This is because this interaction is strongly repulsive. The coverage is

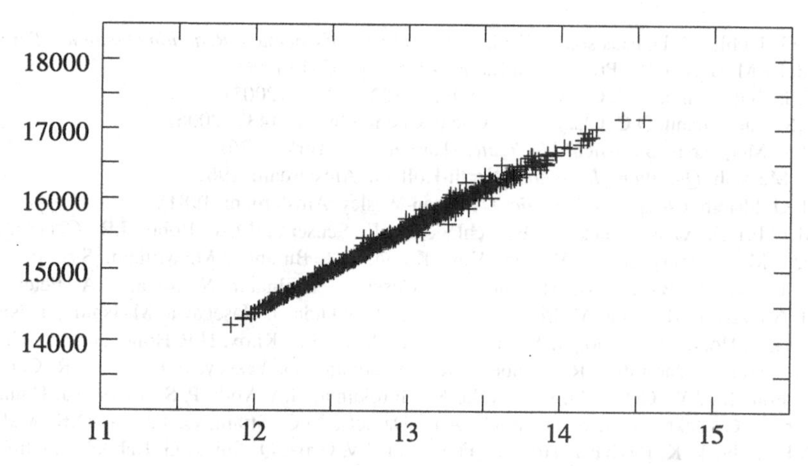


**Fig. 4.6** Results of a fitting procedure to determine the kinetic parameters for desorption of CO molecules at top sites on Rh(100). Each *cross* corresponds to a single set of kinetic parameters for which kMC results have been compared to experimental TPD-spectra during the fitting procedure and for which the fitting function was less than 10. The 10-base logarithm of the prefactor (in 1/s) is plotted on the *horizontal* and the activation energy (divided by $k_B$ in K) is plotted on the *vertical* axis

low so that the system can and does avoid it and it does therefore not affect the TPD spectra. The spectra do depend on the next-nearest-neighbor interaction, which is consequently much better defined. Figure 4.6 shows the region in (prefactor, activation energy) space for which the fitting function is less than 10. Note that this region is very elongated. This means that there is a combination of these parameters that is well defined and another one that is not. The one that is well defined turns out to be the rate constant for desorption.

A follow-up study looked at higher coverages where some CO molecules not only adsorbed at the preferred top but also at bridge sites and where there were more relevant kinetic parameters [72]. This study showed that it is important to have a good model for the interactions in a system to get meaningful values for the interaction parameters. Although the model in this study had double the number of kinetic parameters as the one for low coverages, and the experimental TPD spectra could reasonably well be reproduced, the values of the lateral interactions were quite dubious. More work on this way to use kMC simulations to get kinetic parameters is needed.

## References

1. C.G.M. Hermse, F. Frechard, A.P. van Bavel, J.J. Lukkien, J.W. Niemantsverdriet, R.A. van Santen, A.P.J. Jansen, J. Chem. Phys. **118**, 7081 (2003)
2. R.A. van Santen, J.W. Niemantsverdriet, *Chemical Kinetics and Catalysis* (Plenum, New York, 1995)
3. W.H. Press, B.P. Flannery, S.A. Teukolsky, W.T. Vetterling, *Numerical Recipes. The Art of Scientific Computing* (Cambridge University Press, Cambridge, 1989)

4. D.G. Truhlar, A.D. Isaacson, B.C. Garrett, in *Theory of Chemical Reaction Dynamics, Part IV*, ed. by M. Baer (CRC Press, Boca Raton, 1985), pp. 65–138
5. C.S. Tautermann, D.C. Clary, J. Chem. Phys. **122**, 134702 (2005)
6. C.S. Tautermann, D.C. Clary, Phys. Chem. Chem. Phys. **8**, 1437 (2006)
7. D.A. McQuarrie, *Statistical Mechanics* (Harper, New York, 1976)
8. A. Messiah, *Quantum Mechanics* (North-Holland, Amsterdam, 1961)
9. H. Goldstein, *Classical Mechanics* (Addison-Wesley, Amsterdam, 1981)
10. M.J. Frisch, G.W. Trucks, H.B. Schlegel, G.E. Scuseria, M.A. Robb, J.R. Cheeseman, J.A. Montgomery Jr., T. Vreven, K.N. Kudin, J.C. Burant, J.M. Millam, S.S. Iyengar, J. Tomasi, V. Barone, B. Mennucci, M. Cossi, G. Scalmani, N. Rega, G.A. Petersson, H. Nakatsuji, M. Hada, M. Ehara, K. Toyota, R. Fukuda, J. Hasegawa, M. Ishida, T. Nakajima, Y. Honda, O. Kitao, H. Nakai, M. Klene, X. Li, J.E. Knox, H.P. Hratchian, J.B. Cross, C. Adamo, J. Jaramillo, R. Gomperts, R.E. Stratmann, O. Yazyev, A.J. Austin, R. Cammi, C. Pomelli, J.W. Ochterski, P.Y. Ayala, K. Morokuma, G.A. Voth, P. Salvador, J.J. Dannenberg, V.G. Zakrzewski, S. Dapprich, A.D. Daniels, M.C. Strain, O. Farkas, D.K. Malick, A.D. Rabuck, K. Raghavachari, J.B. Foresman, J.V. Ortiz, Q. Cui, A.G. Baboul, S. Clifford, J. Cioslowski, B.B. Stefanov, G. Liu, A. Liashenko, P. Piskorz, I. Komaromi, R.L. Martin, D.J. Fox, T. Keith, M.A. Al-Laham, C.Y. Peng, A. Nanayakkara, M. Challacombe, P.M.W. Gill, B. Johnson, W. Chen, M.W. Wong, C. Gonzalez, J.A. Pople, Gaussian 03, revision b.04. Gaussian, Inc., Pittsburgh, PA (2003)
11. M. Abramowitz, I.A. Stegun, *Handbook of Mathematical Functions with Formulas, Graphs, and Mathematical Tables* (Dover, New York, 1965)
12. A. Szabo, N.S. Ostlund, *Modern Quantum Chemistry: Introduction to Advanced Electronic Structure Theory* (McGraw-Hill, New York, 1982)
13. D. Frenkel, B. Smit, *Understanding Molecular Simulation: From Algorithms to Applications* (Academic Press, London, 2001)
14. A.R. Leach, *Molecular Modelling. Principles and Applications* (Longman, Singapore, 1996)
15. W. Koch, M.C. Holthausen, *A Chemist's Guide to Density Functional Theory* (Wiley-VCH, New York, 2000)
16. C.J. Cramer, *Essentials of Computational Chemistry* (Wiley, Chichester, 2004)
17. R.A. van Santen, M. Neurock, *Molecular Heterogeneous Catalysis* (Wiley-VCH, Weinheim, 2006)
18. O. Trushin, A. Karim, A. Kara, T.S. Rahman, Phys. Rev. B **72**, 115401 (2005)
19. K. Sastry, D.D. Johnson, D.E. Goldberg, P. Bellon, Phys. Rev. B **72**, 085438 (2005)
20. N. Castin, L. Malerba, Nucl. Instrum. Methods Phys. Res., Sect. B, Beam Interact. Mater. Atoms **267**, 3148 (2009)
21. G. Henkelman, H. Jónsson, J. Chem. Phys. **111**, 7010 (1999)
22. G. Henkelman, G. Jóhannesson, H. Jónsson, in *Progress in Theoretical Chemistry and Physics*, ed. by S.D. Schwarts (Kluwer Academic, London, 2000)
23. G. Mills, H. Jónsson, G.K. Schenter, Surf. Sci. **324**, 305 (1995)
24. C. Popa, W.K. Offermans, R.A. van Santen, A.P.J. Jansen, Phys. Rev. B **74**, 155428 (2006)
25. C. Popa, R.A. van Santen, A.P.J. Jansen, J. Phys. Chem. C **111**, 9839 (2007)
26. D. Curulla, A.P. van Bavel, J.W. Niemantsverdriet, ChemPhysChem **6**, 473 (2005)
27. A.P. van Bavel, Understanding and quantifying interactions between adsorbates: CO, NO, and N- and O-atoms on Rh(100). Ph.D. thesis, Eindhoven University of Technology, Eindhoven (2005)
28. C. Popa, A.P. van Bavel, R.A. van Santen, C.F.J. Flipse, A.P.J. Jansen, Surf. Sci. **602**, 2189 (2008)
29. D.H. Wei, D.C. Skelton, S.D. Kevan, Surf. Sci. **381**, 49 (1997)
30. J.N. Murrell, S. Carter, P. Huxley, S.C. Farantos, A.J.C. Varandas, *Molecular Potential Energy Functions* (Wiley-Interscience, Chichester, 1984)
31. A. van der Walle, G. Ceder, J. Phase Equilibria **23**, 348 (2002)
32. V. Blum, A. Zunger, Phys. Rev. B **69**, 020103(R) (2004)

33. A.P.J. Jansen, W.K. Offermans, in *Computational Science and Its Applications—ICCSA-2005*. LNCS, vol. 3480, ed. by O. Gervasi (Springer, Berlin, 2005)
34. C.G.M. Hermse, A.P.J. Jansen, in *Catalysis*, vol. 19, ed. by J.J. Spivey, K.M. Dooley (Royal Society of Chemistry, London, 2006)
35. Y. Zhang, V. Blum, K. Reuter, Phys. Rev. B **75**, 235406 (2007)
36. D.M. Hawkins, J. Chem. Inf. Comput. Sci. **44**, 1 (2004)
37. A.P.J. Jansen, C. Popa, Phys. Rev. B **78**, 085404 (2008)
38. N.A. Zarkevich, D.D. Johnson, Phys. Rev. Lett. **92**, 255702 (2004)
39. R. Drautz, A. Díaz-Ortiz, Phys. Rev. B **73**, 224207 (2006)
40. D.E. Nanu, Y. Deng, A.J. Böttger, Phys. Rev. B **74**, 014113 (2006)
41. T. Mueller, G. Ceder, Phys. Rev. B **80**, 024103 (2009)
42. E.T. Jaynes, G.L. Bretthorst, *Probability Theory: The Logic of Science* (Cambridge University Press, Cambridge, 2003)
43. A. Gelman, J.B. Carlin, H.S. Stern, D.B. Rubin, *Bayesian Data Analysis* (Chapman & Hall/CRC, Boca Raton, 2003)
44. D. Sivia, J. Skilling, *Data Analysis: A Bayesian Tutorial* (Oxford University Press, Oxford, 2006)
45. B. Hammer, Phys. Rev. B **63**, 205423 (2001)
46. J.N. Brønsted, Chem. Rev. **5**, 231 (1928)
47. M.G. Evans, M. Polanyi, Trans. Faraday Soc. **34**, 11 (1938)
48. P. Hohenberg, W. Kohn, Phys. Rev. **136**, B864 (1964)
49. W. Kohn, L.S. Sham, Phys. Rev. **140**, A1133 (1965)
50. R.G. Parr, W. Yang, *Density-Functional Theory of Atoms and Molecules* (Oxford University Press, New York, 1989)
51. N.G. van Kampen, *Stochastic Processes in Physics and Chemistry* (North-Holland, Amsterdam, 1981)
52. A.P.J. Jansen, Comput. Phys. Commun. **86**, 1 (1995)
53. A.M. de Jong, J.W. Niemantsverdriet, Surf. Sci. **233**, 355 (1990)
54. J.M. Thomas, W.J. Thomas, *Principles and Practice of Heterogeneous Catalysis* (VCH, Weinheim, 1997)
55. G.A. Somorjai, *Introduction to Surface Chemistry and Catalysis* (Wiley, Chichester, 1993)
56. R. Becker, *Theorie der Wärme* (Springer, Berlin, 1985)
57. A. Cassuto, D.A. King, Surf. Sci. **102**, 388 (1981)
58. V.P. Zhdanov, *Elementary Physicochemical Processes on Solid Surfaces* (Plenum, London, 1991)
59. J. Mai, V.N. Kuzovkov, W. von Niessen, Phys. Rev. E **48**, 1700 (1993)
60. J. Mai, V.N. Kuzovkov, W. von Niessen, Physica A **203**, 298 (1994)
61. J. Mai, V.N. Kuzovkov, W. von Niessen, J. Chem. Phys. **100**, 6073 (1994)
62. E.A. Kotomin, V.N. Kuzovkov, *Modern Aspects of Diffusion-Controlled Reactions: Cooperative Phenomena in Bimolecular Processes* (Elsevier, Amsterdam, 1996)
63. O. Kortlüke, V.N. Kuzovkov, W. von Niessen, Chem. Phys. Lett. **275**, 85 (1997)
64. S. Kirkpatric, C.D. Gelatt Jr., M.P. Vecchi, Science **220**, 671 (1983)
65. D.A. Goldberg, *Genetic Algorithms in Search, Optimization, and Machine Learning* (Addison-Wesley, Reading, 1989)
66. H.P. Schwefel, *Evolution and Optimum Seeking* (Wiley, Chichester, 1995)
67. Z. Michalewicz, *Genetic Algorithms + Data Structures = Evolution Programs* (Springer, Berlin, 1999)
68. W. Banzhaf, P. Nordin, R.E. Keller, F.D. Francone, *Genetic Programming: An Introduction* (Morgan Kaufmann, San Francisco, 1998)
69. D. Corne, M. Dorigo, F. Glover, *New Ideas in Optimization* (McGraw-Hill, London, 1999)
70. E. Bonabeau, M. Doriga, G. Theraulaz, *Swarm Intelligence: From Natural to Artificial Systems* (Oxford University Press, New York, 1999)
71. A.P.J. Jansen, Phys. Rev. B **69**, 035414 (2004)
72. M.M.M. Jansen, C.G.M. Hermse, A.P.J. Jansen, Phys. Chem. Chem. Phys. **12**, 8053 (2010)

# Chapter 5
# Modeling Surface Reactions I

**Abstract** The main part of modeling surface reactions is concerned with the description of the processes. For simple systems there is a lattice corresponding to the adsorption sites and the labels of the lattice points describe the occupation of the sites. The labels can however also be used to model steps and other defects and sites on a bimetallic substrates. The lattice points need not necessarily correspond to sites, but can also be used to store other information like the presence of certain structures in the adlayer. Processes need not always correspond to reactions or other actual processes but when they have an infinite rate constant they can be used in a general-purpose code to handle exceptional situations that are normally hard-coded in special-purpose codes. Apart from implementing the processes, modeling also involves specifying system size, length of a simulation, and I/O. This is discussed in relation to reducing the noise in the output of a kinetic Monte Carlo simulation.

## 5.1 Introduction

In this chapter we will start looking at how to model surface processes. This may seem rather trivial. One "just" has to specify which sites are involved in a process, and the occupation of these sites before and after the process. It turns out that this is often indeed all one needs to do to simulate a system, but often there are various ways to model a system, and then the question is which one gives the most efficient simulation. This is typically the case when one has a system with different types of site, processes with very different rate constants, diffusion, and/or lateral interactions.

This chapter is particularly important for people using general-purpose codes to do kinetic Monte Carlo (kMC) simulations, because they have the opportunity to vary their model. If you use a special-purpose code then you do not have that luxury. You are forced to use the model that is hard-coded. However, this chapter is also relevant if you want to write such special-purpose codes, because there may be various ways your model can be coded.

The modeling framework that we present here is based on the assumption that a kMC code can recognize patterns in the occupation of sites, and that it can then change those patterns according to rules that stand for how processes change the site

A.P.J. Jansen, *An Introduction to Kinetic Monte Carlo Simulations of Surface Reactions*, Lecture Notes in Physics 856,
DOI 10.1007/978-3-642-29488-4_5, 

occupations. This framework is explained in Sect. 5.3, and how simple processes are modeled with it in Sect. 5.4.

The framework however is much more flexible and powerful than the straightforward implementation of simple processes. Section 5.5 shows how it can be used to model various types of substrate: e.g., multiple sites in the unit cell, steps and other defects, bimetallic surfaces, and reconstructions.

For some systems the pattern matching approach seems not to be appropriate however. For example, if we have a long reptating chainlike molecule, the number of possible patterns or configurations of the molecule can simply be too large to specify. In a special-purpose code this would be modeled by code that would first move the head of the molecule, then move along its length to find the tail, and finally move the tail. We will show that we can do something similar within our framework with so-called immediate processes. Such processes are also convenient for many other situations. Some examples are shown in Sect. 5.6.

This chapter discusses only the components that can be used to model reaction systems. It gives an overview of the tools that we can use in modeling reaction systems with a general-purpose kMC code. It does not say anything about what would be a good way to model certain processes. This is done in Chap. 6, which shows how these tools can then be used to model specific processes in various ways. Chapter 7 shows examples how reaction systems have actually been modeled. That chapter focuses on the kind of kinetic information one can obtain from kMC simulations.

Apart from specifying the processes, to do a kMC simulation you also need to specify other parameters like the size of the system that you want to simulate, the length of the simulation, information on I/O, and possibly other things. We start this chapter with a discussion of these other parameters and in particular on how to deal with the stochastic nature of kMC. We will see that there are different ways to reduce the noise that is intrinsic in the results of kMC simulations.

## 5.2 Reducing Noise

kMC is a stochastic method. This means that the results will show fluctuations and will be noisy, and one often wants to do something to reduce the noise as it may be unacceptably large. This is often the case because one wants to simulate a surface with macroscopic dimensions, but one can only simulate much smaller systems. In principle, there are three methods that one can use. We will discuss these methods for the Ziff–Gulari–Barshad (ZGB) model of CO oxidation and for Temperature-Programmed Desorption (TPD) spectra of simple desorption. Details of the ZGB model and its implementation can be found in Sect. 7.4.3. Section 5.4.1 has details on TPD and simple desorption. These details are not relevant however for the discussion here on noise reduction.

Figure 5.1 shows how the rate of $CO_2$ production in kMC simulations of the ZGB model changes if we average over several simulations. As the number of simulations increases, the noise decreases. Because the simulations are independent of

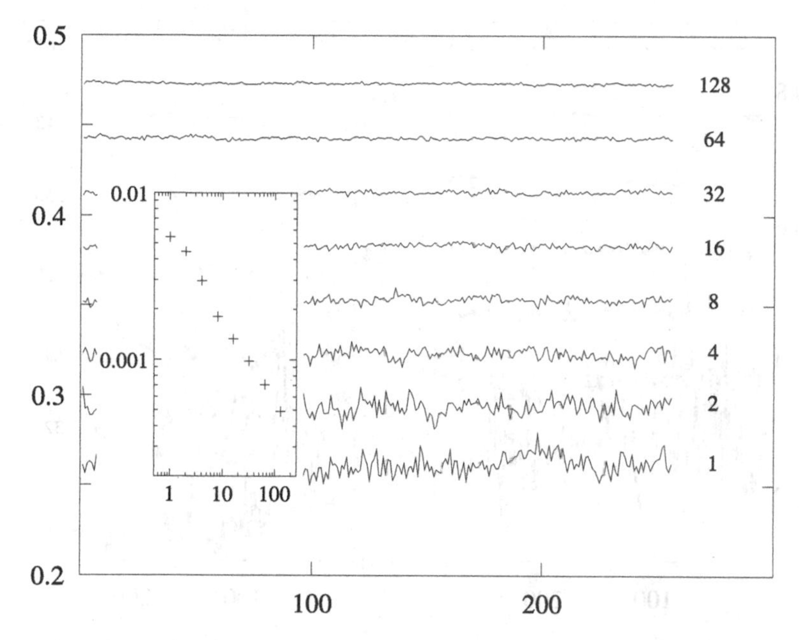


**Fig. 5.1** The reactivity (in reciprocal units of time) defined as the number of reactions forming $CO_2$ per unit cell per unit of time as a function of time in the ZGB model. All simulations were done with a square lattice with $128 \times 128$ lattice points and periodic boundary conditions. The parameter of the model was chosen to be $y = 0.5255$ which is about the value with maximum reactivity. This parameter stands for the fraction of molecules in the gas phase that are CO molecules and it also defines the unit of time as explained in Sect. 7.4.3. The *numbers on the right of the curves* indicate the number of simulations over which the reactivity has been averaged. Only the curve with the result from a single simulation shows the actual reactivity. Each other result has been offset vertically by an additional 0.03 reciprocal units of time. The *inset* shows a log-log plot of the noise, defined as the root-mean-square deviation of the reactivity, as a function as the number of simulations that were used for the averaging. The points lie very close to a line with slope $-\frac{1}{2}$

each other the noise is inverse proportional to the square root of the number of simulations. This way to reduce noise is straightforward. It only requires more computer time the smaller one wants the noise to be.

A second way to reduce noise is to increase the size of the system. That is shown in Fig. 5.2. The effect of this is the same as averaging over multiple simulations. Doubling the linear dimension of a system quadruples the number of sites and gives the same result as averaging over four simulations with the original system size. The only difference may be that changing the system size may cost more computer time. For example, computer time per event in First Reaction Method increases slightly with system size so doubling the linear dimension of a system costs more than four times in computer time (see Sect. 3.5). For other algorithms that need not be the case however.

Figure 5.2 also shows the noise in the coverage of CO. Oxygen shows the same variation. We see that the noise in the coverage relative to the average coverage is somewhat larger than the noise in the reactivity relative to the average reactivity.

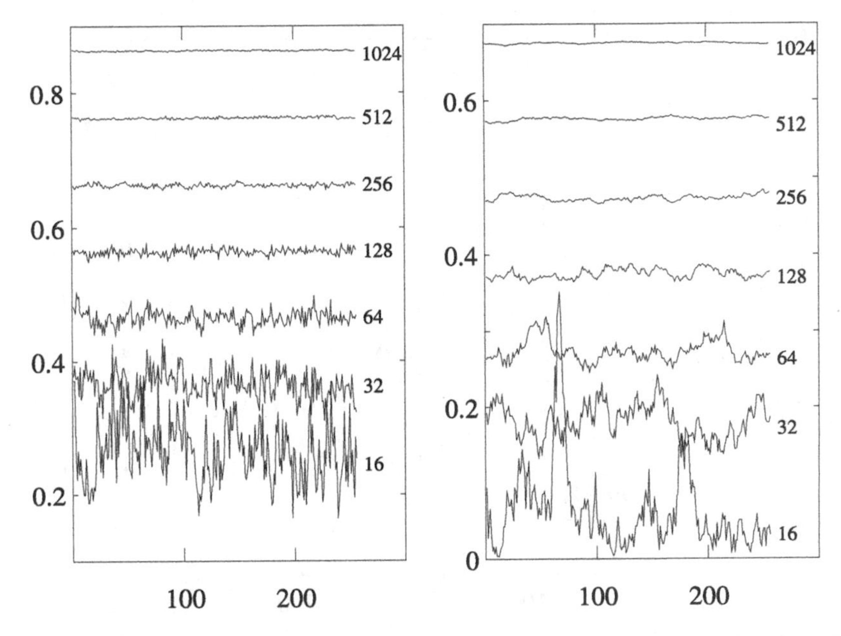


**Fig. 5.2** The reactivity (in reciprocal units of time) on the *left*, defined as the number of reactions forming $CO_2$ per unit cell per unit of time, and the coverage of CO on the *right* as a function of time in the ZGB model. The parameter of the model was chosen to be $y = 0.5255$ which is about the value with maximum reactivity. This parameter stands for the fraction of molecules in the gas phase that are CO molecules and it also defines the unit of time as explained in Sect. 7.4.3. All simulations were done with a square lattice with $L \times L$ lattice points with $L$ given by the *number to the right of the curves* and periodic boundary conditions. Only the curves for $L = 16$ show the actual reactivity and coverage. Each other result has been offset by an additional 0.03 reciprocal units of time for the reactivity and by 0.1 ML for the coverage

This is not always the case. Figure 5.3 shows the desorption rate and the coverage for various system sizes for TPD spectra of simple desorption. We see that the noise in the coverage is much less than the noise in the desorption rate.

Figure 5.4 shows a third way to reduce noise. Instead of averaging over simulation or larger areas one can also average over longer time intervals. The figures of the rates in this section show output at times $t_{\text{start}} + n\Delta t$ with $t_{\text{start}}$ the time at the start of a simulation, $\Delta t$ a time interval, and $n$ a non-negative integer. The rate at time $t_{\text{start}} + n\Delta t$ for a particular $n$ is obtained by counting the number of processes that have occurred between $t_{\text{start}} + (n-1)\Delta t$ and $t_{\text{start}} + n\Delta t$ and then dividing by the size of the system (i.e.: the number of unit cells) and by $\Delta t$. The number of processes that occurs in an interval fluctuates but it is expected that the fluctuation relative to the length of the interval becomes smaller. That is indeed what is seen in the figure.

This last way to reduce noise is the cheapest as the system size can be kept relatively small and only one simulation needs to be done. The drawback is that the amount of data is reduced by averaging over larger and larger intervals. For a steady state this is no problem as for such a situation one is not interested in

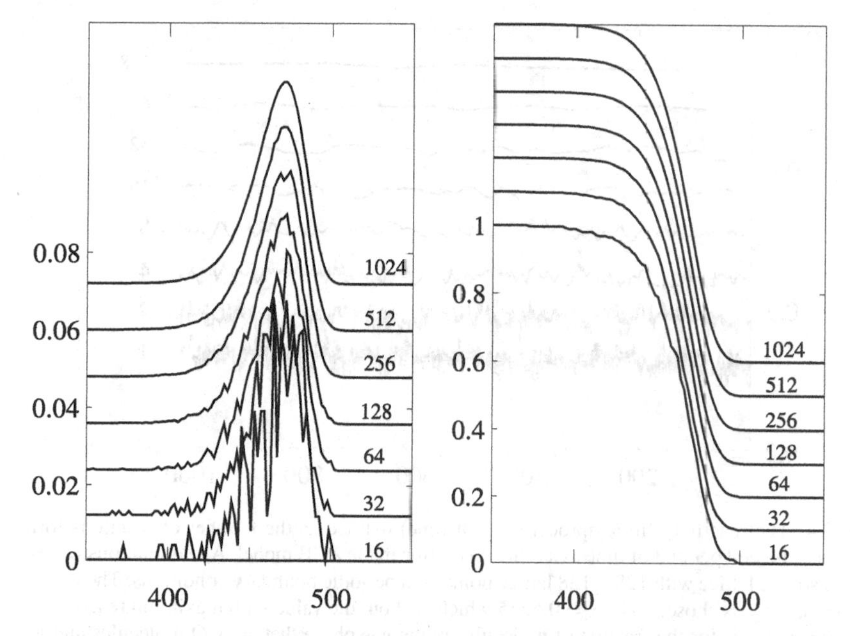


**Fig. 5.3** The desorption rate (in 1/s) on the *left* and the coverage on the *right* as a function of temperature (in K) for a TPD experiment with simple desorption. The activation energy for desorption was 124.7 kJ/mol and the prefactor $10^{13}$ s$^{-1}$. The heating rate was 2 K/s and the initial coverage 1 ML. All simulations were done with a square lattice with $L \times L$ lattice points and periodic boundary conditions with $L$ given by the *number to the right of the curves*. Only the curves for $L = 16$ show the actual desorption rate and coverage. Each other result has been offset by an additional 0.012 s$^{-1}$ for the desorption rate and by 0.1 ML for the coverage

temporal fluctuations anyway and one might as well average over the whole length of a simulation. Note that we are using here the ergodic hypothesis or ergodicity theorem [1]. It states that such average over time is the same as an average over many identical systems. Instead of taking a large time interval $\Delta t$ for steady states to get average coverages and rates, it is also possible to use a small $\Delta t$ and average over the variations as a functions of time after a simulation has been done. This has the advantage that the fluctuations can be used to derive error estimates of the averages.

It is important to realize that we only want to reduce noise if a single simulation shows noise and it is meant to model a system in which this noise is supposed to be absent. There are other situations. For example, if the system that we try to simulate is small, then the coverages and rates may actually fluctuate. The noise in a simulation then represents these fluctuations, and we do not want to average [2, 3].

## 5.3 A Modeling Framework

An implementation of kMC has to have information on the sites of the system that is being simulated and the occupation of these sites. The sites will most likely be

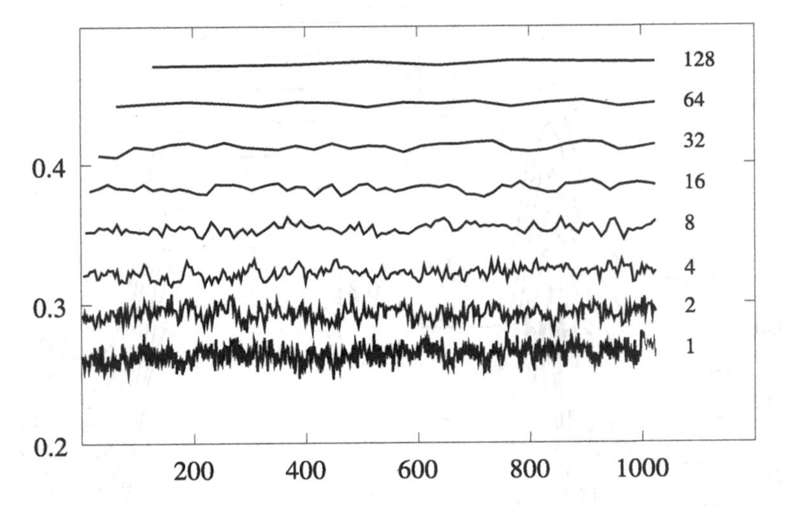


**Fig. 5.4** The reactivity (in reciprocal units of time) defined as the number of reactions forming $CO_2$ per unit cell per unit of time as a function of time in the ZGB model. All simulations were done with a square lattice with $128 \times 128$ lattice points and periodic boundary conditions. The parameter of the model was chosen to be $y = 0.5255$ which is about the value with maximum reactivity. This parameter stands for the fraction of molecules in the gas phase that are CO molecules and it also defines the unit of time as explained in Sect. 7.4.3. The *numbers on the right of the curves* indicate the time interval $\Delta t$ over which the reactivity was averaged. Only the curve with the result with a time interval of 1 time unit shows the actual reactivity. Each other result has been offset by an additional 0.03 reciprocal units of time

implemented as some two-dimensional array of integers, characters, booleans, or strings, which indicates the occupations. As a string can be used to represent an integer, character, or boolean as well, we will use only strings here, but we will call them labels from now on. Such labels have the advantage over integers, characters, and booleans that they can often be given values with a meaning that is obvious.

Apart from the occupation, other information on the sites may be useful as well. Such additional information we will also represent with a label. One may think that a real or floating-point numbers can be useful if one wants to model for example a local temperature. However a continuous change of a property does not fit in the theory as described in Chap. 2. If one wants to model a continuous property of a site, it will have to be discretized, and then a label can be used again.

Our framework has already been discussed briefly in Sects. 2.1.3 and 2.1.4. We can indicate a site by a triplet $(n_1, n_2/s)$. This is the site at position $\mathbf{x}_0^{(s)} + n_1\mathbf{a}_1 + n_2\mathbf{a}_2$ with $\mathbf{a}_1$ and $\mathbf{a}_2$ the primitive vectors of the lattice formed by the sites, and $\mathbf{x}_0^{(s)}$ is the position of site number $s$ within a unit cell. If there is only one site per unit cell, then the "$s$" can be left out, leaving us with $(n_1, n_2)$. Adding a label specifies information on the site. This can most easily be seen using an example. With $(0, 0/0) : \mathrm{CO}$ we mean that there is a CO molecule adsorbed at the site at position $\mathbf{x}_0^{(0)}$. Note however that the meaning of the label is up to the user. For example, $(1, 2/0) : \mathrm{S}$ may indicate a sulfur atom adsorbed on the site at $\mathbf{x}_0^{(0)} + \mathbf{a}_1 +$

$2\mathbf{a}_2$, but that need not be the meaning of the label S at all. It may also be defined by a user to indicate a step site, or something entirely different.

The meaning of a label is really defined implicitly by the way it can change: i.e., by the processes in which it plays a role. For example, if the label S above stands for a sulfur atom then there will probably be reactions between the sulfur atom and other adsorbates, there will be hops from one site to another to model diffusion, et cetera. If S stands for a step site, then there are processes in which it becomes occupied, or it is converted into a terrace site when the substrate changes its structure, et cetera. We therefore need to look at how processes are implemented.

It is in this aspect that general-purpose and special-purpose codes differ most. The processes in a special-purpose code are hard code: i.e., how the occupations or other properties of the sites change is determined by the code. In a general-purpose code the processes are descriptions specified by the user on input that are interpreted by the program.

The way we do this here is as follows. Each process has to be specified as a list of sites involved in the process, the labels of the sites before, and the labels of the sites after the process has taken place. For example, in

$$(0,0),(1,0) : \mathrm{A\ B} \rightarrow * * \tag{5.1}$$

an integer pair $(n_1, n_2)$ indicates as before a unit cell, A and B are reactants, and $*$ the product. This can be interpreted as an molecule A reacting with a molecule B at a neighboring site to the right to form AB which immediately desorbs leaving two vacant sites. This interpretation, however, is based on the meaning of the labels A, B, and $*$ as mentioned before.[1]

The indices specifying the unit cells are relative. The specification above stands not just for a process at (0, 0) and (1, 0), but at $(n_1, n_2)$ and $(n_1 + 1, n_2)$ for any pair of integers $n_1$ and $n_2$. Having more than one site per unit cell is also possible. For example in

$$(0,0/0),(0,0/1) : \mathrm{A\ B} \rightarrow * * \tag{5.2}$$

the reaction takes place on two sites in the same unit cell. The integers after the slash indicate which sites are involved. Note that we do not make any assumptions on distances or angles. So

$$(0,0),(1,0) : \mathrm{A\ B} \rightarrow * * \tag{5.3}$$

---

[1] The point that the meaning of the labels is up to the user can not be emphasized often enough. Suppose we model CO oxidation on some transition metal surface. Then $(0,0),(1,0) : \mathrm{CO\ O} \rightarrow * *$ can be interpreted as a $CO_2$ formation reaction that is followed immediately by desorption of $CO_2$ leaving two vacant sites. This interpretation is obvious as is the interpretation of $(0,0),(1,0)$ : carbonmonoxide oxygenatom $\rightarrow$ vacant vacant for the same process. However, we may also have $(0,0),(1,0)$ : vase gravity $\rightarrow$ shard shard. This is just as correct, although of course, very misleading. For more complicated models the interpretation and the choice of meaningful labels is not always straightforward. For a kMC simulation only the way labels can change as specified by the processes and the rates with which these changes occur is what counts.

may refer to a square, a hexagonal, or any other type of lattice. This too is implicit in the reactions. If $(0,0),(1,0) : \mathrm{A\,B} \rightarrow * *$ stands for a reaction on a square lattice, then there should also be reactions $(0,0),(0,1) : \mathrm{A\,B} \rightarrow * *$, $(0,0),(-1,0) : \mathrm{A\,B} \rightarrow * *$, and $(0,0),(0,-1) : \mathrm{A\,B} \rightarrow * *$. If we have a hexagonal lattice, then we should have the reaction specifications as for a square lattice plus $(0,0),(1,1) : \mathrm{A\,B} \rightarrow * *$, and $(0,0),(-1,-1) : \mathrm{A\,B} \rightarrow * *$ where we assume that the angle between the primitive translations is 120°.

Processes also have a reaction rate constant. That need not concern us here except that for general-purpose codes it may be convenient to have the possibility of having processes with infinite rate constants. Such processes occur immediately when they become possible. We will call such processes immediate processes. In special-purpose codes immediate processes are often not regarded as processes at all, but as situations in which some special action needs to be taken. By extending our concept of processes to these immediate processes, such situations need not be treated differently, and they fit neatly in the theoretical framework that we have described in Chap. 2.

## 5.4 Modeling the Occupation of Sites

### *5.4.1 Simple Adsorption, Desorption, and Unimolecular Conversion*

Modeling simple adsorption, desorption, and unimolecular conversion is essentially the same. The essential part is that only one site is involved. In each case the process can then be written as $\mathrm{A} \rightarrow \mathrm{B}$. For adsorption A is a vacant site and B the site occupied by an adsorbate, for desorption A is the site occupied by an adsorbate and B a vacant site, and for conversion A is the adsorbate that is converted and B the adsorbate into which A is converted. Note that we only look at the site and its occupation. We ignore the fact that prior to adsorption the adsorbate is in the gas phase or dissolved in a solution. For the simulation this is irrelevant. In the following we use $\mathrm{A} \rightarrow \mathrm{B}$ as a generic form for all three cases.

Note that an Eley–Rideal reaction that takes place on a single site is also included here. If the product is a molecule that immediately desorbs, then A is an adsorbate and B is a vacant site. The reaction is then effectively the same as a desorption, but with a rate constant that depends not only on temperature but also on the pressure of the gas-phase reactant. If the product stays on the surface at the same site as the adsorbed reactant, then B is the product. The reaction is then effectively the same as a conversion with the same caveat concerning the rate constant.

Modeling this reaction is very simple. In the notation of our framework (see Sect. 5.3) we have

$$(0,0) : \mathrm{A} \rightarrow \mathrm{B}. \tag{5.4}$$

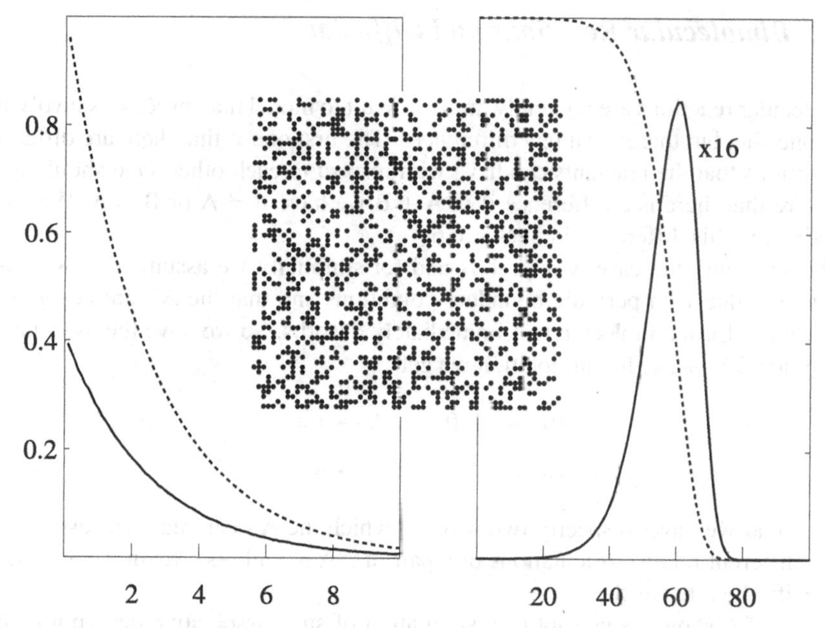


**Fig. 5.5** Change of the coverage (in ML: *dashed curves*) and the desorption rate (in reactions per second per site: *solid curves*) as a function of time (in seconds) for isothermal (*left*) and temperature-programmed (*right*) simple desorption. The lattice size in the kMC simulations was $128 \times 128$ with periodic boundary conditions and each result is an average of 1000 simulations. The initial coverage was 1 ML for all simulations. On the *left* the desorption rate constant is 0.4 $s^{-1}$. On the *right* the activation energy is 124.7 kJ/mol, the prefactor is $10^{13}$ $s^{-1}$, the heating rate is 2 K/s, and the initial temperature is 350 K. The *inset* shows a snapshot of a simulation with a $64 \times 64$ square lattice with the *black circles* depicting the adsorbates. It shows a situation in which about two third of the sites has been vacated

This means that for each lattice point that has a label A, this label can change to B. This is the whole specification of the process except for the rate constant.

Figure 5.5 shows a snapshot of a simulation of simple desorption. A square lattice is used, and about one third of the sites are occupied. Note that the adsorbates are randomly distributed over the lattice. There is no mechanism that can lead to any kind of ordering. Figure 5.5 also shows how the coverage and the desorption rate change in time. The coverage is given in monolayers (ML) which is the fraction of all sites that is occupied. For the isothermal desorption the coverage and the desorption rate are simple exponential decreasing functions of time. For the TPD case we have

$$\theta(t) = \theta(0) \exp\big[\Omega(0) - \Omega(t)\big] \tag{5.5}$$

as was shown in Sect. 4.6.2.

### 5.4.2 Bimolecular Reactions and Diffusion

Bimolecular reactions are not really more difficult to model than processes involving only one site, but there are a few differences. The first one is that there are different orientations that the reactants can have with respect to each other. One should also be aware that there is a difference if in A + B we have B ≠ A or B = A. We have already seen this difference in Sect. 4.6.6.

We start with the case with A + A and for simplicity we assume that we have a square lattice with periodic boundary conditions and that the A's are adsorbates that react to form a molecule that immediately desorbs. So we have the associative desorption $2A \rightarrow 2*$. In our notation we get

$$\begin{aligned} (0,0),(1,0) &: \mathrm{A\,A} \rightarrow *\,*, \\ (0,0),(0,1) &: \mathrm{A\,A} \rightarrow *\,*. \end{aligned} \tag{5.6}$$

We see that we have to specify two ways in which the A's can react corresponding to the different relative orientations of a pair of AA neighbors. Again we also have to specify the rate constant.

Figure 5.6 show a snapshot of a simulation of such associative desorption. We see that we get isolated A's. If we have no diffusion then these will remain on the surface indefinitely. Figure 5.6 also shows how the coverage and the desorption rate change in time.

If we include diffusion all the A's will eventually react. Figure 5.7 shows that indeed the coverage then goes to zero for large time $t$. We model the diffusion by

$$\begin{aligned} (0,0),(1,0) &: \mathrm{A}\,* \rightarrow *\,\mathrm{A}, \\ (0,0),(0,1) &: \mathrm{A}\,* \rightarrow *\,\mathrm{A}, \\ (0,0),(-1,0) &: \mathrm{A}\,* \rightarrow *\,\mathrm{A}, \\ (0,0),(0,-1) &: \mathrm{A}\,* \rightarrow *\,\mathrm{A}. \end{aligned} \tag{5.7}$$

There are four processes because an A can hop to one of four neighboring sites if vacant. If the diffusion is so fast that the particles are randomly mixed then we have

$$\frac{d\theta}{dt} = -ZW_{\text{des}}\theta^2. \tag{5.8}$$

This was shown in Sect. 4.6.6. For the isothermal case this yields

$$\theta(t) = \frac{\theta(0)}{1 + ZW_{\text{des}}\theta(0)t}, \tag{5.9}$$

and for the TPD case

$$\theta(t) = \frac{\theta(0)}{1 + Z[\Omega(t) - \Omega(0)]\theta(0)} \tag{5.10}$$

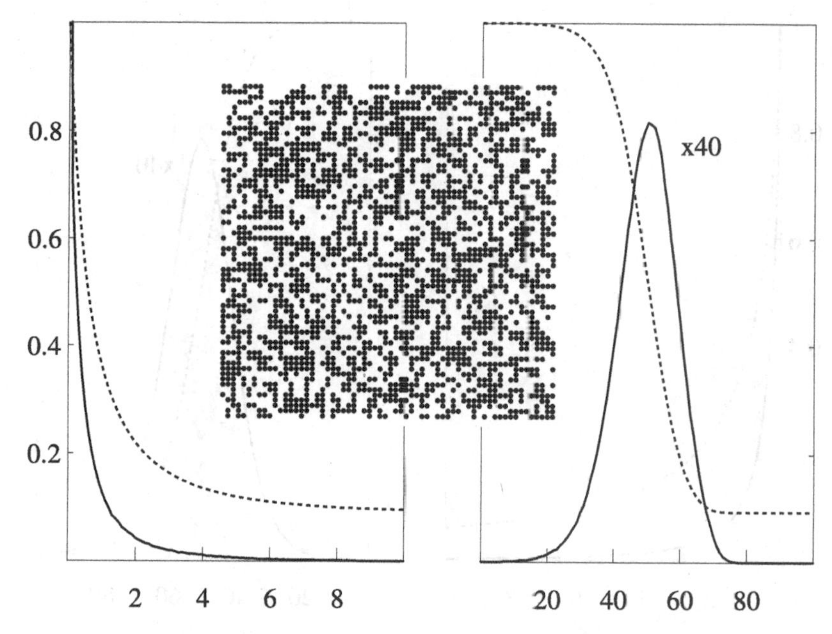


**Fig. 5.6** Change of the coverage (in ML: *dashed curves*) and the desorption rate (in reactions per second per site: *solid curves*) as a function of time (in seconds) for isothermal (*left*) and temperature-programmed (*right*) associative desorption. The lattice size in the kMC simulations was $128 \times 128$ with periodic boundary conditions and each result is an average of 1000 simulations. The initial coverage was 1 ML for all simulations. On the *left* the desorption rate constant is $0.4\ \mathrm{s}^{-1}$. On the *right* the activation energy is 124.7 kJ/mol, the prefactor is $10^{13}\ \mathrm{s}^{-1}$, the heating rate is 2 K/s, and the initial temperature is 350 K. The *inset* shows a snapshot of a simulation with a $64 \times 64$ square lattice with the black circles depicting the adsorbates. It shows a situation in which about half of the sites has been vacated. We get horizontal and vertical rows of adsorbates, because the adsorbates desorb in pairs. Note that there is no diffusion in the simulations

with $\Omega$ given by Eq. (4.85). The coverages in Fig. 5.7 vary according to Eqs. (5.9) and (5.10) because the diffusion is fast.

Next we deal with the case with $\mathrm{B} \neq \mathrm{A}$ and for simplicity we again assume that we have a square lattice with periodic boundary conditions and that A and B react to form a molecule that immediately desorbs so that we have the associative desorption $\mathrm{A} + \mathrm{B} \rightarrow 2*$. We get

$$\begin{aligned}
(0,0),(1,0) &: \mathrm{A}\,\mathrm{B} \rightarrow **, \\
(0,0),(0,1) &: \mathrm{A}\,\mathrm{B} \rightarrow **, \\
(0,0),(-1,0) &: \mathrm{A}\,\mathrm{B} \rightarrow **, \\
(0,0),(0,-1) &: \mathrm{A}\,\mathrm{B} \rightarrow **.
\end{aligned} \tag{5.11}$$

We see that we have to specify four ways in which an A can react with a B corresponding to the different relative orientations of a pair of AB neighbors. There are two more than for $2\mathrm{A} \rightarrow 2*$, because AB and BA are of course the same when A

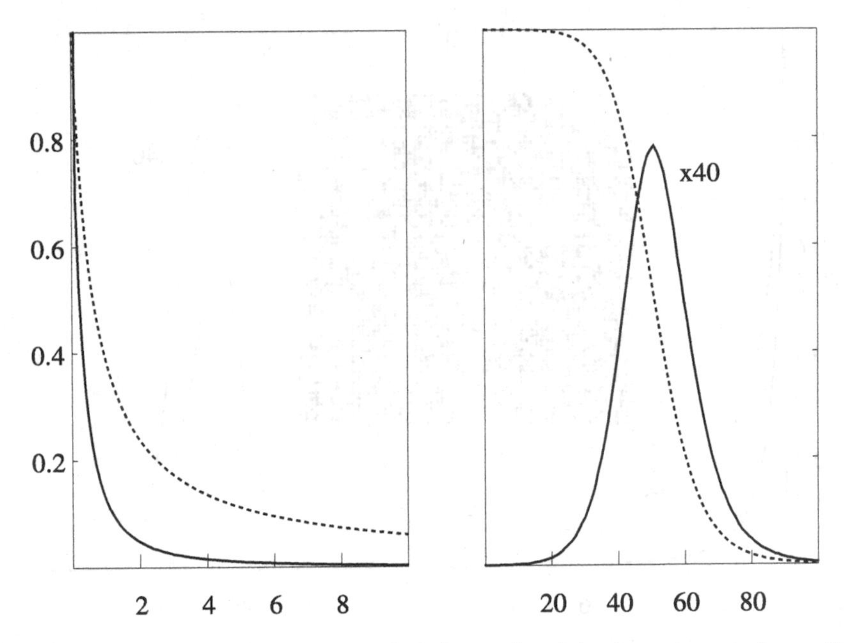


**Fig. 5.7** Change of the coverage (in ML: *dashed curves*) and the desorption rate (in reactions per second per site: *solid curves*) as a function of time (in seconds) for isothermal (*left*) and temperature-programmed (*right*) associative desorption with diffusion with a rate constant that is 100 times the one of desorption. The lattice size in the kMC simulations was 128 × 128 with periodic boundary conditions and each result is an average of 1000 simulations. The initial coverage was 1 ML for all simulations. On the *left* the desorption rate constant is 0.4 $s^{-1}$. On the *right* the activation energy is 124.7 kJ/mol, the prefactor is $10^{13}$ $s^{-1}$, the heating rate is 2 K/s, and the initial temperature is 350 K

and B are the same. We can also include diffusion of A and B, which is modeled as in Eq. (5.7).

Figure 5.8 show a snapshot of a simulation of such a model. The initial configuration corresponds to a random mixture of equal numbers of A's and B's. It can be noted that there are areas that have almost no B's and others that have almost no A's. The reason is that locally the number of A's and the number of B's are not the same. After some time the particles in the minority have reacted and the only particles of the other type are left [4]. The size of the areas with only A's or only B's depends on the ratio between the rate constant of the reaction and the hopping rate constant. This size increases as $\sqrt{t}$ with time [5]. Figure 5.8 also shows that the coverages decrease as $1/\sqrt{t}$ for large $t$ [4, 6, 7]. Initially, however, the particles are still randomly mixed, and the coverage decreases according to the macroscopic rate equations which yields $\theta(0)/[1 + ZW_{\mathrm{rx}}\theta(0)t]$.

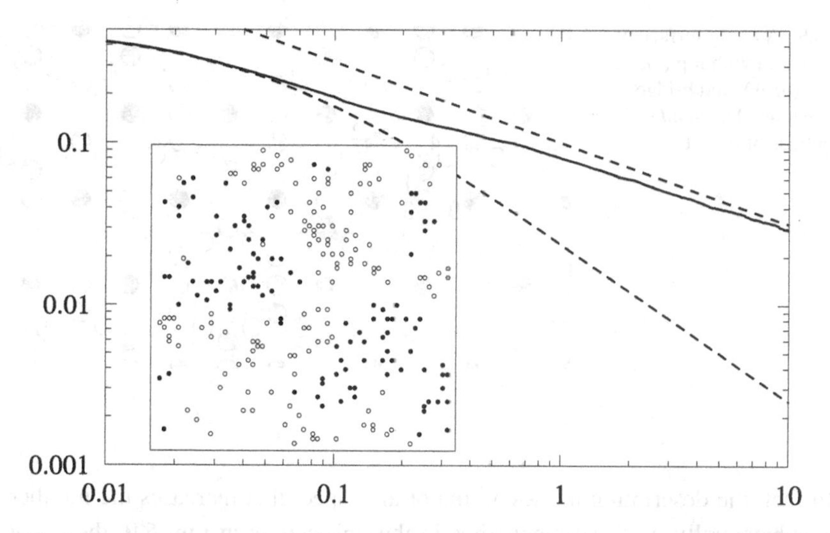


**Fig. 5.8** Change of the coverage (*solid line*) as a function of time (in seconds) for the reaction $A + B \to 2*$ on a $128 \times 128$ square lattice with periodic boundary conditions. The initial configuration was a fully occupied lattice with equal numbers of A's and B's. The rate constant for the reaction is $W_{rx} = 10\ s^{-1}$ and for diffusion $W_{hop} = 1\ s^{-1}$. The *lower dashed line* depicts the change in coverage obtained from the macroscopic rate equations: i.e., when the adsorbates would be randomly distributed at all times. The *upper dashed line* is proportional to $t^{-1/2}$, and shows the dependence of the coverage at long times. The *inset* shows a snapshot of an area of $64 \times 64$ sites at the end of the simulation. A's are *closed* and B's are *open circles*

## 5.5 Modeling Adsorption Sites

### 5.5.1 Using the Sublattice Index

The previous sections have shown how to model processes involving one or two sites on a surface with just one site per unit cell. If there is more than one site per unit cell, then we can use the third index in our notation to model the processes. Figure 5.9 shows a (111) surface of an fcc metal (e.g., Pt). A study of the dehydrogenation of $NH_3$ on Pt(111) showed that $NH_3$ prefers to adsorb at top sites, $NH_2$ prefers bridge sites, NH and N prefer fcc sites, and H does not really have a preference [8, 9]. To model the dehydrogenation of $NH_3$ ($NH_3 \to N_2 + H_2$) we can choose primitive translations $\mathbf{a}_1$ and $\mathbf{a}_2$ making an angle of 60°, the top site at $\mathbf{0}$ with index 0, the fcc site at $\frac{1}{3}\mathbf{a}_1 + \frac{1}{3}\mathbf{a}_2$ with index 1, the hcp site at $\frac{2}{3}\mathbf{a}_1 + \frac{2}{3}\mathbf{a}_2$ with index 2, the bridge sites at $\frac{1}{2}\mathbf{a}_1$ with index 3, at $\frac{1}{2}\mathbf{a}_2$ with index 4, and at $\frac{1}{2}\mathbf{a}_1 + \frac{1}{2}\mathbf{a}_2$ with index 5. The dehydrogenation can then be written as

$$(0,0/0), (-1,0/3), (0,0/1) : NH_3 \ * \ * \to * \ NH_2 \ H \tag{5.12}$$

with $*$ a vacant site. We see that the site preference of the adsorbates necessitates a process in which three sites are involved even though it is just a unimolecular reaction. This is quite usual for realistic models of reaction systems.

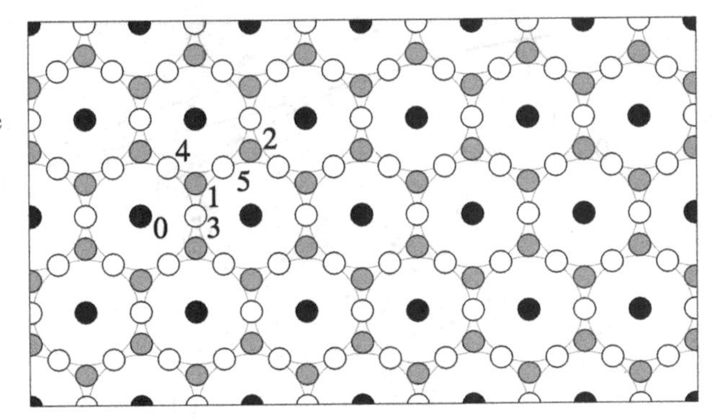


**Fig. 5.9** A (111) surface of an fcc metal with top (*black*), hollow (*gray*), and bridge sites (*white*). The *numbers* are the indices of the sites

In fact, the description misses an important aspect that increases the number of sites substantially. With so many sites in the unit cell as in Fig. 5.9, the distance between the sites is quite small. As a consequence there is a very strong repulsion between adsorbates at two different sites in the unit cell. In fact, in the Density-Functional Theory study mentioned above it was found that the adsorbates did not stay at their preferred site during a geometry optimization when two of them were placed together in one unit cell [8, 9]. One way to deal with this is to say that this repulsion is so strong that an adsorbate at some site prevents other adsorbates from occupying some of the neighboring sites. We can also say that these sites are blocked. For the dissociation of $NH_3$ this means that it can only take place if the target sites for $NH_2$ and H are not blocked. Suppose that an adsorbate blocks all sites within a circle with a radius equal to the lattice cell parameter of the substrate. The description above then has to be replaced by

$$\begin{aligned}
&(0,0/0),(-1,0/3),(0,0/1),\\
&(-2,0/2),(-2,0/5),(-1,0/0),(-1,0/1),(-1,0/2),(-1,0/4),(-1,0/5),\\
&(0,0/2),(0,0/3),(0,0/4),(0,0/5),(1,0/0),(1,0/4),\\
&(-1,-1/1),(-1,-1/2),(-1,-1/4),(-1,-1/5),\\
&(0,-1/1),(0,-1/2),(0,-1/4),(0,-1/5),(1,-1/4):\\
&\mathrm{NH_3} * * * * * * * * * * * * * * * * * * * * * * * *\\
&\to * \,\mathrm{NH_2}\,\mathrm{H} * * * * * * * * * * * * * * * * * * * * * *
\end{aligned} \tag{5.13}$$

We see that only three sites actually change their occupation, but a further 22 have to be included in the description to make sure that sites $(-1,0/3)$ and $(0,0/1)$ are not blocked.

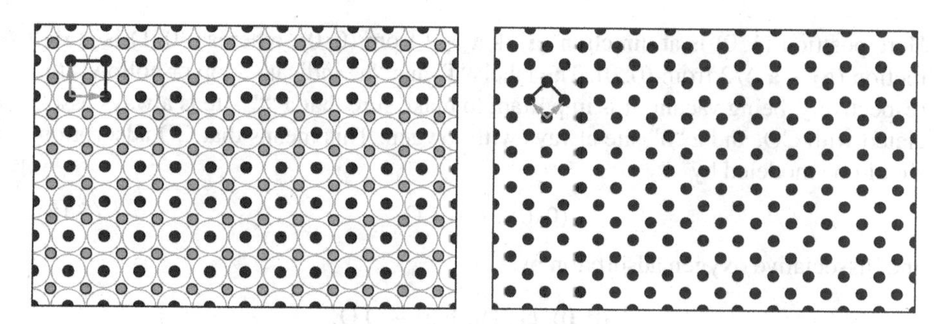

**Fig. 5.10** A (100) surface of an fcc metal with top (*black*) and hollow sites (*gray*) indicated on the *left*. On the *right* it is shown that a simple lattice with the indicated unit cell is obtained if we ignore the difference between the sites

## 5.5.2 Using Labels to Distinguish Sublattices

Using an index to distinguish between the various sites in a unit cell is not the only way to model sublattices. It is often also possible to do this with labels. Whether one or the other method is more efficient depends on the substrate and the processes. It is sometimes also a matter of personal preference.

Suppose that we are dealing with a (100) surface of an fcc metal and that we have processes involving the top (1-fold) and the hollow (4-fold) site (see Fig. 5.10). Such system has two sites per unit cell. However, we can ignore this difference as is shown in the right part of Fig. 5.10. The main advantage is that we can work with a smaller unit cell with just one site. This can be seen as follows. If $\mathbf{a}_1$ and $\mathbf{a}_2$ are the primitive vectors which span the unit cell of the (100) surface (see the left part of Fig. 5.10), then the sites in the unit cell can be given positions $\mathbf{0}$ and $(\mathbf{a}_1+\mathbf{a}_2)/2$. If we can ignore the difference between the two sites, then these sites form a simple lattice with primitive vectors $(\mathbf{a}_1+\mathbf{a}_2)/2$ and $(\mathbf{a}_2-\mathbf{a}_1)/2$. The advantage of working with such a simple lattice is that in general we may have fewer processes to specify because we do not distinguish between the two sites, and it will be easier (and computationally cheaper) to do the calculations of position on the surface where a process will take place. Whether or not this is useful depends on the processes.

Even if we need to distinguish between the two sites, then it is still possible to work with the lattice with just one site per unit cell. We need, however, to use labels that distinguish the sites. For example, suppose we are dealing with CO oxidation with CO adsorbing on top sites and atomic oxygen on 4-fold sites. If the reaction to form $CO_2$ occurs when a CO and oxygen are at neighboring sites, then we can model this with

$$
\begin{aligned}
(0,0),(1,0) &: \mathrm{CO\ O} \to \mathrm{t\ h},\\
(0,0),(0,1) &: \mathrm{CO\ O} \to \mathrm{t\ h},\\
(0,0),(-1,0) &: \mathrm{CO\ O} \to \mathrm{t\ h},\\
(0,0),(0,-1) &: \mathrm{CO\ O} \to \mathrm{t\ h}.
\end{aligned}
\tag{5.14}
$$

Here position $(1, 0)$ is at direction $(\mathbf{a}_1 + \mathbf{a}_2)/2$ from $(0, 0)$, whereas $(0, 1)$ is at direction $(\mathbf{a}_2 - \mathbf{a}_1)/2$ from $(0, 0)$. The labels "t" and "h" indicate a top or hollow site, respectively, being vacant. It is important to make sure that a "t" site is always associated with CO, and a "h" site always with oxygen. This means that CO adsorption should be modeled by

$$(0,0) : \mathrm{t} \rightarrow \mathrm{CO}, \tag{5.15}$$

and dissociative oxygen adsorption by

$$\begin{aligned} (0,0), (1,1) &: \mathrm{h\,h} \rightarrow \mathrm{O\,O}, \\ (0,0), (1,-1) &: \mathrm{h\,h} \rightarrow \mathrm{O\,O}. \end{aligned} \tag{5.16}$$

The relative positions $(1, 1)$ and $(1, -1)$ here correspond to the translations $\mathbf{a}_2$ and $\mathbf{a}_1$, respectively. It is also important to make sure that the initial configuration corresponds to a checkerboard pattern of "t" or CO and "h" or O. This kind of bookkeeping has the obvious drawback of being error-prone.

With a unit cell with two sites we get for the CO adsorption

$$(0, 0/0) : * \rightarrow \mathrm{CO}, \tag{5.17}$$

with the third "0" in $(0, 0/0)$ referring to the top site in the unit cell. The dissociative adsorption of oxygen becomes

$$\begin{aligned} (0, 0/1), (1, 0/1) &: *\,* \rightarrow \mathrm{O\,O}, \\ (0, 0/1), (0, 1/1) &: *\,* \rightarrow \mathrm{O\,O}. \end{aligned} \tag{5.18}$$

The "1" in $(0, 0/1)$ indicates the hollow site. Note that we can now use the same label "$*$" for any vacant site. The oxidation reaction becomes

$$\begin{aligned} (0, 0/0), (0, 0/1) &: \mathrm{CO\,O} \rightarrow *\,*, \\ (0, 0/0), (-1, 0/1) &: \mathrm{CO\,O} \rightarrow *\,*, \\ (0, 0/0), (0, -1/1) &: \mathrm{CO\,O} \rightarrow *\,*, \\ (0, 0/0), (-1, -1/1) &: \mathrm{CO\,O} \rightarrow *\,*. \end{aligned} \tag{5.19}$$

Whether it is better to use labels to distinguish the sites or to work with a unit cell with two sites depends on the processes and the substrate. Also coding may play a role. Calculations of the positions of where processes can take place can often be done more efficient when the lattice sizes are powers of two using shift operators [10]. If we use labels, then this might not be possible. For example, suppose we have a (111) surface of an fcc metal and we are dealing with top and the two hollow sites. These form again a simple lattice, but if we use this simple lattice and periodic boundary conditions then we have to work with system sizes that are multiples of three.

Sometimes extra sites need to be introduced if we want to use labels to distinguish sublattices. Suppose that we have a (111) surface of an fcc metal. Suppose

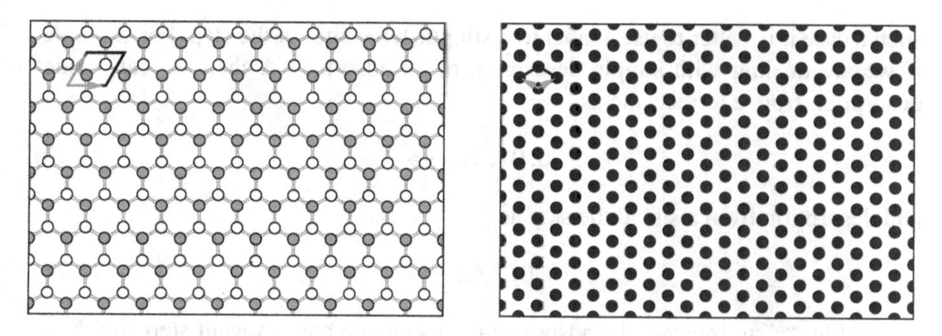

**Fig. 5.11** A (111) surface of an fcc metal with only hollow sites indicated on the *left*. These sites form a honeycomb lattice. If we add top sites we get a simple lattice as shown on the *right*

also that all adsorbates prefer the 3-fold sites: both the fcc and hcp sites. These sites together form the honeycomb structure shown in Fig. 5.11. If we ignore the difference between the hcp and fcc sites, we do not get a simple lattice. So it seems not possible to work with a smaller unit cell and use labels to distinguish between the sublattices. That impression is incorrect however.

The trick is to add sites to make a simple lattice. The right part of Fig. 5.11 shows that if we add a site in the middle of each hexagon formed by the 3-fold sites, we do get a simple lattice. The extra site is at the position of a top site, but that is irrelevant as we will not use the site anyway. With the extra sites we can work with a smaller unit cell, but we need labels to distinguish between three types of sites.

It is clear that we are paying a price when we add extra sites. Although we do not use them in the processes, we still need to store them. That price is however very modest. It depends of course on the size of the system, but the number of sites rarely exceeds 1 000 000, which means that our trick costs only a few Mb extra storage at most.

### 5.5.3 Systems Without Translational Symmetry

The kMC simulations discussed here always assume that the sites form a lattice, but this does not mean that we can only model systems with translational symmetry. The previous section on multiple sites in the unit cell has shown that by adding labels specifying extra properties of a site we can modify the lattice. We can use this to model a system without translational symmetry.

Suppose we want to model a stepped surface. We can model such a surface with a large unit cell and multiple sites. This may not be a good idea, however. Suppose we have parallel steps separated by wide terraces. We will then be dealing with a large unit cell with many sites, and we will have to specify for each site its processes. This will generally lead to a long list of processes, even if the different sites on the terraces have the same properties, and only the sites at the steps behave differently. Moreover, we can not define such a unit cell if the width of the terraces varies. In such a case it is better to use a label to distinguish the sites at the step. For example, if we are dealing with simple desorption of an adsorbate A then we can model desorption from a terrace site as

$$(0,0) : \mathrm{A} \rightarrow *, \tag{5.20}$$

and desorption from a site at the step as

$$(0,0) : \mathrm{As} \rightarrow *\mathrm{s}. \tag{5.21}$$

We add an "s" to indicate the adsorbate on a step site and a vacant step site. Note that there is a difference with the example of CO oxidation on a (100) surface of the section on multiple sites in Sect. 5.5.2. There the label "CO" already implied one type of site and "O" another. Here the same adsorbate can be found on both types of site. We also need to specify different rate constants for the desorption for both types of site, because otherwise distinguishing them would not be meaningful. The precise position of the steps has to be specified in the initial configuration.

For a unimolecular reaction on a step we need to specify just two reactions. If we have a bimolecular reaction we need to specify reactions for all possible combinations of occupations of step and terrace sites. So if we have a reaction $\mathrm{A} + \mathrm{B} \rightarrow 2*$, then we will have $\mathrm{A\,B} \rightarrow *\,*$, $\mathrm{As\,B} \rightarrow *\mathrm{s}\,*$, $\mathrm{A\,Bs} \rightarrow *\,*\mathrm{s}$, and $\mathrm{As\,Bs} \rightarrow *\mathrm{s}\,*\mathrm{s}$. Diffusion can also be regarded as a bimolecular reaction, and we need to specify $\mathrm{A} * \rightarrow * \mathrm{A}$, $\mathrm{As} * \rightarrow *\mathrm{s\,A}$, $\mathrm{A} *\mathrm{s} \rightarrow * \mathrm{As}$, and $\mathrm{As} *\mathrm{s} \rightarrow *\mathrm{s\,As}$ for the diffusion of A. Needless to say that all these possibilities will have in general different rate constants.

In a realistic model of a surface it may not suffice to just change labels to indicate step sites. Figure 5.12 shows a so-called (111) step of an fcc(111) surface. The figure also shows the relevant sites if we assume that all adsorbates prefer top sites. If we draw only the sites, we see that we have no translational symmetry in the direction perpendicular to the step. We can try to get translational symmetry by introducing extra sites as we have done with using labels to distinguish different sublattices (see Sect. 5.5.2). That is possible as is shown in the figure, but there is a much simpler approach.

It is important to realize that the only geometric information that is used in kMC simulations is the translational symmetry of the lattice points. There is no information on distances between the lattice points and their relative position with respect to each other. All that information is implicitly contained in the processes and their rate constants. We have already seen that in the discussion of the modeling framework in Sect. 5.3 where we saw the difference in the associative desorption on a square and a hexagonal lattice.

The relevance of these remarks on the geometric information is that we are allowed to shift sites and draw them as shown on the right in Fig. 5.12. We see that we have restored the translational symmetry in two directions. The way we can implement the step is now as follows. We label the sites at the step on the upper terrace using "u" (for "up"), those at the step on the lower terrace using "d" (for "down"), and those on the terraces using "t" (for "terrace"). The processes use these labels to

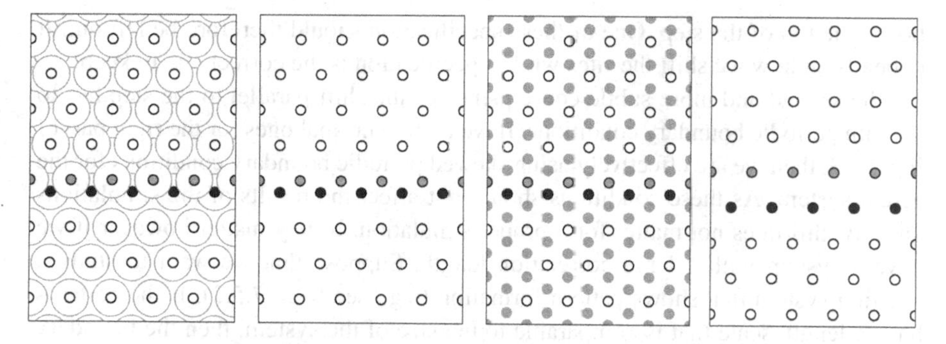

**Fig. 5.12** A (111) step on a (111) surface of an fcc metal on the *left*, the top sites in the *middle-left*, top sites with extra sites in *gray* to form a simple lattice in the *middle-right*, and the simple lattice that is formed after the top sites of one terrace have been shifted on the *right*

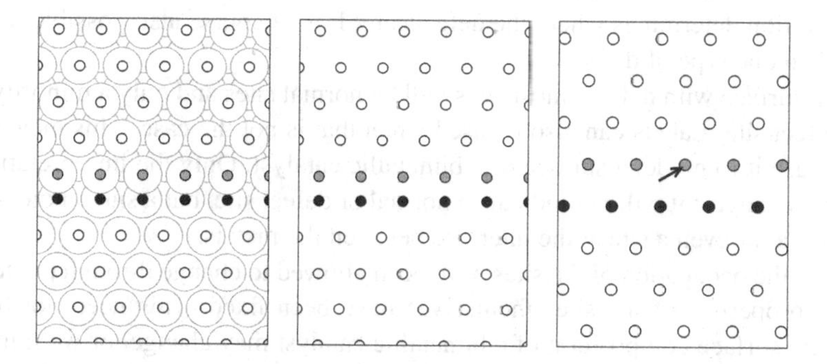

**Fig. 5.13** A (100) step on a (111) surface of an fcc metal on the *left*, the top sites in the *middle*, and the simple lattice that is formed after the top sites of one terrace have been shifted as indicated by the *arrow* on the *right*

distinguish between terrace sites and those at the step just as before. That there is no translational symmetry perpendicular to the step at the step is irrelevant.

Extra care is sometimes necessary. Figure 5.13 shows a (100) step of an fcc(111) surface. Again the top sites show only translational symmetry in two directions if we shift the sites of the upper terrace with respect to those of the lower terrace. But the shift is not perpendicular to the step in this case. There is also a component parallel to the step. This has two consequences. The first and most important one has to do with the specification of the processes. Suppose we have a reaction between an adsorbate A at the bottom of the step and a neighboring adsorbate B at the top of the step. This will give us a specification of the form Ad Bu $\rightarrow \ldots$, where we have added "u" and "d" as defined above for a (111) step. The problem is the indexing of the sites. The right part of Fig. 5.13 suggests that we should have $(0,0), (0,1)$ : Ad Bu $\rightarrow \ldots$ and $(0,0), (-1,1)$ : Ad Bu $\rightarrow \ldots$ ($\mathbf{a}_1$ is then parallel to the step and $\mathbf{a}_2$ makes an angle of 60° with $\mathbf{a}_1$), but the middle part of the figure clearly shows that a top site at the bottom of the step has only one top site as neighbor at the top of the step. One of these specification should therefore be left out. It depends on how we shift the sites which specification is the correct one to keep.

The second and more subtle consequence of the shift parallel to the step has to do with periodic boundary conditions. If we use the normal ones for the right part of Fig. 5.13, then we are effectively using skewed periodic boundary conditions for the actual system. As these conditions should not affect the results of the simulations anyway, this does not matter for a proper simulation. It may matter however if we have a system with a large correlation length. Suppose that we want to study a reaction system that shows pattern formation (e.g., see Sect. 7.5.2). If the patterns have a length scale that is comparable to the size of the system, then the boundary conditions can affect the results. Of course, one can argue that the system is then chosen too small, but it may not be practical to increase the size of the system.

Using labels to indicate defects will be unavoidable if we have point defects that are not regularly distributed over the surface. The specification of the initial configuration determines where the defects are. It is of course also possible to have more than one type of defect.

On a surface with defects most sites will be normal sites and only a minority will be a defect site. Labels can also be used when that is not the case. This means that we can use it to model a surface of a bimetallic catalyst. Only the interpretation of the label changes: it will not indicate a normal or defect site, but a site on one or the other metal or even a site at the interface between the metals.

So far the occupation of the sites have been allowed to change through processes, but the properties of the sites themselves have been fixed. This need not be the case. The surface composition of a bimetallic catalyst may change, or we might be dealing with a reconstructing surface. For a reconstructing surface we can introduce a label that specifies to which phase of the substrate a site belongs. A change of surface composition of a bimetallic catalyst or a reconstruction can be modeled with processes that specify the changes in the substrate. For example, there have been many studies of CO oxidation on reconstructing platinum surfaces. In one of the simplest models of this process there are process for the growth of the phases [11–13]. One phase, called the $\alpha$ phase, growths if there are no adsorbates. This can be modeled by

$$(0,0),(1,0) : *\alpha \ *\beta \rightarrow *\alpha \ *\alpha. \tag{5.22}$$

The $\beta$ phase growths in an area with sites occupied by CO.

$$(0,0),(1,0) : \mathrm{CO}\beta \ \mathrm{CO}\alpha \rightarrow \mathrm{CO}\beta \ \mathrm{CO}\beta. \tag{5.23}$$

It should be realized that there are restrictions in what one can do with a changing substrate in kMC simulations. One does need to be able to put everything on a lattice. Reconstructions that lead, for example, to surface structure with a different density can only be modeled if this change is ignored [14]. Also it may not be possible to model changes of the local point group symmetry.

### 5.5.4 Bookkeeping Sites

All the lattices that we have seen so far correspond to the actual adsorption sites of a substrate. That need not be. Often it is convenient to introduce an extra sublattice (i.e., extra sites in a unit cell) to store information. Suppose we have a Pt-Ru alloy that is used as an electrode for electrocatalytic CO oxidation [15, 16]. Also suppose for simplicity that the sites form a simple lattice, but some sites are on Pt and others on Ru. We have CO and OH as adsorbates. The latter is formed when water dissociatively adsorbs. This forms OH on the electrode, a proton that dissolves, and an electron that is taken up by the electrode. CO and OH react to form $CO_2$ that desorbs immediately, a proton, and an electron. To distinguish between adsorbates on Pt or Ru we can use, for example, the labels CO(Pt), CO(Ru), OH(Pt), OH(Ru), $*$(Pt), and $*$(Ru) with obvious meaning. This would be the approach used in Sect. 5.5.3.

Another way to model this CO oxidation however is to use two sublattices or to have two lattice points in the unit cell that both correspond to the same adsorption site. The idea of these two lattice points is that one has information on the occupation (CO, OH, or vacant) and the other on the metal of the substrate (Pt or Ru). The formation of $CO_2$ in the former approach would be represented by

$$(0,0),(1,0) : \text{CO(Pt) OH(Ru)} \rightarrow *\text{(Pt)} *\text{(Ru)} \tag{5.24}$$

and similar reactions. In the model with two sublattices this becomes

$$(0,0/0),(0,0/1),(1,0/0),(1,0/1) : \text{CO Pt OH Ru} \rightarrow * \text{ Pt} * \text{ Ru}. \tag{5.25}$$

The drawbacks of this approach are obvious. The description of the processes are more extensive, and there is a higher memory use. The advantage is that the labels become simpler, because we decouple the occupation from the type of site. Sometimes it also makes the process description simpler. Suppose we have a diffusion of CO that is the same on Pt and Ru. In such a case

$$(0,0/0),(1,0/0) : \text{CO} * \rightarrow * \text{ CO} \tag{5.26}$$

is a good description. There is no reference to the substrate. This makes not only the description simpler, but it is also more efficient, because no checking is needed for the labels of the other sublattice.

It is clear that splitting the information on a site over two or more sublattices can also be used to model steps and reconstruction. But that is not the only way such extra lattice points can be used. We will see later that it can be convenient for storing all kinds of useful information. This information can refer to one site or to a group of sites. We will call such extra lattice points bookkeeping sites, because of the nature of the information that is often stored in them.

## 5.6 Using Immediate Processes

We have not given much attention in this chapter to the rate constants of the processes. We have only said that there is such a rate constant and from Chap. 3 we know that this means that there is a probability distribution for the time that the process will take place. It turns out to be very convenient to allow the rate constant to become infinite and even to distinguish between different orders of infinity. Processes with an infinite rate constant occur immediately when they become possible (i.e., enabled in the terminology of Chap. 3) and we call them therefore immediate processes.

### *5.6.1 Very Fast Processes*

The most trivial use of immediate processes is to model a very fast process: i.e., a process with a rate constant that is so large that it is almost certain that the process will occur before any other process. In such a situation it is advantageous to use an immediate process. The first advantage is that one need not bother about determining the precise value of the rate constant. The second, and more important, advantage is that the time that the process occurs is the time that it has become possible, and no random number needs to be generated and a time determined. All computations concerning Eq. (3.11) or Eq. (3.25) are thus avoided.

There are many examples in the literature of such a use of immediate processes, in particular in studies of simple models directed at understanding certain kinetic aspects and the development of kinetic concepts. For example, in the ZGB model (see Sect. 5.2) the formation of $CO_2$ is assumed to be infinitely fast. This, and a few other assumptions, allows the reduction of the number of kinetic parameters to just one: the fraction of gas molecules that are CO [17].

Using immediate processes does not mean that the generation of random numbers is completely avoided for such processes. Suppose that in the ZGB model just mentioned a CO molecule adsorbs next to two oxygen atoms. Because $CO_2$ formation is an immediate process, CO will react with one of the oxygen atoms, but which one is arbitrary and we need to generate a random number to decide which process will actually take place.

Suppose that the CO molecule after adsorption could also react differently: e.g., it could also dissociate. The question is then what the molecule will actually do. If the other process has a finite rate constant, then the immediate process ($CO_2$ formation) will occur. But what if the other process is also an immediate process. For example, there is no CO diffusion in the original ZGB model, but diffusion is generally a very fast process. So if we would have diffusion as an immediate process as well, then the situation is not clear. We might simply choose one of the processes, but that may be unrealistic.

For such a case we distinguish between different orders of infinite rate constants: i.e., we give the immediate processes priorities. An immediate process with a high

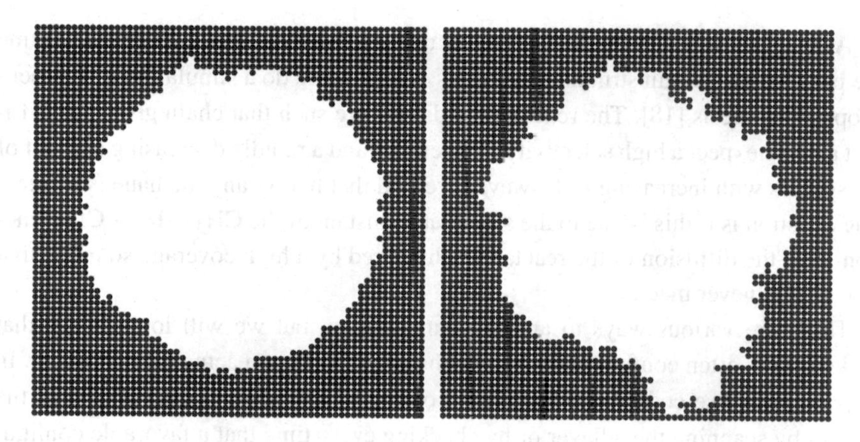

**Fig. 5.14** Snapshots of holes in an oxygen layer that have been formed from reactions with CO. In both cases the initial situation was a surface completely covered with oxygen except for one site. Diffusion of CO and reaction with oxygen are infinitely fast, but on the *left* the reaction is infinitely faster than the diffusion (i.e., it has a higher priority) and on the *right* it is the other way around

priority will occur before an immediate process with a low priority. This can be important. Figure 5.14 shows an oxygen layer that started with a small hole after exposure to CO. CO adsorbs, reacts with oxygen possibly after diffusing over the surface, and then forms $CO_2$ which will desorb from the surface, thus enlarging the hole in the oxygen layer. The shape of the hole depends on whether diffusion or $CO_2$ formation is the faster of the two processes.

The ZGB model uses a square lattice representing a (100) surface of an fcc metal. A CO molecule on such a lattice has a probability of reacting with a neighboring oxygen atom no matter on which neighboring site the oxygen atom sits. This should not be the case for a rectangular lattice that represents a (110) surface of an fcc metal. There should be a different rate constant for a reaction with a neighboring oxygen atom in one direction than for the direction perpendicular to it. It is possible in this situation too to have different priorities for the infinitely fast reactions. Suppose a CO molecule has an oxygen atom neighbor in direction $[1\,\bar{1}\,0]$ and an oxygen atom neighbor in direction $[1\,0\,0]$. It will react with both of them infinitely fast, but it prefers to react with the one in direction $[1\,\bar{1}\,0]$ because it is nearer. In such a situation one should give the reaction with the neighbor in direction $[1\,\bar{1}\,0]$ a higher priority.

### *5.6.2 Flagging Structural Elements*

Very fast processes are certainly not the most important use of immediate processes. They are more important in situations where a special-purpose code would have a piece of code to handle special situations. We will show here that the same can often be done with immediate processes. This has the advantage that the theoretical framework of Chap. 2 still applies.

As a first application we will look at using immediate processes to determine the presence of certain structural elements. Suppose we do a simulation of Fischer–Tropsch synthesis [18]. The reaction conditions are such that chain growth is difficult and we expect a high selectivity for methane and a rapidly decreasing amount of $C_n$ species with increasing $n$. However, we find that hardly any methane is formed. The question is if this is due to the small rate constant of the $CH_3 + H \rightarrow CH_4$ reaction, or if the diffusion of the reactants is hindered by a high coverage so much that $CH_3$ and H never meet.

There are various ways to answer that question, but we will look at one that looks at how often configurations in which the reactants can actually react occur. In a special-purpose code this would be accomplished by code that would check the system by scanning the adlayer or by checking every time that a favorable configuration might have formed. Immediate processes do however as well. Apart from the processes we need bookkeeping sites to store information.

Let's suppose for simplicity that we have a square lattice, and the $CH_4$ can be formed if $CH_3$ and H are at neighboring sites. For each pair of neighboring sites we create one bookkeeping site. Per unit cell we then have one real and two bookkeeping sites. Site $(n_1, n_2/1)$ is the bookkeeping site for the pair of sites $(n_1, n_2/0)$ and $(n_1 + 1, n_2/0)$, and $(n_1, n_2/2)$ for $(n_1, n_2/0)$ and $(n_1, n_2 + 1/0)$. The bookkeeping sites have a label "set" or "unset". We start with all labels for the bookkeeping sites equal to "unset". A site becomes "set" via one of the next immediate processes.

$$\begin{aligned}
&(0,0/0), (1,0/0), (0,0/1) : \mathrm{CH_3\ H\ unset} \rightarrow \mathrm{CH_3\ H\ set}, \\
&(0,0/0), (1,0/0), (0,0/1) : \mathrm{H\ CH_3\ unset} \rightarrow \mathrm{H\ CH_3\ set}, \\
&(0,0/0), (0,1/0), (0,0/2) : \mathrm{CH_3\ H\ unset} \rightarrow \mathrm{CH_3\ H\ set}, \\
&(0,0/0), (0,1/0), (0,0/2) : \mathrm{H\ CH_3\ unset} \rightarrow \mathrm{H\ CH_3\ set}.
\end{aligned} \tag{5.27}$$

These processes allow us to determine the number of $CH_3$–H pairs in a configuration.

The procedure above is all we need if we have a given configuration and want to count the number of $CH_3$–H pairs in that configuration. The immediate processes are then also the only processes that we need in a simulation: i.e., we have a separate simulation for counting the number of $CH_3$–H pairs. This is straightforward, but also quite cumbersome if we want to know how the number of $CH_3$–H pairs varies in a kMC simulation. What we rather have is a procedure that keeps track of the number of $CH_3$–H pairs during the Fischer–Tropsch simulation. For that we can still use the immediate processes above, but we also need processes that unset the bookkeeping sites if processes occur that remove $CH_3$–H pairs.

If we have only $CH_3$ or H as adsorbates on the surface, then we only need

$$\begin{aligned}
(0,0/0),(1,0/0),(0,0/1) &: CH_3 * \text{set} \rightarrow CH_3 * \text{unset},\\
(0,0/0),(1,0/0),(0,0/1) &: * CH_3 \text{ set} \rightarrow * CH_3 \text{ unset},\\
(0,0/0),(0,1/0),(0,0/2) &: CH_3 * \text{set} \rightarrow CH_3 * \text{unset},\\
(0,0/0),(0,1/0),(0,0/2) &: * CH_3 \text{ set} \rightarrow * CH_3 \text{ unset}
\end{aligned} \tag{5.28}$$

where $*$ is a vacant site, to handle situations where hydrogen is removed from the site. A similar set of immediate processes is needed for the case where $CH_3$ is removed. This method becomes rather cumbersome if $CH_3$ or H can also react to form other adsorbates. For example, $CH_3$ may split off another hydrogen atom and form $CH_2$. We can of course extend the set of immediate processes, but another approach may be easier.

We define another bookkeeping site $(n_1, n_2/3)$. This site normally has a label "idle". For each process that involves $CH_3$ or H at site $(n_1, n_2/0)$ the label of $(n_1, n_2/3)$ is changed to "reset". This triggers the following immediate processes, which change the older bookkeeping sites.

$$\begin{aligned}
(0,0/3)(0,0/1) &: \text{reset set} \rightarrow \text{reset unset},\\
(0,0/3)(0,0/2) &: \text{reset set} \rightarrow \text{reset unset},\\
(0,0/3)(-1,0/1) &: \text{reset set} \rightarrow \text{reset unset},\\
(0,0/3)(0,-1/2) &: \text{reset set} \rightarrow \text{reset unset}.
\end{aligned} \tag{5.29}$$

We don't want the label of $(n_1, n_2/3)$ to remain "reset" so we also need

$$(0,0/3) : \text{reset} \rightarrow \text{idle}. \tag{5.30}$$

This gives us all the necessary immediate processes, except that we also need some control that makes sure that process (5.30) does not occur before all appropriate "set" labels are changed to "unset". So as long as processes (5.29) are possible, process (5.30) should not take place. This can be accomplished by giving (5.29) a higher priority than (5.30).

Suppose we have a process that swaps $CH_3$ and H when they are neighbors. We do not want to change the "set" label corresponding to this pair, but because $CH_3$ and H might be part of other $CH_3$–H pairs, we do change the "idle" labels of these adsorbates to "reset". This causes all neighboring "set" labels to "reset", including the one of the $CH_3$–H pair that is swapped. This is not what we want, but it is not really a problem, because (5.27) changes it back to "set" again. However, this works only if (5.27) is the last immediate process to occur. So (5.27) should be given a lower priority than (5.30) and (5.29).

We can now keep track of the number of $CH_3$–H pairs in the Fischer–Tropsch simulation. There are two ways in which sublattices 1 and 2 can be used. One way is simply to look at the "coverage" of the "set" labels. This is equal to the number of $CH_3$–H pairs. This method might not work if this pairs exist only very briefly and the labels of sublattices 1 and 2 are always all "unset". It is then better to count the number of times (5.27) occurs minus the number of times (5.29) occurs. There should be no "set" labels in the initial configuration with this approach. This method may not work however if $CH_3$–H pairs are very often formed and removed. The reason is that one tries to determine a small difference between two large numbers, which is hard to do accurately.[2]

The method described above can be used to detect all other kind of structures. It can be transferred to the detection of any pair correlations as is, but also the way to set up the detection of more complicated structures is straightforward. The idea is always to have an immediate process that is triggered when the structure that we are looking for is formed. This should change the label of a bookkeeping site. The difficult part is to change this label back when the structure is broken up. A careful assignment of priorities is required as incorrect priorities can easily lead to infinite loops of immediate processes.

### 5.6.3 Counting

Sometimes one wants to count how often some structural element is found at a certain position on the surface. This is really an extension of setting a flag to indicate that a structural element is present or not. We will see in Sect. 6.3 how the result of such counts can be used.

Suppose we have a site and we want to know how many of the neighboring sites are occupied. For simplicity we assume that we have a square lattice and that there is only one type of adsorbate which we call "A". To store the count we need bookkeeping sites: one for each real site. So we use a second sublattice. We need labels that indicate how many of the neighboring sites are occupied. It is natural to use the labels "0", "1", "2", "3", and "4". Note however that they look like integers, but in our formalism they are labels. In particular, we can not use them for arithmetic.

[2]Yet another way to determine the number of $CH_3$–H pairs would be to have a normal process that takes place if there is a $CH_3$–H pair, but that does not change the configuration. The rate of such a process is equal to its rate constant times the coordination number of the lattice times the probability that a pair of neighboring sites is occupied by a $CH_3$–H pair. As we know the coordination number and can choose the rate constant, the rate allows us to calculate the probability of a $CH_3$–H pair. This method has however one drawback. The result is approximate, because the process is a stochastic process. We can make the result more accurate by increasing the rate constant, but that would increase the number of times that the process occurs and slow down the simulation. The method with the immediate processes does not have this drawback, but it does require more computer memory because of the bookkeeping sites and it is also more complex.

If a site becomes occupied, then we need to increase the counts of the neighboring sites by 1. Just having processes

$$\begin{aligned} &(0,0/0),(1,0/1) : \mathrm{A}\ 0 \rightarrow \mathrm{A}\ 1,\\ &(0,0/0),(1,0/1) : \mathrm{A}\ 1 \rightarrow \mathrm{A}\ 2,\\ &(0,0/0),(1,0/1) : \mathrm{A}\ 2 \rightarrow \mathrm{A}\ 3,\\ &(0,0/0),(1,0/1) : \mathrm{A}\ 3 \rightarrow \mathrm{A}\ 4, \end{aligned} \tag{5.31}$$

and symmetry-related processes won't do, because these processes will continue to occur until all counts have become "4". To avoid this we need to change the labels "0", "1", "2", "3", and "4" to labels for which there are no processes that lead to a further increase of the count. At some stage however we want to change these labels back to the labels "0", "1", "2", "3", and "4" again without a further increase of the count. This means that we also need to use other labels for the adsorbate.

A solution to the problem of runaway counts is the following. If there is an adsorption, this does not create a new label "A" at the site of adsorption, but a label "new". We have the following processes for that label.

$$\begin{aligned} &(0,0/0),(1,0/1) : \mathrm{new}\ 0 \rightarrow \mathrm{new}\ \mathrm{t1},\\ &(0,0/0),(1,0/1) : \mathrm{new}\ 1 \rightarrow \mathrm{new}\ \mathrm{t2},\\ &(0,0/0),(1,0/1) : \mathrm{new}\ 2 \rightarrow \mathrm{new}\ \mathrm{t3},\\ &(0,0/0),(1,0/1) : \mathrm{new}\ 3 \rightarrow \mathrm{new}\ \mathrm{t4}. \end{aligned} \tag{5.32}$$

Note that instead of "1", "2", "3", and "4" we have labels "t1", "t2", "t3", and "t4" for the counts. Because there are no processes for these labels, there is no runaway effect.

After all the counts have been updated we change the "new" label.

$$(0,0/0) : \mathrm{new} \rightarrow \mathrm{A}. \tag{5.33}$$

This creates the proper label for an adsorbate. There are no processes containing "A" that change the counts, so they retain there correct value. After the "new" label has been removed we have

$$\begin{aligned} &(0,0/1) : \mathrm{t0} \rightarrow 0,\\ &(0,0/1) : \mathrm{t1} \rightarrow 1,\\ &(0,0/1) : \mathrm{t2} \rightarrow 2,\\ &(0,0/1) : \mathrm{t3} \rightarrow 3,\\ &(0,0/1) : \mathrm{t4} \rightarrow 4. \end{aligned} \tag{5.34}$$

This changes the labels of the counts to their correct values.

Of course, the specification above calls for a specific order in which the processes should occur. This means that these processes should all be immediate processes with a certain priority. Processes (5.32) should have the highest priority, process (5.33) should have a lower priority, and processes (5.34) should have the lowest priority.

In a similar way we can model desorption. The site from which an adsorbate has just desorbed gets a label "gone". Then there is a set of processes with the same priority as processes (5.32), which are giving by

$$\begin{aligned}
&(0,0/0),(1,0/1) : \text{gone } 1 \rightarrow \text{gone t0},\\
&(0,0/0),(1,0/1) : \text{gone } 2 \rightarrow \text{gone t1},\\
&(0,0/0),(1,0/1) : \text{gone } 3 \rightarrow \text{gone t2},\\
&(0,0/0),(1,0/1) : \text{gone } 4 \rightarrow \text{gone t3}.
\end{aligned} \tag{5.35}$$

Then there is

$$(0,0/0) : \text{gone} \rightarrow *, \tag{5.36}$$

with $*$ standing for a vacant site, which has the same priority as process (5.33). Finally, processes (5.34) occur which change the labels of the counts to their correct values.

Diffusion can be implemented using

$$(0,0/0),(1,0/0) : \text{A} * \rightarrow \text{gone new}, \tag{5.37}$$

but note that this works for a square lattice, but not for an hexagonal lattice. The reason why this works for a square lattice is that the neighbors of $(0,0/0)$ and $(1,0/0)$ form disjoint sets, and counts stored at $(0,0/1)$ and $(1,0/1)$ are handled correctly. If we have a hexagonal lattice, then there are two sites that are neighbors of both $(0,0/0)$ and $(1,0/0)$. The process above gives incorrect results for these sites. The simplest way to handle this problem is to have an immediate processes like

$$(0,0/0),(1,0/0),(1,1/1) : \text{gone new } 1 \rightarrow \text{gone new t1}, \tag{5.38}$$

et cetera with $(1,1/1)$ a neighboring site of $(0,0/0)$ and $(1,0/0)$. This, and similar processes, should have a priority that is higher than any of the processes above.

Other processes that create or remove the adsorbate can be handled in the same way. The approach can also be extended to systems with different types of adsorbate.

### *5.6.4 Decomposing the Implementation of Processes*

When many sites are involved in a process they can be included directly in the description, but sometimes it is better to split it into a number of processes. Because the combination of these processes must have a probability distribution for the time that the process takes place that is the same of that of the description as a single process, all but the first of these processes to occur must be immediate processes. This can be seen as follows. Suppose that we have just two processes with rate constants $W_1$ and $W_2$. The process with rate constant $W_2$ takes place after the one with rate constant $W_1$. If we start at time $t = 0$, then the second process will take place at a time $t$ with probability distribution

$$\int_0^t dt' W_2 e^{-W_2(t-t')} W_1 e^{-W_1 t'} = \frac{W_1 W_2}{W_2 - W_1} \left[ e^{-W_1 t} - e^{-W_2 t} \right]. \tag{5.39}$$

This probability distribution differs from the usual exponential distribution unless $W_1$ or $W_2$ is infinite. Because we want both processes to take place at the same time, we need $W_2 \to \infty$.

Suppose that we have an adsorbate A with a strong repulsion so that the neighboring sites can not be occupied. There may be several reasons for this. The adsorbate may be rather bulky, neighboring sites are very close, there is a strong through-metal interaction, et cetera [19]. We might try to model this using a description that involves the adsorption site and its neighbors as is discussed in Sect. 6.2, but there is another option. The adsorption we model simply with

$$(0,0) : * \to \mathrm{A} \tag{5.40}$$

with $*$ a vacant site. This process has a finite rate constant. The blocking of the neighboring sites is modeled with

$$(0,0),(1,0) : \mathrm{A}\ * \to \mathrm{A\ blocked}. \tag{5.41}$$

This process, and the symmetry-related ones, are infinitely fast. This seems quite straightforward, but things become a bit more tricky when the adsorbate can also diffuse and desorb. We look at desorption. Just

$$(0,0) : \mathrm{A} \to * \tag{5.42}$$

does not work, because this leaves "blocked" labels for sites that are not blocked any longer.

$$(0,0),(1,0) : \mathrm{A\ blocked} \to *\ * \tag{5.43}$$

doesn't work either, because this process can occur only once for each adsorbate. This means only one label "blocked" is changed back to $*$, but the adsorbate may have been blocking more than one site. What does work is first

$$(0,0) : \mathrm{A} \to \mathrm{vacated}, \tag{5.44}$$

The label "vacated" indicates that an adsorbate has just desorbed from the site. It is used to remove the "blocked" labels by

$$(0,0),(1,0) : \mathrm{vacated\ blocked} \to \mathrm{vacated}\ * \tag{5.45}$$

and symmetry-related immediate processes. Note that the label "vacated" stays so that it can remove all "blocked" labels. To get rid of the "vacated" label we finally have another immediate process

$$(0,0) : \text{vacated} \rightarrow *. \tag{5.46}$$

It is clear that this process should only occur after all "blocked" labels have been removed, so it should have a lower priority than the previous process. The last thing that we now need to do is to give the original blocking process (5.41) an even lower priority. If we would not do that we could get the following infinite loop

$$\begin{array}{llllllll} \text{blocked} & \text{A} & \text{blocked} & \text{A} & \rightarrow & \text{blocked} & \text{vacated} & \text{blocked} \;\; \text{A} \\ & & & \rightarrow & \text{blocked} & \text{vacated} & * & \text{A} \\ & & & \rightarrow & \text{blocked} & \text{vacated} & \text{blocked} & \text{A} \\ & & & \rightarrow & \text{blocked} & \text{vacated} & * & \text{A} \\ & & & \rightarrow & \ldots & & & \end{array} \tag{5.47}$$

(We show only the sites along one line for convenience.) With the blocking process having the lowest priority, sites only become blocked again after the "vacated" label has been removed.

### *5.6.5 Implementing Procedures*

The real strength of immediate processes becomes apparent when we model large molecules with many configurations and possibly with different and/or variable sizes. For such systems it is not possible to list all patterns. Let's assume we want to model linear polymer chains on a surface. We assume that we can model such molecules with a square lattice with the monomers forming the polymer chains adsorbed on neighboring sites horizontally or vertically. Moreover, we assume that monomers that are not linked to each other can not be at neighboring sites either horizontally, vertically, or diagonally. We show how immediate processes can be used to model reptation for such molecules.

We use three labels to model a polymer chain: "head", "body", and "tail" (see Fig. 5.15). The labels "head" and "tail" indicate the monomers at the ends. The other monomers are labeled "body". We will show how to implement a reptation "head"-first, but we will also show how to extend the model to have the polymer reptate in the opposite direction.

The first process, and the only one that is not an immediate process, moves the head to a new position.

$$\begin{array}{ccc} & (0,1) & (1,1) \\ (-1,0) & (0,0) & (1,0) \\ & (0,-1) & (1,-1) \end{array} : \begin{array}{ccc} & * & * \\ \text{head} & * & * \\ & * & * \end{array} \rightarrow \begin{array}{ccc} & * & * \\ \text{fh} & \text{head} & * \\ & * & * \end{array}. \tag{5.48}$$

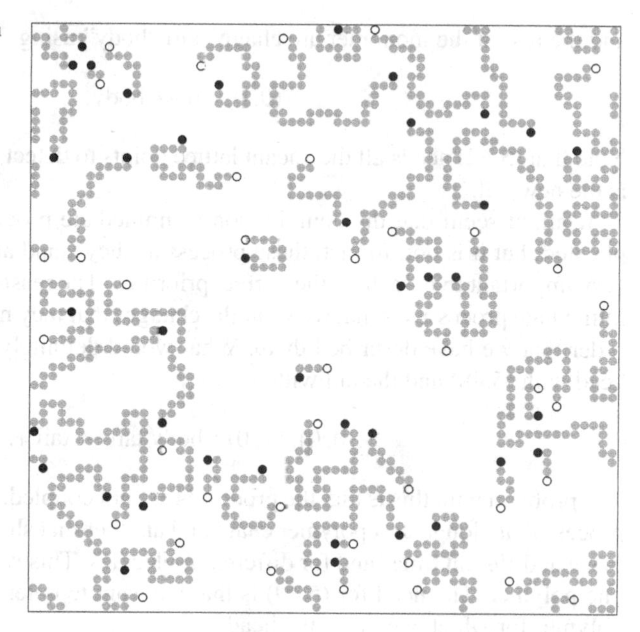

**Fig. 5.15** Square lattice with an example of polymer chains. The labels "head", "tail", and "body" are depicted by *black*, *white*, and *gray circles*, respectively

Note all the extra lattice points that need to be vacant for the head to move. This is to make sure that only linked monomers can occur at neighboring sites. The label "fh" stands for "follow head", and indicates the monomers that move after the head. Note also that we only give one process that moves the head. There are three others that are symmetry-related and that correspond to reptation in other directions. We give also just one representative description for all subsequent processes.

This "fh" label initiates a chain of immediate processes. The process above only changes the monomer that used to be the head. To change the other monomers we have the immediate process

$$(0,0),(1,0): \text{fh body} \rightarrow \text{fh fh}. \tag{5.49}$$

This changes all the "body" labels to "fh". To retract the tail we need

$$(0,0),(1,0): \text{fh tail} \rightarrow \text{ft} *. \tag{5.50}$$

We see that the lattice point with the tail becomes vacant, and the new tail is labeled "ft". This new label initiates another chain of immediate processes that transform the monomers again.

$$(0,0),(1,0): \text{fh ft} \rightarrow \text{ft ft}. \tag{5.51}$$

Then the new tail is given its proper label using

$$\begin{array}{ccccccc} & (0,1) & & & * & & * \\ (-1,0) & (0,0) & (1,0) & : & \text{ft ft} * & \rightarrow & \text{ft tail} * \\ & (0,-1) & & & * & & * \end{array} \tag{5.52}$$

and the rest of the monomer are changed to "body" using

$$(0,0) : \text{ft} \rightarrow \text{body}. \tag{5.53}$$

Note that (5.52) needs all the vacant lattice points to detect the proper "ft" label that is the new tail.

It might seem that the combination of immediate processes is needlessly complicated, but it is not. In fact, these process as they stand are not complete yet. It is very important to give them the correct priorities. The reason why there are so many immediate processes is that we want the changes that they make to occur in the same order that we have described them. What would definitely not work is moving the head with (5.48) and the tail with

$$(0,0), (1,0) : \text{body tail} \rightarrow \text{tail} *. \tag{5.54}$$

The problem with this is that the processes are not coupled. We would then have one process that elongates a polymer chain and another that shortens it but they can take place at different times and for different molecules. This would not be a reptation of one polymer. The need for (5.49) is that we want to determine the tail of the same polymer for which we move the head.

To see why we need (5.50) and (5.51) is more subtle. Why not have

$$(0,0), (1,0) : \text{fh tail} \rightarrow \text{tail} * \tag{5.55}$$

instead of (5.50) and

$$(0,0) : \text{fb} \rightarrow \text{body ?} \tag{5.56}$$

The problem with this is that we get an infinite loop of immediate processes. As soon as the last process takes place, process (5.49) becomes possible again and changes "body" back to "fb". The reason for changing "fb" to "ft" is to avoid such infinite loops.

For processes (5.48) to (5.53) to work we need to assign priorities to avoid the same kind of infinite loop. This means that we should not change "ft" too soon to "body". Changing "fh" to "ft", process (5.51), and making the new tail, process (5.52), should have a higher priority than the other processes. However, (5.51) and (5.52) can be given the same priority, and also the other processes can be given the same, but lower, priority. Note that although some processes need to occur before others, this does not necessarily imply that they need to have a higher priority. For example, (5.50) will always occur after (5.49) independent on the priorities.

The implementation above distinguishes between the two ends of the polymer, and reptation is only possible in one direction. If you want the polymer to reptate in both directions there are at least two possibilities. The easiest is to add a set of processes that is identical to the ones above except that the roles of "head" and "tail" are reversed. (The names "head" and "tail" are then not appropriate anymore, and should be changed.) It is advisable not to use "fh" and "ft" again, however, but use other labels. This is to avoid triggering any of the processes above.

The alternative is to add a set of processes that swap the labels "head" and "tail". The drawback of this approach is that reptation in the opposite direction becomes a combination of two processes. This makes it slower and the distribution function for the time of occurrence is not an exponential function anymore (see Eq. (5.39)).

## References


1. D.A. McQuarrie, *Statistical Mechanics* (Harper, New York, 1976)
2. D.T. Gillespie, J. Comput. Phys. **22**, 403 (1976)
3. D.T. Gillespie, J. Phys. Chem. **81**, 2340 (1977)
4. V. Privman, *Nonequilibrium Statistical Mechanics in One Dimension* (Cambridge University Press, Cambridge, 1997)
5. S. Redner, in *Nonequilibrium Statistical Mechanics in One Dimension*, ed. by V. Privman (Cambridge University Press, Cambridge, 1996), pp. 3–27
6. A.A. Ovchinnikov, Y.B. Zeldovich, Chem. Phys. **28**, 215 (1978)
7. D. Toussaint, F. Wilczek, J. Chem. Phys. **78**, 2642 (1983)
8. W.K. Offermans, A.P.J. Jansen, R.A. van Santen, Surf. Sci. **600**, 1714 (2006)
9. W.K. Offermans, A.P.J. Jansen, R.A. van Santen, G. Novell-Leruth, J.M. Ricart, J. Pérez-Ramirez, J. Phys. Chem. C **111**, 17551 (2007)
10. B. Stroustrup, *The C++ Programming Language* (Addison-Wesley, Reading, 1991)
11. V.N. Kuzovkov, O. Kortlüke, W. von Niessen, Phys. Rev. Lett. **83**, 1636 (1999)
12. O. Kortlüke, V.N. Kuzovkov, W. von Niessen, Phys. Rev. Lett. **83**, 3089 (1999)
13. V.N. Kuzovkov, O. Kortlüke, W. von Niessen, Phys. Rev. E **66**, 011603 (2002)
14. R.J. Gelten, A.P.J. Jansen, R.A. van Santen, J.J. Lukkien, J.P.L. Segers, P.A.J. Hilbers, J. Chem. Phys. **108**, 5921 (1998)
15. M.T.M. Koper, J.J. Lukkien, A.P.J. Jansen, R.A. van Santen, J. Phys. Chem. B **103**, 5522 (1999)
16. M.T.M. Koper, N.P. Lebedeva, C.G.M. Hermse, Faraday Discuss. **121**, 301 (2002)
17. R.M. Ziff, E. Gulari, Y. Barshad, Phys. Rev. Lett. **56**, 2553 (1986)
18. B.H. Davis, M.L. Occelli, *Fischer–Tropsch Synthesis, Catalysts and Catalysis*. Studies in Surface Science and Catalysis, vol. 163 (Elsevier, Amsterdam, 2006)
19. C.G.M. Hermse, A.P.J. Jansen, in *Catalysis*, vol. 19, ed. by J.J. Spivey, K.M. Dooley (Royal Society of Chemistry, London, 2006)

# Chapter 6
# Modeling Surface Reactions II

**Abstract** The way to model many processes for kinetic Monte Carlo simulations is straightforward. There are however also processes that one encounters regularly and for which there are more modeling options and for which the best is not always clear. We discuss here several of them. We look at how to handle site blocking by large adsorbates and other cases with strong repulsion. We show several ways to implement finite lateral interactions. Fast diffusion and other fast processes are shown to be not necessarily a hindrance for efficient simulations. Some fast processes can even be combined with slower processes in one effective process. Tagging adsorbates is introduced to simulate isotope experiments and to obtain information on diffusion. Our two-dimensional modeling framework is shown to be capable to deal with simulating reactions on nanoparticles. Non-physical processes are shown to be useful to create the initial configuration of a kinetic Monte Carlo simulation.

## 6.1 Introduction

Chapter 5 has introduced the tools that are available within the theoretical framework of Chap. 2 to model processes on surfaces. In this chapter we discuss how these tools can be used to model specific chemical and physical aspects that one meets when one starts implementing such processes. The distinction between the topics discussed here and those in Chap. 5 are not completely unambiguous. For example, the discussion on how to implement multiple sites in the unit cell of a surface (Sect. 5.5.2) and the various ways to model steps (Sect. 5.5.3) would also fit well in this chapter. The difference between the chapters is really a point of view. In Chap. 5 we discussed modeling tools and showed what could be done with them. Here we will start with the question of how to model some process and then show how to do that.

This chapter is also mainly important if you are using a general-purpose code for your kinetic Monte Carlo (kMC) simulations, but it also gives an impression of the various ways you can hardcode some model in special-purpose codes. In fact, because the starting point in this chapter is some chemical or physical problem, it brings the fact that there are often several ways to implement a model better to the fore.

A.P.J. Jansen, *An Introduction to Kinetic Monte Carlo Simulations of Surface Reactions*, Lecture Notes in Physics 856,
DOI 10.1007/978-3-642-29488-4_6, © Springer-Verlag Berlin Heidelberg 2012

## 6.2 Large Adsorbates and Strong Repulsion

Suppose that we have adsorption of a somewhat bulky adsorbate A. The adsorbate occupies not only a particular site, but also makes it impossible for other adsorbates to occupy neighboring sites. We have already seen how this can be done with infinitely fast processes in Sect. 5.6.4. The adsorption we model simply with

$$(0,0) : * \to \mathrm{A} \tag{6.1}$$

with $*$ a vacant site. This process has a finite rate constant. The blocking of the neighboring sites is modeled with the immediate process

$$(0,0), (1,0) : \mathrm{A}\ * \to \mathrm{A}\ \text{blocked}, \tag{6.2}$$

and symmetry-related processes. For desorption we have

$$(0,0) : \mathrm{A} \to \text{vacated}. \tag{6.3}$$

The label "vacated" indicates that an adsorbate has just desorbed from the site. It is used to remove the "blocked" labels by

$$(0,0), (1,0) : \text{vacated blocked} \to \text{vacated}\ *, \tag{6.4}$$

which is an another immediate process. Note that the label "vacated" stays so that it can remove all "blocked" labels. To get rid of the "vacated" label we finally have yet another immediate process

$$(0,0) : \text{vacated} \to *. \tag{6.5}$$

These immediate processes should be given the right priorities as discussed in Sect. 5.6.4: i.e., (6.4) should have the highest priority, (6.5) a lower one, and (6.2) should have the lowest.

We have to take care if we want to extend this approach to diffusion. We might try

$$(0,0), (1,0) : \mathrm{A}\ \text{blocked} \to \text{vacated}\ \mathrm{A}. \tag{6.6}$$

The "vacated" label removes the "blocked" labels for the neighbors of the old adsorption site of A. However, this is not always correct, because site $(1,0)$ might have other A's at neighboring sites besides the one originally at $(0,0)$. The process would then put two, or more, A's at neighboring sites.

One way to deal with this is to check the neighboring sites of $(1,0)$. If we have a square lattice this might look as follows. Instead of the process above, we have

$$\begin{array}{cccccccccc} & (1,1) & & & & \text{notA} & & & \text{notA} & \\ (0,0) & (1,0) & (2,0) & : & \mathrm{A} & \text{blocked} & \text{notA} \to & \text{vacated} & \mathrm{A} & \text{notA}\,. \\ & (1,-1) & & & & \text{notA} & & & \text{notA} & \end{array} \tag{6.7}$$

Here "notA" stands for either "blocked" or "*". So there are really eight different processes, plus symmetry-related ones, that we have to include to model the diffusion in this way. An alternative approach would be to undo a diffusion process that should not have occurred. We do use (6.6), but also add the immediate process

$$(0,0),(1,0),(2,0) : \text{vacated A A} \rightarrow \text{A blocked A} \tag{6.8}$$

with a priority higher than any of the other immediate processes. This process puts the A that has just jumped back on its original site when it gets another A as its neighbor. We also need to add two more processes to test for A's at sites $(1,1)$ and $(1,-1)$ instead of $(2,0)$.

Of course, undoing a process is inefficient. There may also be an inefficiency related to the "vacated" label. The "blocked" labels on neighbors of "vacated" are changed to "*", but this is not always what we want, because these sites may be neighbors of other adsorbates. These adsorbates will change the appropriate "*" labels back to "blocked", but only after "vacated" has been changed to "*". For this reason another way to model these processes might be considered.

The advantage of using immediate processes is that the number of sites involved in the description of a process can be kept low. Pattern matching will be efficient. Another way to model the same processes above is to check always the neighbors of a site that becomes occupied. If we have a square lattice the adsorption becomes

$$\begin{array}{ccc} & (0,1) & \\ (-1,0) & (0,0) & (1,0) \\ & (0,-1) & \end{array} : \begin{array}{ccc} & * & \\ * & * & * \\ & * & \end{array} \rightarrow \begin{array}{ccc} & * & \\ * & \text{A} & * \\ & * & \end{array}. \tag{6.9}$$

The desorption is very simple, because it involves no changes in labels except to one of the adsorption site.

$$(0,0) : \text{A} \rightarrow *. \tag{6.10}$$

Diffusion is very similar to adsorption. We get

$$\begin{array}{ccc} & (0,1) & \\ (-1,0) & (0,0) & (1,0) \\ & (0,-1) & \end{array} : \begin{array}{ccc} & * & \\ \text{A} & * & * \\ & * & \end{array} \rightarrow \begin{array}{ccc} & * & \\ * & \text{A} & * \\ & * & \end{array} \tag{6.11}$$

for a square lattice. For diffusion there are three other symmetry-related processes. We see that with this approach we do not have immediate processes, but adsorption and diffusion involve more than two sites. Note also that we need to specify explicitly all neighbors of a site. The description of adsorption and diffusion would be different for a hexagonal lattice.

Although we have started this section with stating that the adsorbate A is bulky, the descriptions above are also valid for other adsorbates that prevent neighboring sites from becoming occupied. This is the case for example when an adsorbate changes the electronic structure of the substrate in such a way that adsorption on neighboring sites becomes unfavorable. Other mechanisms might be operable as well: see [1]. Suppose that the adsorbate is really large and not only prevents other adsorbates from occupying neighboring sites, but actually occupies not just one site but also the neighboring sites. The following implementation then seems obvious. For adsorption we use

$$\begin{array}{ccc} & (0,1) & \\ (-1,0) & (0,0) & (1,0) \\ & (0,-1) & \end{array} : \begin{array}{ccc} & * & \\ * & * & * \\ & * & \end{array} \rightarrow \begin{array}{ccc} & \mathrm{A} & \\ \mathrm{A} & \mathrm{A} & \mathrm{A} \\ & \mathrm{A} & \end{array}, \tag{6.12}$$

for desorption

$$\begin{array}{ccc} & (0,1) & \\ (-1,0) & (0,0) & (1,0) \\ & (0,-1) & \end{array} : \begin{array}{ccc} & \mathrm{A} & \\ \mathrm{A} & \mathrm{A} & \mathrm{A} \\ & \mathrm{A} & \end{array} \rightarrow \begin{array}{ccc} & * & \\ * & * & * \\ & * & \end{array}, \tag{6.13}$$

and for diffusion

$$\begin{array}{cccc} & (0,1) & (1,1) & \\ (-1,0) & (0,0) & (1,0) & (2,0) \\ & (0,-1) & (1,-1) & \end{array} : \begin{array}{cccc} & \mathrm{A} & * & \\ \mathrm{A} & \mathrm{A} & \mathrm{A} & * \\ & \mathrm{A} & * & \end{array} \rightarrow \begin{array}{cccc} & * & \mathrm{A} & \\ * & \mathrm{A} & \mathrm{A} & \mathrm{A} \\ & * & \mathrm{A} & \end{array}. \tag{6.14}$$

There is one big problem with this implementation, however, and if that problem is solved then there is still one important difference with the two preceding implementations.

The big problem is that the implementation is incorrect. When several adsorbates are located close to each other, it is no longer clear which label "A" belongs to which adsorbate. This makes it possible that processes take place that involve some A's from one adsorbate and some from another. This is obviously not what we want. We need to distinguish between the different parts of the adsorbate. The adsorption could for example be modeled by

$$\begin{array}{ccc} & (0,1) & \\ (-1,0) & (0,0) & (1,0) \\ & (0,-1) & \end{array} : \begin{array}{ccc} & * & \\ * & * & * \\ & * & \end{array} \rightarrow \begin{array}{ccc} & \mathrm{a} & \\ \mathrm{a} & \mathrm{A} & \mathrm{a} \\ & \mathrm{a} & \end{array}. \tag{6.15}$$

An adsorbate is now well-defined, because it is an "A" surrounded by "a"'s.

This implementation however is not equivalent to the previous two. Only neighboring sites were blocked for adsorption in the previous two implementations, but in the last implementation more sites are blocked. This can most easily be seen for a square lattice. An adsorbate on site $(0,0)$ blocks only sites $(1,0)$, $(0,1)$, $(-1,0)$, and $(0,-1)$ in the first two implementations. A site like $(1,1)$ is available for adsorption. This is not the case in the last implementation. The reason is that an adsorbate at $(1,1)$ in the last implementation would also occupy the neighboring sites, among them $(1,0)$ and $(0,1)$. But these are already occupied. In the first two implementation the maximum coverage for a square lattice is 0.5 ML and the adlayer then has a translational symmetry given by the primitive vectors $a(1,1)$ and $a(1,-1)$ with $a$ the distance between neighboring sites. In last implementation the maximum coverage for a square lattice is 0.2 ML and the adlayer then has a translational symmetry given by the primitive vectors $a(2,1)$ and $a(1,-2)$ (see Fig. 6.1).

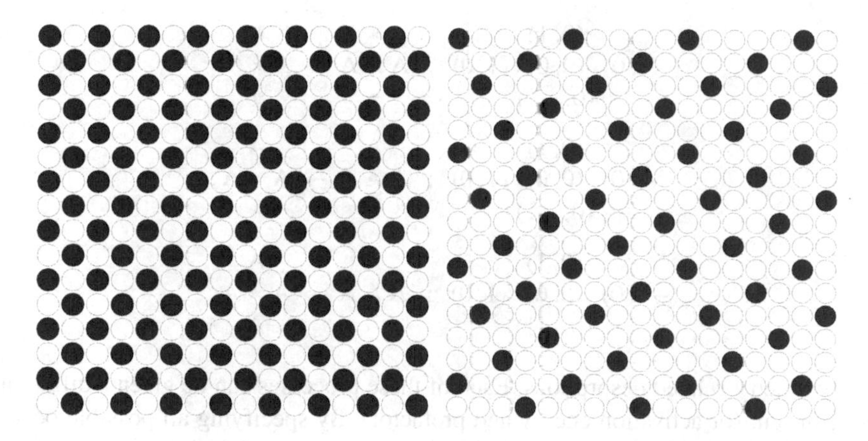

**Fig. 6.1** A square lattice with an adsorbate that blocks neighboring sites for adsorption (*left*) and an adsorbate that occupies a central site and the four neighboring sites (*right*). In both cases an adlayer is shown with a maximum coverage. The *black circles* show the adsorbates (*left*) or the central sites occupied by the adsorbates (*right*). The *white circles* indicate all other sites

## 6.3 Lateral Interactions

Lateral interactions are interactions between adsorbates. It has been well-known for a long time that these interactions lead to structured or ordered adlayers at low temperatures. The realization that the kinetics of surface reactions can be substantially affected by these interactions even at high temperatures is more recent. Relatively little is known about the form of these interactions and even less about the strength of them [1]. In this section we present a number of ways to implement these interactions. In all cases we assume that the lateral interactions are short range.

The most general method to model lateral interactions consists of specifying the process and the occupation of the sites that may have adsorbates that will affect the process. For example, if we have a simple desorption of an adsorbate A on a square lattice with a rate constant that depends on the occupation of the four neighboring sites, then we have

$$
\begin{array}{ccc} & (0,1) & \\ (-1,0) & (0,0) & (1,0) \\ & (0,-1) & \end{array} : \begin{array}{ccc} & * & \\ * & \mathrm{A} & * \\ & * & \end{array} \rightarrow \begin{array}{ccc} & * & \\ * & * & * \\ & * & \end{array},
$$

$$
\begin{array}{ccc} & (0,1) & \\ (-1,0) & (0,0) & (1,0) \\ & (0,-1) & \end{array} : \begin{array}{ccc} & * & \\ * & \mathrm{A} & \mathrm{A} \\ & * & \end{array} \rightarrow \begin{array}{ccc} & * & \\ * & * & \mathrm{A} \\ & * & \end{array},
$$

$$
\begin{array}{ccc} & (0,1) & \\ (-1,0) & (0,0) & (1,0) \\ & (0,-1) & \end{array} : \begin{array}{ccc} & \mathrm{A} & \\ * & \mathrm{A} & \mathrm{A} \\ & * & \end{array} \rightarrow \begin{array}{ccc} & \mathrm{A} & \\ * & * & \mathrm{A} \\ & * & \end{array}, \tag{6.16}
$$

$$\begin{array}{ccccc} & (0,1) & & & * \quad\quad * \\ (-1,0) & (0,0) & (1,0) & : & \mathrm{A\,A\,A} \to \mathrm{A} * \mathrm{A}, \\ & (0,-1) & & & * \quad\quad * \end{array}$$

$$\begin{array}{ccccc} & (0,1) & & & \mathrm{A} \quad\quad \mathrm{A} \\ (-1,0) & (0,0) & (1,0) & : & \mathrm{A\,A\,A} \to \mathrm{A} * \mathrm{A}, \\ & (0,-1) & & & * \quad\quad * \end{array}$$

$$\begin{array}{ccccc} & (0,1) & & & \mathrm{A} \quad\quad \mathrm{A} \\ (-1,0) & (0,0) & (1,0) & : & \mathrm{A\,A\,A} \to \mathrm{A} * \mathrm{A}, \\ & (0,-1) & & & \mathrm{A} \quad\quad \mathrm{A} \end{array}$$

and symmetry-related desorptions. Each of these processes can be given a different rate constant (or activation energy and prefactor). By specifying all possible occupations of the neighboring sites explicitly any dependence of the lateral interactions on the occupation of neighboring sites can be modeled. The disadvantage should also be clear. The list of processes can be quite long. With $Z$ neighboring sites that may be occupied by an adsorbate affecting a process and $A$ possible occupations of each of these sites (including no adsorbate) there are $A^Z$ processes to specify.

Suppose that the rate constant for the desorption above depends only on the number of neighbors of the adsorbate that desorbs. This is for example the case if we have only interactions between pairs of adsorbates that change the activation energy: i.e., the activation energy can be written as $E_{\text{act}}^{(0)} - n\varphi$ with $E_{\text{act}}^{(0)}$ the activation energy in the absence of neighbors, $n$ the number of neighbors, and $\varphi$ the interaction parameter that is negative for attractive and positive for repulsive interactions. Instead of (6.16) we have

$$\begin{aligned} &(0,0/0),(0,0/1) : \mathrm{A}\ 0 \to \text{gone}\ 0, \\ &(0,0/0),(0,0/1) : \mathrm{A}\ 1 \to \text{gone}\ 1, \\ &(0,0/0),(0,0/1) : \mathrm{A}\ 2 \to \text{gone}\ 2, \\ &(0,0/0),(0,0/1) : \mathrm{A}\ 3 \to \text{gone}\ 3, \\ &(0,0/0),(0,0/1) : \mathrm{A}\ 4 \to \text{gone}\ 4 \end{aligned} \tag{6.17}$$

each with a different rate constant. We have introduced a bookkeeping site $(0,0/1)$ that holds the number of neighbors of the desorbing adsorbate at $(0,0/0)$. How to count these neighbors was shown in Sect. 5.6.3. The label "gone" was introduced there also. It is used to update the counts of the neighbors of $(0,0/0)$.

The advantage of this implementation of lateral interactions is that it requires fewer process descriptions. Instead of $2^Z$ we only need $Z+1$. On a hexagonal lattice ($Z=6$) this reduces the number of process descriptions by almost an order of magnitude. If there are adsorbates with different lateral interaction parameters, we need a sublattice of bookkeeping sites for each type of adsorbate. The number of process descriptions changes then from $A^Z$ to $(Z+1)^{A-1}$. We see that this approach works best if $A \ll Z$. It can however also be extended to include more complicated interactions by including more bookkeeping sites: e.g., three-particle interactions, et cetera.

It is clear from these expressions for the number of processes that lateral interactions involve many processes in the formalism that we have used so far. The reason is fundamental. There can be a very large number of ways that neighboring sites can be occupied, which may lead to an equally large number of rate constants, and hence an equally large number of processes. We need to extend our formalism if we want to avoid this.

One way of doing this is to partition the sites involved in a process in two groups. The first group has the sites with a changing occupation when the process takes place. The second group contains the sites that do not change their occupation, but the adsorbates on these sites affect the rate constant. For the first group we can use the formalism as before. For the second group we need a description, rule, or function that yields a rate constant given a particular occupation of its sites. It is not possible to say more than this for general lateral interactions. A popular form for the lateral interactions is the cluster expansion [1–12]. The adsorption energy $E_{\text{ads}}$ of an adsorbate in a particular adlayer structure is then written as

$$E_{\text{ads}} = \sum_{m} c_m V_m \tag{6.18}$$

with $V_m$ the value of the interaction of type $m$, and $c_m$ the number of interactions of type $m$ per adsorbate. The interactions $V_m$ stand for the interaction of the adsorbate with the substrate (adsorption energy of an isolated adsorbate), pair interactions between adsorbates at various distances, all possible three-particle interactions, four-particle interactions, et cetera. This expansion can be made to reproduce the calculated adsorption energy as accurately as one wants, and therefore can describe any form of lateral interaction [2]. It also depends linearly on its parameters $c_m$, which makes them relatively easy to determine. The number of terms can become very large however. Moreover, it may lead to overfitting: i.e, the cluster expansion will not only describe the interactions, but also the errors one makes in the calculations of the adsorption energies. To avoid this one needs to truncate the cluster expansion (see Sects. 4.5.3 and 4.5.4) [3, 4, 8–13]. Other functional forms for the lateral interactions that are based on the physical mechanism that causes the lateral interactions exist as well [1], but, as there are a number of such mechanisms, none of them can be used universally.

The Carlos/Kinetix code uses such a partitioning of the sites to describe the lateral interactions, but implements only pair interactions [14]. This may seem rather limited, but by using immediate processes and bookkeeping sites more complicated interactions can be modeled as well. An example of this is the model to implement the effect of stress in the top layer of a Cu(110) surface caused by adsorption of (R,R)-bitartrate [15]. This adsorbate forms diagonal rows in the $\langle 1\bar{1}4 \rangle$ direction at maximum coverage. Three of these rows are close together and form a bundle, but there is a trough separating these bundles. The reason is that the rows of Cu atoms in the $\langle 1\bar{1}0 \rangle$ direction are stressed. This needs to be relieved by interspersing the Cu atoms with an adsorbate attached by Cu atoms without adsorbate. The effect can be regarded as an elastic lateral interaction, but it is clearly not a pair interaction [1].

## 6.4 Diffusion and Fast Reversible Reactions

Diffusion is often mentioned when shortcomings of kMC are discussed. This is not quite appropriate. It is even less appropriate to say that kMC has problems when there are processes with very different rate constants. This is indeed a problem when an algorithm with a fixed step size is used, because the step size should be small enough so that the fastest process is simulated correctly. For the slower processes a small step size is, however, inefficient. But there is no problem with algorithms with variable step sizes. Diffusion is often much faster than the other processes. kMC will happily simulate this fast diffusion, but most computer time is then spent on diffusion and only a very small fraction on the other process, which are often the ones one is really interested in. This is not really a shortcoming of kMC, however, but reflects an intrinsic property of the system one is studying. Any method that simulates all processes that take place will have to spend most time on the diffusion simply because most events are adsorbates moving from site to site. A similar situation arises also when one has a very fast adsorption-desorption equilibrium: most events are then adsorption and desorption processes. Nevertheless, there are some ways to reduce the fraction of computer time that one has to spend on such fast processes.

It is important to understand what the problem really is. A fast process in itself is not a problem. Suppose we have NO at high coverage on a Rh(111) surface [16]. On this surface NO readily dissociates, provided that the coverage is low enough so that there are vacant sites for the oxygen that is formed. At high coverage these sites may not be available, and then the dissociation is suppressed. Increasing the temperature causes some NO to desorb. This creates vacant sites for oxygen, and the NO remaining on the surface will then dissociate. The rate constant for he dissociation at these elevated temperatures is many orders of magnitude larger than those of all other processes. This is however no problem at all for kMC, because the process is irreversible. It would not even be a problem if the reverse process would not be fast. There is only a problem if we have processes that lead to an infinite chain of events taking place on a very short time scale. This occurs if both a process and its reverse have a high rate constant, or if there is a whole set of fast consecutive processes as for diffusion.

Mason et al. have discussed a method in which a number of subsequent process are treated analytically and then have modeled them as one single effective process taking place on a longer time scale [17]. We discuss this method in Sect. 8.1.4. It is often possible to use a much simpler approach, which we first discuss here for diffusion and then for other processes.

The idea is based on the fact that the main effect of diffusion is to bring the adlayer to steady state or equilibrium, and that realistic rate constants do this at a much shorter time scale than the time scale of all other processes. This means that if we reduce the rate constant for diffusion, but not too much, the adlayer is still brought to steady state or equilibrium on a time scale short compared to that of other processes. As a consequence the kinetics does not change, but the number of diffusional hops in a kMC simulation can be reduced: in practice often by orders of magnitude.

Finding by how much the rate constants can be reduced requires doing several simulations, which can be short however. The simulations need to be done with different values for the rate constant. The times the simulations take can be minimized by starting with low values of the rate constants, increase them, and look when the results converge. This yields minimal values for the rate constants and fastest simulations with good results for the kinetics. The same approach can not only be used for diffusion, but also for other fast processes with a fast reverse process that rapidly reach equilibrium.

The precise value that can be used for the rate constant for diffusion depends on the system size. The displacement of a particle through diffusion increases with the square of time. If the system is small, then it will rapidly have moved through the whole system. If the system is large, then this will take longer. This means that for small systems the diffusion rate can be reduced more than for large systems.

Sometimes it also suffices if there is a possibility for the chain of fast processes to break off. Figure 5.14 shows the results of fast CO diffusion followed by an even faster $CO_2$ formation. Because the latter is irreversible, the chain of diffusional hops will end when a CO meets an oxygen atom with which it can react. Section 8.1 has more on advanced methods to deal with fast processes.

## 6.5 Combining Processes

Infinitely fast processes can sometimes be removed altogether. Suppose we have the Ziff–Gulari–Barshad (ZGB) model of Sects. 5.6.1 and 7.4.3. Straightforward modeling as in that section yields

$$(0,0) : * \to \mathrm{A} \tag{6.19}$$

for the adsorption, and

$$(0,0),(1,0) : \mathrm{A}\,\mathrm{B} \to *\,* \tag{6.20}$$

plus symmetry-related expressions for the infinitely fast formation of AB. We can now combine these two processes. For example an adsorption on a site with vacant neighboring sites can be modeled as

$$\begin{array}{ccccccc} & (0,1) & & & * & & * \\ (-1,0) & (0,0) & (1,0) & : & *\,*\,* & \to & *\,\mathrm{A}\,* . \\ & (0,-1) & & & * & & * \end{array} \tag{6.21}$$

(For simplicity we have assumed that we have a square lattice.) The rate constant for this process equals the rate constant for adsorption. More interesting is the adsorption on a site with one B neighbor. We then have

$$\begin{array}{ccc} & (0,1) & \\ (-1,0) & (0,0) & (1,0) \\ & (0,-1) & \end{array} : \begin{array}{ccc} & * & \\ * & * & \mathrm{B} \\ & * & \end{array} \rightarrow \begin{array}{ccc} & * & \\ * & * & * \\ & * & \end{array} . \qquad (6.22)$$

The adsorption takes place at $(0, 0)$, but there is an immediate process with the B so that the sites at $(0, 0)$ and $(1, 0)$ become vacant again. On a square lattice there are three other symmetry-related processes. The rate constant for each of them is again the rate constant for adsorption. If there are two B neighbors we get

$$\begin{array}{ccc} & (0,1) & \\ (-1,0) & (0,0) & (1,0) \\ & (0,-1) & \end{array} : \begin{array}{ccc} & \mathrm{B} & \\ * & * & \mathrm{B} \\ & * & \end{array} \rightarrow \begin{array}{ccc} & \mathrm{B} & \\ * & * & * \\ & * & \end{array} \qquad (6.23)$$

and symmetry-related processes or

$$\begin{array}{ccc} & (0,1) & \\ (-1,0) & (0,0) & (1,0) \\ & (0,-1) & \end{array} : \begin{array}{ccc} & * & \\ \mathrm{B} & * & \mathrm{B} \\ & * & \end{array} \rightarrow \begin{array}{ccc} & * & \\ \mathrm{B} & * & * \\ & * & \end{array} \qquad (6.24)$$

and symmetry-related processes. In both processes there are two possibilities for A to react. This affects the rate constant. Suppose we have at time $t$ the situation

$$\begin{array}{ccc} & * & \\ \mathrm{B} & * & \mathrm{B} \\ & * & \end{array} \qquad (6.25)$$

and an A adsorbs in the middle. This adsorption takes on average a time $W_{\mathrm{ads}}^{-1}$, where $W_{\mathrm{ads}}$ is the rate constant for adsorption. So after adsorption we are on average at time $t + W_{\mathrm{ads}}^{-1}$ and have the situation

$$\begin{array}{ccc} & * & \\ \mathrm{B} & \mathrm{A} & \mathrm{B} \\ & * & \end{array} . \qquad (6.26)$$

Next the A reacts with one of the B's to give

$$\begin{array}{ccc} & * & \\ * & * & \mathrm{B} \\ & * & \end{array} , \quad \text{or} \quad \begin{array}{ccc} & * & \\ \mathrm{B} & * & * \\ & * & \end{array} \qquad (6.27)$$

with equal probability, and we are still at time $t + W_{\mathrm{ads}}^{-1}$. If we use process (6.24) to model the adsorption of A followed by the immediate formation and desorption of AB, we need to be careful in defining the rate constant for (6.24). If the rate constant is $W_{2\mathrm{B}}$, then on average the process leading to one of the two possibilities of (6.27) takes place at time $t + (2W_{2B})^{-1}$. We get a factor 2, because in the initial situation there are two processes possible. The total rate constant for that situation is the sum of the rate constants of all possible processes: i.e., $2W_{2B}$. To get the same result as before we therefore must have $W_{2B} = W_{ads}/2$. In general, if we have $N$ B neighbors in the initial situation then the direct processes should have a rate constant that is equal to the rate constant for adsorption divided by $N$.

Such a combination of processes as above may seem artificial, but it is actually quite common for dissociative adsorption and associative desorption. In the ZGB model we have A's and B's reacting with each other immediately when they become neighbors. The A's stand for CO molecules and the B's for oxygen atoms. They form $CO_2$ which subsequently desorbs from the surface. Neither process is always fast. Often the desorption is fast, but the $CO_2$ formation is not. In that case we can model the combined process as a slow reaction between CO and O that leads to two vacant sites.

Another example is methane adsorption, which is often a difficult process [18]. The molecule is either scattered or it dissociates upon adsorption. So instead of having adsorption of $CH_4$ and then a dissociation of the adsorbed $CH_4$, we can describe this as a single process that leads directly to adsorbed $CH_3$ and H [19, 20]. The remarks concerning the rate constants made above apply to such description and also the description for the $CO_2$ formation.

## 6.6 Isotope Experiments and Diffusion

Isotope experiments are often done to resolve questions on reaction mechanisms [21]. By using different isotopes of an element for different reactants or in different positions in a reactant it is often possible to determine which atoms in reactants end up at which places in the products. The isotopes can also be used to study kinetics. By switching in an experiment to another isotope it might be possible to make statements on the rate of certain processes.

It is very easy to simulate isotope experiments. The only thing that needs to be done is to copy each process that involves isotopes. For example, suppose we have CO oxidation with $^{12}CO$ and $^{13}CO$. For the adsorption of CO we then might have

$$(0,0) : * \rightarrow 12CO \tag{6.28}$$

and

$$(0,0) : * \rightarrow 13CO. \tag{6.29}$$

If we would neglect the isotope effect, we would not even have to give the processes different rate constants.

Figure 6.2 shows how the coverage of $^{12}CO$ and $^{13}CO$ varies in a simulated isotope experiment based on the ZGB model (see Sects. 5.6.1 and 7.4.3). CO is switched every $\Delta t = 40$ unit of time from one isotope to another. This is about the time to replace all CO on the surface. The coverage of vacancies is $\theta_* = 0.50$ ML,

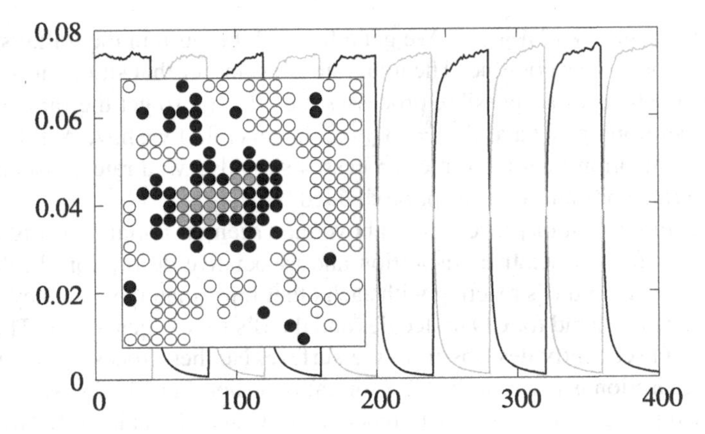


**Fig. 6.2** Variations in the coverage of the $^{12}CO$ and $^{13}CO$ isotopes during CO oxidation simulated using the ZGB-model. Isotopes were switched every 40 units of time. The parameter of the model was chosen to be $y = 0.5255$ which is about the value with maximum reactivity. This parameter stands for the fraction of molecules in the gas phase that are CO molecules and it also defines the unit of time as explained in Sect. 7.4.3. The simulation was done with a square lattice with $1024 \times 1024$ lattice points and periodic boundary conditions. The *inset* shows a snapshot of one isotope (*gray circles*) being replaced by another (*black circles*). The *white circles* depict oxygen atoms

so the number of CO molecules that adsorb per site during these 40 units of time is $y\theta_* \Delta t = 10.5$. The CO coverage is $\theta_{CO} = 0.075$ ML, which means that there are about $y\theta_* \Delta t/\theta_{CO} = 140$ adsorptions of CO necessary per adsorbed CO to replace all CO on the surface. Figure 6.2 also shows why so many are needed. The CO molecules form islands, and only those at the edge can react with oxygen and are removed from the surface. To get to the center of the islands the $CO_2$ formations that break down the islands should be faster locally than adsorption of CO which makes the islands grow.

The same technique that is used to simulate isotope experiments can also be used to get information on diffusion of adsorbates. Suppose we have an adlayer with just one type of adsorbate, the only process is hopping from one site to a neighboring one, and we want to know how the diffusion constant depends on coverage. The hopping can be modeled as

$$(0,0),(1,0) : \mathrm{A} * \rightarrow * \mathrm{A} \tag{6.30}$$

and symmetry-related processes for a simple lattice with A an adsorbate and $*$ a vacant site. We now tag one adsorbate: i.e., we have

$$(0,0),(1,0) : \mathrm{T} * \rightarrow * \mathrm{T} \tag{6.31}$$

and symmetry-related processes with T being a single tagged adsorbate. Figure 6.3 shows the square of the difference of the number of hops to the right and the number of hops to the left plus the square of the difference of the number of hops upward and the number of hops downward as a function of time. If we multiply this by the

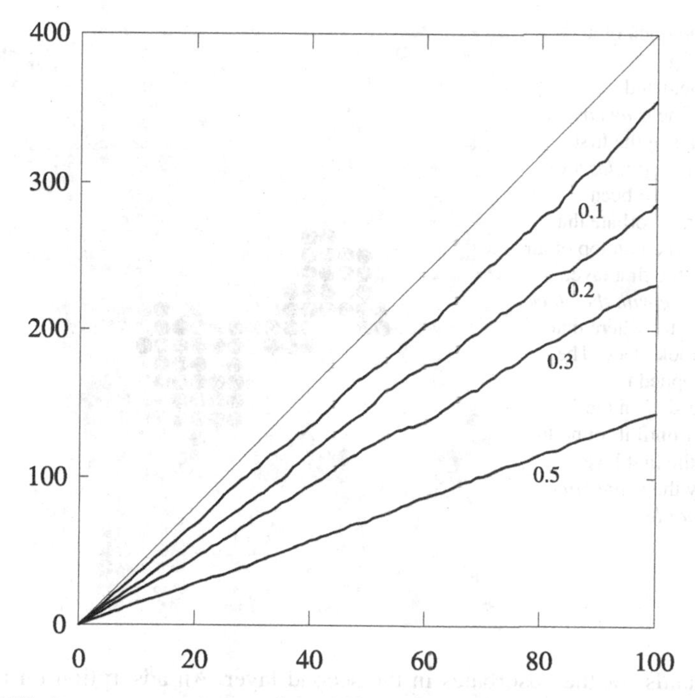


**Fig. 6.3** Diffusion on a square lattice of size 128 × 128 modeled as hops for various coverages. Shown is the squared displacement (in Å$^2$) as a function of time (in s) averaged over 1000 simulations. Distance between lattice points 2.72 Å and rate constant for hopping is 0.1352 s$^{-1}$. The *line* shows the analytical result in the limit of zero coverage (see Sect. 4.6.5). The *curves* are for the coverages indicated next to each curve

square of the distance between the sites (which is actually done for the results in the figure) we get the displacement squared of the tagged particle as a function of time. Averaged over many adsorbates this should give a straight line through the origin with slope $4D$ with $D$ the diffusion coefficient. As we follow only one adsorbate in a simulation, we would get quite bad statistics. The determination of the diffusion coefficient from a single simulation would not be very reliable.

It is possible to improve the statistics by using more tagged particles, but one should make sure that they are distinguishable, because otherwise it is not possible to determine the displacements of the individual adsorbates. This is the reason why we need to tag an adsorbate in the first place. It is also possible to average over many simulations as was done in the figure, or take different moments during a single simulation to start looking at the displacement of the tagged adsorbate from the position it is at these moments.

Another way such tagged adsorbate can be used is to have it adsorb and then monitor its diffusion until it reacts in some way. Figure 6.4 for example shows the sites that an adsorbate visits during a simulation of precursor-mediated adsorption. The most straightforward way to model this is to have two lattice points in the unit cell. The first stands for the adsorption site and the adsorbates in the first layer. The

**Fig. 6.4** The track of a particle during precursor-mediated adsorption. The *gray circles* are adsorbates in the first layer. The *black circles* are the sites that have been visited by an adsorbate that initially adsorbed on top of an adsorbate in the first layer. The *gray circle with the black border* indicates where that adsorption took place. The adsorbate hopped to neighboring sites in the second layer until it found the vacancy in the first layer indicated by the *white circle with black border*

second stands for the adsorbates in the second layer. An adsorption on top of an adsorbate in the first layer can then be model by

$$(0,0/0),(0,0/1) : \mathrm{A} * \to \mathrm{A}\ \mathrm{T} \tag{6.32}$$

with A the adsorbate in the first layer, $*$ a vacant site in the second layer, and T the particle adsorbing in the second layer: i.e., as a precursor.

The particle will then start moving around in the second layer as follows.

$$(0,0/0),(0,0/1),(1,0/0),(1,0/1) : \mathrm{A}\ \mathrm{T}\ \mathrm{A} * \to \mathrm{At} * \mathrm{A}\ \mathrm{T}. \tag{6.33}$$

At is an adsorbate in the first layer on a site that has been visited by the diffusing T. We introduce this label so that we can see how T has moved. If this is of no interest, then the process might be simplified to

$$(0,0/1),(1,0/0),(1,0/1) : \mathrm{T}\ \mathrm{A} * \to * \mathrm{A}\ \mathrm{T}. \tag{6.34}$$

If we decide however to keep track of where T has been, then we have to take care of the possibility that T revisits sites. This means we need

$$\begin{aligned} &(0,0/0),(0,0/1),(1,0/0),(1,0/1) : \mathrm{A}\ \mathrm{T}\ \mathrm{A} * \to \mathrm{At} * \mathrm{A}\ \mathrm{T}, \\ &(0,0/0),(0,0/1),(1,0/0),(1,0/1) : \mathrm{At}\ \mathrm{T}\ \mathrm{A} * \to \mathrm{At} * \mathrm{A}\ \mathrm{T}, \\ &(0,0/0),(0,0/1),(1,0/0),(1,0/1) : \mathrm{A}\ \mathrm{T}\ \mathrm{At} * \to \mathrm{At} * \mathrm{At}\ \mathrm{T}, \end{aligned} \tag{6.35}$$

and

$$(0,0/0),(0,0/1),(1,0/0),(1,0/1) : \mathrm{At}\ \mathrm{T}\ \mathrm{At} * \to \mathrm{At} * \mathrm{At}\ \mathrm{T} \tag{6.36}$$

to deal with all combinations of A's and At's and the two sites involved.

A desorption of T before it adsorbs in the first layer can be modeled by

$$(0,0/0),(0,0/1): \mathrm{A\,T} \rightarrow \mathrm{At}\,* \tag{6.37}$$

and

$$(0,0/0),(0,0/1): \mathrm{At\,T} \rightarrow \mathrm{At}\,*. \tag{6.38}$$

If we want to remember where the desorption took place, then the At on the right should be replaced by another label.

A final adsorption in the first layer can be modeled by

$$(0,0/0),(0,0/1),(1,0/0): \mathrm{A\,T}\,* \rightarrow \mathrm{At}\,*\,\mathrm{T} \tag{6.39}$$

and

$$(0,0/0),(0,0/1),(1,0/0): \mathrm{At\,T}\,* \rightarrow \mathrm{At}\,*\,\mathrm{T}. \tag{6.40}$$

The T on the right is not changed to A or At, so we can see where the final adsorption has taken place.

Note that we have not shown the symmetry-related processes above. We have also assumed above that the diffusion in the second layer is much faster than diffusion in the first layer so that we can neglect the latter. If we want to include diffusion in the first layer, we need to make sure that the "t" in "At" is transferred to the vacancy that is formed when the At hops to another site. We have also assumed that there is only one particle in the second layer. If that is not the case, then the processes above do not distinguish between the tracks formed by the different particles in that layer. The processes also model only an extrinsic precursor: i.e., there is always an adsorbate in the first layer below an adsorbate in the second layer. A hop of T to a position with a vacancy in the first layer always results in an adsorption in the first layer. It would of course be possible to extend the model to allow for an intrinsic precursor. The number of hops of T in the second layer is useful to know as it tells us how difficult it was to find a vacancy in the first layer.

It is also possible to model the precursor-mediated adsorption with a single site in the unit cell. This is very easy, because we did not include an intrinsic precursor. For adsorption on top of an adsorbate in the first layer we have

$$(0,0): \mathrm{A} \rightarrow \mathrm{T} \tag{6.41}$$

with A the adsorbate in the first layer and T now two adsorbates on top of each other. The hops in the second layer become

$$(0,0),(1,0): \mathrm{T\,A} \rightarrow \mathrm{At\,T} \tag{6.42}$$

and

$$(0,0),(1,0): \mathrm{T\,At} \rightarrow \mathrm{At\,T}. \tag{6.43}$$

A desorption of T before it adsorbs in the first layer can be modeled by

$$(0,0) : \mathrm{T} \to \mathrm{At}, \tag{6.44}$$

and a final adsorption in the first layer can be modeled by

$$(0,0),(1,0) : \mathrm{T}\ast \to \mathrm{At\ Tt}. \tag{6.45}$$

The Tt on the right is used so we can see where the final adsorption has taken place. We see that this way to model the process is conceptually a little bit more difficult, but the processes are simpler.

## 6.7 Simulating Nanoparticles and Facets

We have already seen in Sect. 5.5.3 that the surface need not be flat. Our modeling framework allows steps, but, as we will see here, also the simulation of reactions and other processes on nanoparticles. In fact, there are several ways to do this.

If the nanoparticles are very small so that the number of adsorption sites is also very small, then we can simply model these sites by using a unit cell with the same number of sites. A single unit cell then models the whole nanoparticle. For example, suppose that we have a tetrahedral metal particle of just four atoms, and suppose that we have just one adsorption site per atom. We can then model the processes on such a particle with four sites in a unit cell. Simple adsorption then becomes

$$\begin{aligned}
(0,0/0) &: \ast \to \mathrm{A},\\
(0,0/1) &: \ast \to \mathrm{A},\\
(0,0/2) &: \ast \to \mathrm{A},\\
(0,0/3) &: \ast \to \mathrm{A},
\end{aligned} \tag{6.46}$$

with A an adsorbate and $\ast$ a vacant site. A dissociation $\mathrm{AB} \to \mathrm{A} + \mathrm{B}$ will be

$$\begin{aligned}
(0,0/0),(0,0/1) &: \mathrm{AB}\ast \to \mathrm{A\ B},\\
(0,0/0),(0,0/2) &: \mathrm{AB}\ast \to \mathrm{A\ B},
\end{aligned} \tag{6.47}$$

and possibly other combinations of sites on which the dissociation is possible.

This way of modeling a nanoparticle does not use the translational symmetry that is built in our modeling framework. Each unit cell corresponds to one nanoparticle, and by having a large system we simulate simply a large number of nanoparticles. This has the advantage that we can get good statistics with even just a single simulation. We should note however that it is questionable if kMC simulations are really the best way to study the kinetics on such small particle. When the number of ways in which the sites can be occupied is not too large, it might be possible to solve

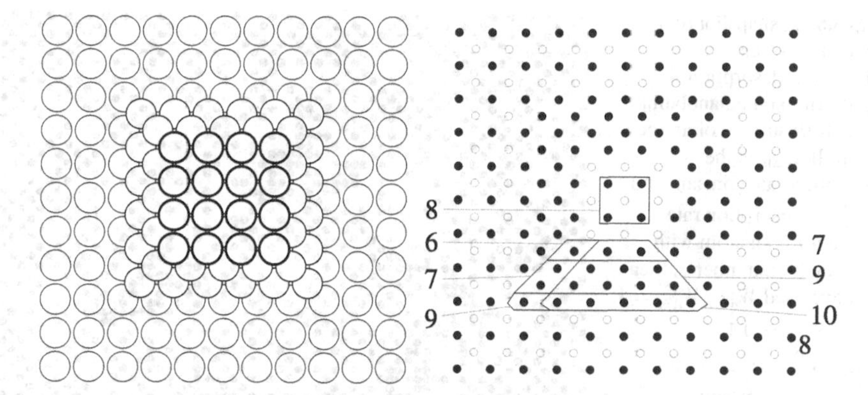


**Fig. 6.5** A nanobump on a (100) surface of an fcc metal on the *left*. We can model the top sites with the lattice on the *right*. The *open circles* are not used as adsorption sites. The *numbers* indicate coordination numbers of the sites in the enclosed areas. The atoms forming the facets on the sides of the bump form (111) surfaces and have therefore coordination number *9*. The sites on the edges have all coordination number *7*. The sites at the vertices have coordination number *6*

the probabilities in the master equation explicitly. This can either be done numerically by using a method to solve sets of ordinary differential equations [22], or by diagonalizing the matrix $\mathbf{W}$ in Eq. (2.7) and using Eqs. (2.12)–(2.14).

If the nanoparticle becomes too large for the approach above, then the whole lattice needs to be used to simulate the particle. There are three problems that we need to address. The first is that a nanoparticle has various facets that might have different surface structures. The second is that a nanoparticle is a three-dimensional object, whereas our framework is two-dimensional. The third is that we have to make sure that periodic boundary conditions of our two-dimensional lattice does not put sites next to each other that are far apart on the nanoparticle. We start with looking at the facets.

Figure 6.5 shows a (100) surface of an fcc metal with a nanobump. The bump has five facets. The top one has also the structure of a (100) surface, but the other ones have the structure of a (111) surface. Moreover, there are different edges and corner atoms. The coordination numbers of the exposed metal atoms vary from 6 to 10.

Now suppose that we want to model processes on such a surface with a bump, that the processes only involve top sites, and that there is one top site per exposed metal atom. The sites are then the ones depicted by the black circles on the right of Fig. 6.5. They do not form a lattice, but by adding the white circles we do get a lattice. This can be regarded as taking the top sites of the original lattice of the (100) surface and then adding a second site in each unit cell at the hollow position. Alternatively, we can take the top and hollow positions, disregard their difference, which gives us a simple square lattice, and use labels to indicate what kind of site a lattice point corresponds to. This last approach seems simpler, as we will want to use labels anyway to indicate the coordination number of the metal atom at a site. If we take the primitive vectors horizontally and vertically in both cases, then the two

**Fig. 6.6** A snapshot of a simulation of an adsorption-desorption equilibrium on a nanobump on a (100) surface of an fcc metal. If $W_{ads}$ is the adsorption rate constant and $W_{des}^{(z)}$ the desorption rate constant for an atom with coordination number $z$, then we have used $W_{ads}/W_{des}^{(8)} = 4$, $W_{des}^{(6)}/W_{des}^{(8)} = 0.1$, $W_{des}^{(7)}/W_{des}^{(8)} = 0.2$, $W_{des}^{(9)}/W_{des}^{(8)} = 5$, and $W_{des}^{(10)}/W_{des}^{(8)} = 10$ to account for the variation in adsorption energy with coordination number

approaches yield the same system but rotated with respect to each other over 45°. This can be seen when comparing Figs. 6.5 and 6.6.

This second approach then looks as follows. We take a simple square lattice. We have a first sublattice with labels that indicate the occupation of the adsorption sites. We add a second sublattice that functions as an overlay to indicate the facets. Its labels have the coordination numbers. So we have $(n_1, n_2/1) : c$, which means that the site at $n_1[\frac{1}{2}(\mathbf{a}_1 + \mathbf{a}_2)] + n_2[\frac{1}{2}(\mathbf{a}_1 - \mathbf{a}_2)]$ is on top of an atom with coordination number $c$. Here $\mathbf{a}_1$ and $\mathbf{a}_2$ are the primitive vectors of the (100) surface, and we use a special value, say $c = 0$, to indicate a lattice point that does not correspond to an actual site. These lattice points are the white circles on the right of Fig. 6.5. Figure 6.6 shows the result of a simple adsorption-desorption equilibrium where the desorption rate constant depends on the coordination number. The atoms forming the facets with (111) structure have a higher coordination number than the atoms forming the surface and the facet with (100) structure. The adsorbate is therefore weaker bounded, the rate constant for desorption is higher, and the coverage lower for the (111) facets. The same effect is even more apparent where the (111) facets meet the (100) surface. There is a row of atoms with coordination number 10. The probability to have an adsorbate there is very low. There are indeed four lines in the figure with vacant sites.

The second problem with nanoparticles is that they are really three-dimensional objects whereas our modeling framework is two-dimensional. This does not make it impossible to use our framework. The surface of a nanoparticle is after all also two-dimensional and it is quite possible to model that surface within our framework. As an example we look at a cubic particle and we assume that we can model the sites on each of its six facets with a simple square lattice. We take the processes from

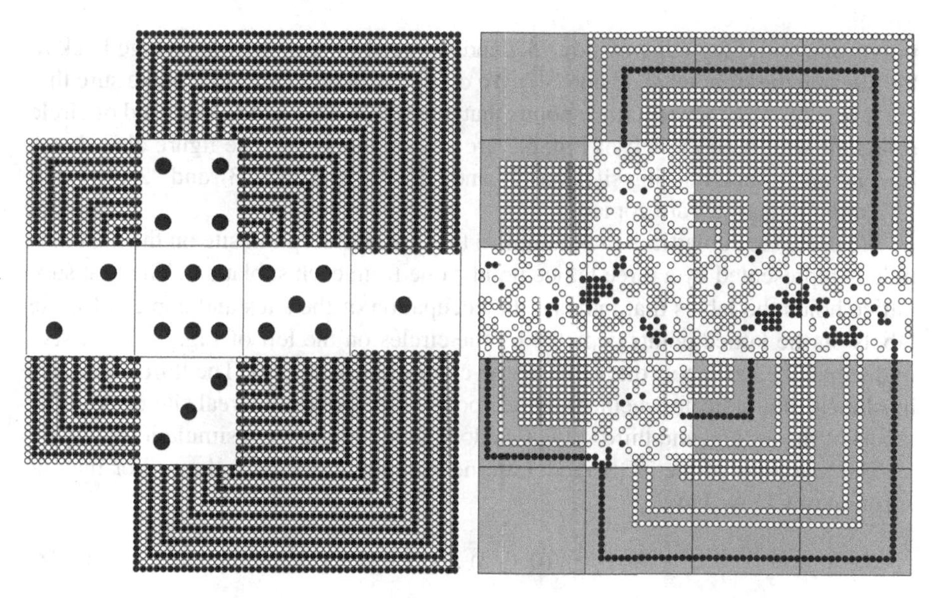

**Fig. 6.7** A snapshot of a simulation of the ZGB model on a cube that is modeled with a simple square lattice. The *squares on the left with the big spots* indicate the facets. If we imagine the cube to be a regular die, then the number of the spots indicate the six facets. *Neighboring circles on the left with the same color* indicate lattice points on the *right* that are forced to have the same label. *Black circles on the right* depict CO molecules and *white circles* oxygen atoms

the ZGB model. We assume for simplicity that the oxygen adsorption and $CO_2$ formation can also occur at on two neighboring sites that straddle the edge of two adjacent facets.

We start with mapping the facets onto a simple square lattice. We do this as is shown on the left of Fig. 6.7. We imagine the nanoparticle to be a normal six-sided die with opposite sides having a total number of spots equal to seven. The problem with this mapping is that sites on different facets that are neighbors are no longer neighbors with this mapping. The mapping retains the neighborhood relations between the sites on sides "2" and "6", "4" and "6", "3" and "6", "5" and "6" and "5" and "1". (We mean with side "X" the side with X spots in the figure.) The neighborhood relation between sides "1" and "2" is also retained because we can use periodic boundary conditions in horizontal direction. The neighborhood relation is lost however for the other six pairs of neighboring sides. The size of the two-dimensional lattice in vertical direction is made large enough so that periodic boundary conditions in that direction do not accidentally cause sites to become neighbors that are really not.

The sites on the right edge of side "4" and those on the top edge of side "5" are neighbors on the nanoparticle, but not on the two-dimensional map in Fig. 6.7. To allow reactions on one site of side "4" and a neighboring one on side "5" we copy the labels of the sites on the right edge of side "4" to the lattice points above the top edge of side "5". As a consequence the reactions now become possible on sites on opposite sides of the top edge of side "5". This will change labels of the lattice points above the top edge of side "5", and we need to copy such a change back to the sites on the right edge of side "4". We do this with processes that make sure that the labels on neighboring lattice points that are depicted with the same kind of circle (black or white) on the left of Fig. 6.7 are always the same. The figure also shows how labels for neighboring sites on "5" and "2", "3" and "5", "3" and "2", "1" and "5", and "1" and "2" are copied.

We use three sublattices to accomplish this. Each adsorption site on the nanoparticle is represented by a three lattice points: one from each sublattice. The first sublattice holds the labels that indicate the occupation of the sites and copies of these labels on the lattice points depicted by the circles on the left of Fig. 6.7. The second sublattice has labels that indicate the color of these circles. The third sublattice has labels that indicate whether a lattice point corresponds to a real site or not. The labels of the second and third sublattice do not change during a simulation.

The reactions of the ZGB model are now implemented as follows. For the adsorption of CO we have

$$(0,0/0),(0,0/2): \mathrm{V\ S} \rightarrow \mathrm{COt\ S}. \tag{6.48}$$

The label S for the lattice point of the sublattice 2 indicates that the lattice point is a real site on the nanoparticle. Labels V and COt of sublattice 0 indicate that a vacant site becomes a site that is occupied by CO. The reason for the "t" in COt is that COt is a temporary label. The site where the adsorption takes place may be one of the sites for which we need to copy the label.

For the oxygen adsorption we have

$$(0,0/0),(1,0/0),(0,0/2),(1,0/2): \mathrm{V\ V\ S\ S} \rightarrow \mathrm{Ot\ Ot\ S\ S} \tag{6.49}$$

and symmetry-related descriptions. The "t" in Ot again indicates that we have a temporary label that may need to be copied. The lattice points in this description are both actual sites, but that need not always be the case. The adsorption of oxygen can also occur on two sites that are neighbors but on different facets of the nanoparticle, but that are not neighbors on the two-dimensional lattice. We therefore also have

$$(0,0/0),(1,0/0),(0,0/2),(1,0/2): \mathrm{V\ V\ S\ X} \rightarrow \mathrm{Ot\ Ot\ S\ X} \tag{6.50}$$

and

$$(0,0/0),(1,0/0),(0,0/2),(1,0/2): \mathrm{V\ V\ X\ S} \rightarrow \mathrm{Ot\ Ot\ X\ S}, \tag{6.51}$$

and symmetry-related descriptions, with X indicating a lattice point that does not correspond to an actual site. This allows for example adsorption on oxygen on a lattice point below the top edge of side "5" and on a lattice point above that edge. The latter corresponds to a site at the right edge of side "4". Note that we do not have

$$(0,0/0),(1,0/0),(0,0/2),(1,0/2): \mathrm{V\ V\ X\ X} \rightarrow \mathrm{Ot\ Ot\ X\ X}. \tag{6.52}$$

For the $CO_2$ formation we have

$$
\begin{aligned}
&(0,0/0),(1,0/0),(0,0/2),(1,0/2) : \text{CO O S S} \to \text{Vt Vt S S},\\
&(0,0/0),(1,0/0),(0,0/2),(1,0/2) : \text{CO O S X} \to \text{Vt Vt S X},\\
&(0,0/0),(1,0/0),(0,0/2),(1,0/2) : \text{CO O X S} \to \text{Vt Vt X S}
\end{aligned} \tag{6.53}
$$

and symmetry-related descriptions. This is very similar to oxygen adsorption. The "t" in Vt is again to indicate that the label is temporary.

To copy labels we need some immediate processes. The actual copying is done with

$$
\begin{aligned}
&(0,0/0),(1,0/0),(0,0/1),(1,0/1) : \text{COt CO B B} \to \text{COt COt B B},\\
&(0,0/0),(1,0/0),(0,0/1),(1,0/1) : \text{COt O B B} \to \text{COt COt B B},\\
&(0,0/0),(1,0/0),(0,0/1),(1,0/1) : \text{COt V B B} \to \text{COt COt B B},
\end{aligned} \tag{6.54}
$$

and symmetry-related descriptions. These processes take care of copying COt along the black circles, hence the label B, on the left of Fig. 6.7. If we replace all B's by W's we take care of copying along the white circles. Copying Ot and Vt is accomplished with the same processes but with COt replaced by Ot and Vt, respectively. These processes make clear why we need the "t" at the end of COt, Ot, and Vt. It makes clear what needs to be copied.

At the end of the copying process the labels COt, Ot, and Vt have to be changed to the normal form CO, O, and V. Having immediate processes that simply do that does not work. The reason is that either the normal forms are immediately changed back to one of the forms with a "t" at the end, or the copying above does not work. It depends on the priority which of these possibilities actually occurs. Instead we use the immediate processes

$$
\begin{aligned}
&(0,0/0) : \text{COt} \to \text{COtt},\\
&(0,0/0) : \text{Ot} \to \text{Ott},\\
&(0,0/0) : \text{Vt} \to \text{Vtt},
\end{aligned} \tag{6.55}
$$

and

$$
\begin{aligned}
&(0,0/0) : \text{COtt} \to \text{CO},\\
&(0,0/0) : \text{Ott} \to \text{O},\\
&(0,0/0) : \text{Vtt} \to \text{V},
\end{aligned} \tag{6.56}
$$

with processes (6.54) having a higher priority than (6.55), and (6.56) having the lowest priority. The reason for the labels ending in "tt" is that this prevents (6.54) to occur when the normal forms of the labels are restored. The formation of $CO_2$ is also an immediate process, but it should have a priority that is even lower than the one of (6.56).

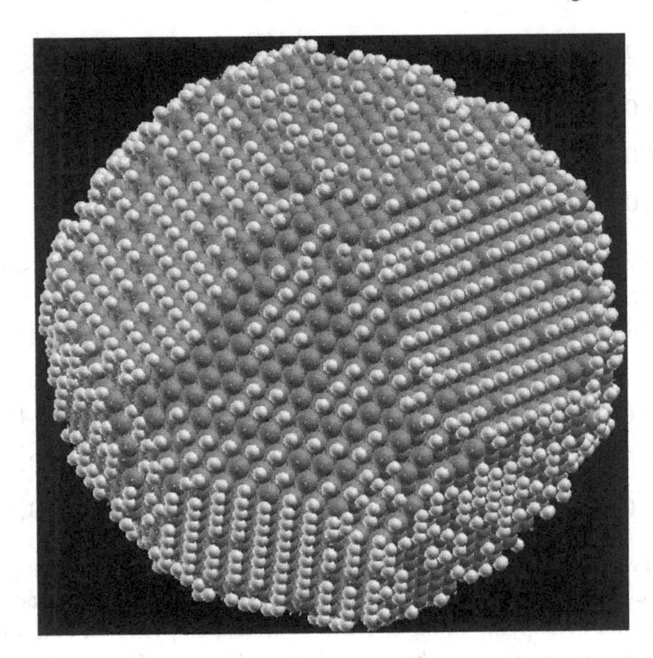

**Fig. 6.8** A snapshot of a simulation of an adsorption-desorption equilibrium for the top sites on a nanocluster of about 10 000 platinum atoms

The result of a simulation with these processes is shown on the right of Fig. 6.7. The facets show distribution of the adsorbates that is characteristic for the ZGB model, but also see how the labels are copied from the edge from one facet to another.

A much more flexible and simpler approach is to extend our modeling framework to three spatial dimensions. Figure 6.8 shows an example. The nanoparticle and the adsorbates shown in that figure were modeled as follows. We used a three-dimensional lattice with a cubic unit cell. We defined four sublattices by taking four lattice points in the unit cell at positions $(\mathbf{0}, \mathbf{0}, \mathbf{0})$, $(\frac{1}{2}\mathbf{a}_1, \frac{1}{2}\mathbf{a}_2, \mathbf{0})$, $(\frac{1}{2}\mathbf{a}_1, \mathbf{0}, \frac{1}{2}\mathbf{a}_3)$, and $(\mathbf{0}, \frac{1}{2}\mathbf{a}_2, \frac{1}{2}\mathbf{a}_3)$. This gave us an fcc lattice. The four sublattices were used to construct the nanoparticle. About 10 000 Pt atoms were put together on lattice positions. (We used only pair interactions between neighboring atoms with a value that gave the proper formation energy for bulk platinum.) The atoms were allowed to hop to neighboring lattice points at 2200 K to equilibrate. The particle was then slowly cooled down to room temperature. This yielded the particle in Fig. 6.8 with (111) and (100) facets, but also with steps and kinks on these facets.

After the nanoparticle was created, we added 12 more sublattices corresponding the tetrahedral and octahedral interstitial positions of the fcc lattice. Tetrahedral position with only one neighboring Pt atom were used to define top sites on (111) facets. Octahedral positions with only one neighboring Pt atom were used to define top sites on (100) facets. These positions were then used to simulate a simple adsorption-desorption equilibrium. The desorption from tetrahedral positions was treated as the desorption from a (111) facets of the nanobump shown in Fig. 6.6. The desorption from octahedral positions was treated as the desorption from the (100) facets. The adsorption rate constant was the same for all sites.

## 6.8 Making the Initial Configuration

We have focused in this chapter, and also in the preceding ones, almost exclusively on the processes. We need however also some initial configuration to do a kMC simulation. Creating such an initial configuration is often fairly straightforward, but not always. It is possible to make it by hand or to use a special-purpose tool or code. We will look here however how to use our modeling framework to create various types of initial configurations.

Kinetic experiments often start with the preparation of the system. One might try to simulate this preparation to get the initial configuration for the simulation of the kinetic experiment itself. That is indeed an option, but it may be not a convenient one. The preparation often consists of processes that may be hard to simulate or it may not be completely clear what the mechanisms of the processes are that take place during the preparation. For example, there often is adsorption involved. Simple adsorption causes no problems for a simulation, but if there is an precursor involved, then it might be difficult to find out which processes take place and what the rate constants are of all these processes. As the desired end result is just an adlayer with a specific coverage, there are much easier ways to get the initial configuration.

We will assume that it is a trivial matter to create a configuration in which all lattice points have the same label. This already takes care of many situations. For example, kinetic experiments in catalysis often start with an empty surface of the catalyst. Suppose however that we want to create an adlayer with some fractional coverage. The easiest way to get this would be to implement a simple adsorption-desorption equilibrium. Assuming for simplicity that we have just one site per unit cell, then

$$(0,0) : * \to \mathrm{A} \tag{6.57}$$

with rate constant $W_{\mathrm{ads}}$ and

$$(0,0) : \mathrm{A} \to * \tag{6.58}$$

with rate constant $W_{\mathrm{des}}$, and A an adsorbate and $*$ a vacant site give a coverage of $\theta = \lambda$ with

$$\lambda = \frac{W_{\mathrm{ads}}}{W_{\mathrm{ads}} + W_{\mathrm{des}}}. \tag{6.59}$$

Because a kMC simulation is a stochastic simulation, the coverage is not exactly $\lambda$. It will fluctuate around that value. The standard deviation of the occupation of a single site equals $\sqrt{\lambda(1-\lambda)}$. The adsorptions and desorptions at different sites are independent of each other. The standard deviation of the coverage over the whole surface is then $\sqrt{\lambda(1-\lambda)/S}$ with $S$ the number of sites. It is important to realize that the adsorption and desorption need not correspond to processes that actually take place or are even actually possible. If the adsorbates would be oxygen atoms, then the species in the gas phase would be dioxygen. Using $O_2(\text{gas}) \rightleftharpoons 2O(\text{ads})$ to get a certain coverage is possible, but the relation between the rate constants and the coverage will not be as simple as the one above.

We use simple adsorption and desorption also to make adlayers of more than one adsorbate. We must have for each adsorbate X

$$W^{(X)}_{\text{ads}}\theta_* = W^{(X)}_{\text{des}}\theta_X \tag{6.60}$$

with $\theta_*$ the fraction of vacant sites. This expression gives us rate constants to get given coverages. The use of processes that are not actually possible also opens up other possibilities. Suppose we want to create an adlayer with adsorbates A and B with coverage $\theta_A$ and $\theta_B$, respectively. We can then also use a two-step approach as follows. We first create an adlayer with only A's and coverage $\theta_A + \theta_B$ as described above. The second step consists of the conversions $A \to B$ and $B \to A$ with rate constants $W_{A\to B} = W\theta_B/(\theta_A + \theta_B)$ and $W_{B\to A} = W\theta_A/(\theta_A + \theta_B)$ with $W$ some arbitrary rate constant.

The processes mentioned so far yield adlayer structures that consist of adsorbates that are randomly distributed over the surface. Often the adsorbates form a superstructure or islands. Such structures can be created simply as follows. After an adlayer is made with a given coverage, a second simulation is done in which the adsorbates are allowed to diffuse. The desired structure is obtained by including the proper lateral interactions in the second simulation. The same approach can also be used to construct bimetallic surfaces. Instead of labels standing for occupied and vacant sites, we need to use labels denoting the two different metals. Instead of diffusion, we have exchange of substrate atoms.

The drawback of the approach with diffusion or exchange is that the resulting structure will show defects. Instead of an adlayer showing a perfect superstructure, we will get domains each showing the same superstructure but displaced or rotated with respect to the superstructures in neighboring domains. To obtain a perfect adlayer immediate processes form a better approach.

Suppose we have a square lattice and an adsorbate that blocks neighboring sites. We want to construct an adlayer with a perfect checkerboard structure. We start with a lattice with all labels equal to "T" that stands for "temporary". We then have a simple adsorption

$$(0,0) : \mathrm{T} \to \mathrm{A} \tag{6.61}$$

with A the adsorbate. This is a normal process with a finite rate constant. The checkerboard structure is then made by the immediate processes

$$\begin{aligned}(0,0),(1,0) &: \mathrm{A}\,\mathrm{T} \to \mathrm{A}\,*,\\ (0,0),(1,0) &: *\,\mathrm{T} \to *\,\mathrm{A},\end{aligned} \tag{6.62}$$

and symmetry-related processes. These processes start after a single occurrence of T $\rightarrow$ A and change the whole system to a checkerboard structure. There will not be a second adsorption.

The reason for the label "T" is the following. Suppose we start with all labels equal to "$*$" and adsorption

$$(0,0) : * \rightarrow \mathrm{A}. \tag{6.63}$$

We might try to construct the checkerboard structure using

$$(0,0),(1,1) : \mathrm{A} * \rightarrow \mathrm{A}\;\mathrm{A} \tag{6.64}$$

and symmetry-related processes as immediate processes. This will create indeed the checkerboard structure initially, but then there will be a second adsorption followed by more immediate processes and we will end up with a surface with all sites occupied by A's. The label "T" is used to prevent this. It allows for adsorption, but it is removed when the checkerboard structure is created which prevents new adsorptions. This is however not the only way to prevent a second adsorption. It is possible to use (6.64) and only "A" and "$*$" as labels by using

$$(0,0),(1,0) : * * \rightarrow \mathrm{A} * \tag{6.65}$$

as adsorption. Here a second adsorption is prevented because after the checkerboard structure is formed there is no vacant site anymore with a neighboring site that is also vacant.

Immediate processes can also be used to create other adlayers with a defect-free structure or substrates with a perfect structure: e.g., a substrate with parallel steps separated by terraces with a given width. Note also that this approach also enables us to create adlayers and substrate structures without long-range order but with a well-defined ratio of different labels. For example, in the A + B model (see Sect. 8.2) we want to start with an adlayer that is a random mixture of A's and B's except that their numbers should be exactly the same. To get such a structure we can start with constructing a checkerboard structure of A's and B's and then use exchange processes A B $\rightleftharpoons$ B A to randomize the adlayer.

## References


1. C.G.M. Hermse, A.P.J. Jansen, in *Catalysis*, vol. 19, ed. by J.J. Spivey, K.M. Dooley (Royal Society of Chemistry, London, 2006)
2. J.N. Murrell, S. Carter, P. Huxley, S.C. Farantos, A.J.C. Varandas, *Molecular Potential Energy Functions* (Wiley-Interscience, Chichester, 1984)
3. A. van der Walle, G. Ceder, J. Phase Equilibria, **23**, 348 (2002)
4. V. Blum, A. Zunger, Phys. Rev. B **69**, 020103(R) (2004)
5. A.P.J. Jansen, W.K. Offermans, in *Computational Science and Its Applications—ICCSA-2005*. LNCS, vol. 3480, ed. by O. Gervasi (Springer, Berlin, 2005)
6. Y. Zhang, V. Blum, K. Reuter, Phys. Rev. B **75**, 235406 (2007)
7. D.M. Hawkins, J. Chem. Inf. Comput. Sci. **44**, 1 (2004)

8. A.P.J. Jansen, C. Popa, Phys. Rev. B **78**, 085404 (2008)
9. N.A. Zarkevich, D.D. Johnson, Phys. Rev. Lett. **92**, 255702 (2004)
10. R. Drautz, A. Díaz-Ortiz, Phys. Rev. B **73**, 224207 (2006)
11. D.E. Nanu, Y. Deng, A.J. Böttger, Phys. Rev. B **74**, 014113 (2006)
12. T. Mueller, G. Ceder, Phys. Rev. B **80**, 024103 (2009)
13. T. Mueller, G. Ceder, Phys. Rev. B **82**, 184107 (2010)
14. Carlos is a general-purpose program, written in C by J.J. Lukkien, for simulating reactions on surfaces that can be represented by regular lattices: an implementation of the First Reaction Method, the Variable Step Size Method, and the Random Selection Method. http://www.win.tue.nl/~johanl/projects/Carlos/
15. C.G.M. Hermse, A.P. van Bavel, A.P.J. Jansen, L.A.M.M. Barbosa, P. Sautet, R.A. van Santen, J. Phys. Chem. B **108**, 11035 (2004)
16. C.G.M. Hermse, F. Frechard, A.P. van Bavel, J.J. Lukkien, J.W. Niemantsverdriet, R.A. van Santen, A.P.J. Jansen, J. Chem. Phys. **118**, 7081 (2003)
17. D.R. Mason, R.E. Rudd, A.P. Sutton, Comput. Phys. Commun. **160**, 140 (2004)
18. J.H. Larsen, I. Chorkendorff, Surf. Sci. Rep. **35**, 163 (2000)
19. I.M. Ciobîcã, F. Frechard, A.P.J. Jansen, R.A. van Santen, in *Studies in Surfaces Science and Catalysis*, vol. 133, ed. by G.F. Froment, K.C. Waugh (Elsevier, Amsterdam, 2001), pp. 221–228
20. I.M. Ciobîcã, F. Frechard, C.G.M. Hermse, A.P.J. Jansen, R.A. van Santen, in *Surface Chemistry and Catalysis*, ed. by A.F. Carley, P.R. Davies, G.J. Hutchings, M.S. Spencer (Kluwer Academic/Plenum, New York, 2001)
21. J.M. Thomas, W.J. Thomas, *Principles and Practice of Heterogeneous Catalysis* (VCH, Weinheim, 1997)
22. W.H. Press, B.P. Flannery, S.A. Teukolsky, W.T. Vetterling, *Numerical Recipes. The Art of Scientific Computing* (Cambridge University Press, Cambridge, 1989)

# Chapter 7
# Examples

**Abstract** We discuss a number of complete surface reaction systems and show the benefits of using kinetic Monte Carlo simulations for modeling them. The following topics will be discussed: lateral interactions and the link with the occupation of unfavorable sites and the suppression of processes, the effect of the structure of the substrate on kinetics, the role of equilibrium and non-equilibrium phase transitions and symmetry breaking, and non-linear kinetics.

## 7.1 Introduction

This chapter is somewhat similar to Chaps. 5 and 6. It also discusses how to model reactions. However, the emphasis here is more on the information one can get from kinetic Monte Carlo (kMC) simulations: i.e., the reason why one wants to do kMC simulations instead of using conventional macroscopic rate equations. The discussions in this chapter focus more on aspects of kinetics than on technical details on how to implement processes. Also whole reactions systems are discussed instead of individual processes. The discussion on the derivation of macroscopic rate equations in Sect. 4.6 has shown that rate equations are approximate equations except for very simple systems. This means that there always will be a quantitative difference between the results from rate equations and those from kMC simulations. The examples in this chapter are chosen because these differences are very large or the behavior predicted by rate equations and by kMC simulations is even qualitatively different. This is not to say that it is hard to find such systems. On the contrary, especially for systems for which lateral interactions are important (and those include all systems with high coverages) rate equations can at best be regarded as descriptive. Their drawback is then that the rate constants have no physical meaning as has been pointed out in Chap. 1.

## 7.2 NO Reduction on Rh(111)

One of the most important reasons for doing kMC simulations is to include the effect of lateral interactions on the kinetics. It is still not widely appreciated how

A.P.J. Jansen, *An Introduction to Kinetic Monte Carlo Simulations of Surface Reactions*, Lecture Notes in Physics 856,
DOI 10.1007/978-3-642-29488-4_7, 

important these interactions can be. It has been argued that much of catalysis takes place at high temperatures where the adsorbates are randomly distributed over the sites, and that lateral interactions, if present at all, can be incorporated in an effective rate constant. This however underestimates how strong lateral interactions can be, and that adsorbates are rarely, if at all, randomly distributed. Moreover, even if they are so distributed, the lateral interactions lead to a very complicated dependence on the coverage that can not just be represented by an effective rate constant.

Lateral interactions can easily be just as large as the interactions between adsorbate and substrate [1]. This holds in particular when they are repulsive. Kinetic experiments however do not always show these large interactions. Let's look at CO on Rh(100) as an example [2]. Temperature-Programmed Desorption (TPD) experiments for initial coverages of up to 0.5 ML show hardly any effect of lateral interactions. This is because two CO molecules that prefer top sites never need to occupy neighboring sites and thus strong repulsion is avoided. The TPD spectra show only variation in the energetics comparable to the thermal energy $k_B T$. The fact that the system avoids two CO molecules at neighboring sites indicates that the interactions between such molecules must be well above the thermal energy. A Density-Functional Theory (DFT) calculation of NO molecules on Rh(111) at the same distance gives lateral interactions between 15 and 20 kJ/mol depending on the adsorption site [3]. However, the distance between a preferred fcc site of Rh(111) and the nearest hcp site is much smaller, and a substantial higher repulsion is expected if both of these sites were to be occupied. CO/Rh(100) with coverage above 0.5 ML still doesn't show the high lateral interaction of CO's on neighboring top sites. Instead, a smaller increase of energy is seen that results from moving CO molecules to bridge sites. To avoid the large fcc-hcp interaction in NO/Rh(111) some NO molecules are moved to top sites, which have an adsorption energy that is 55 kJ/mol less than fcc sites, and 64 kJ/mol less than hcp sites. This means that the lateral interactions that are avoided are larger than these numbers.

Even if a system avoids large repulsive interactions, they can still have an effect on the kinetics although that effect will be subtle. Let's look a bit closer at CO/Rh(100). At very low coverage the top sites already occupied and the four neighboring top sites are not available for adsorption. This means that the macroscopic rate equation for adsorption

$$\frac{d\theta_{\mathrm{CO}}}{dt} = W_{\mathrm{ads}}(1 - 5\theta_{\mathrm{CO}}) \tag{7.1}$$

is a good approximation for small $\theta_{\mathrm{CO}}$. At coverages just below 0.5 ML when we assume that the CO's can not become neighbors, each CO that is removed leads to just one new possibility for adsorption. We therefore get

$$\frac{d\theta_{\mathrm{CO}}}{dt} = W_{\mathrm{ads}}\left(\frac{1}{2} - \theta_{\mathrm{CO}}\right) \tag{7.2}$$

(see Fig. 7.1). We see that the lateral interactions do not seem to affect the kinetics, but the rate at intermediate coverages has a complicated dependence on the coverage. The structure of the adlayers is determined by the configurational entropy. For $\theta_{\mathrm{CO}} < 0.369$ ML this results in a disordered structure. At $\theta_{\mathrm{CO}} = 0.369$ ML there is

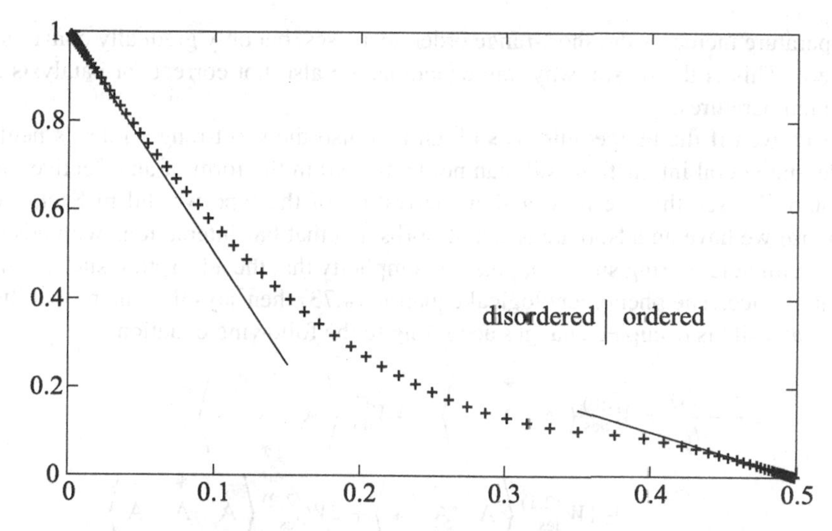


**Fig. 7.1** The rate for adsorption divided by the rate constant (i.e., only the coverage dependence) as a function of coverage for the situation of a square lattice and repulsion between adsorbates at neighboring sites that is so strong that the system avoids it (e.g., CO/Rh(100)). This leads to a order-disorder phase transition at 0.369 ML. The *tangents* show that at low coverage the rate is proportional to $1 - 5\theta$ and at high coverage to $1/2 - \theta$. Each *cross* is the results of a kMC simulation with a $512 \times 512$ lattice, a fixed adsorption rate constant and a desorption rate constant between 0.001 and 1000 times the adsorption rate constant

a phase transition to a $c(2 \times 2)$ structure [4]. Even below that coverage some CO molecules form that structure so that the others have more room to move. This gives a high entropy.

The effects of lateral interactions at low temperatures have at least been known for as long as diffraction techniques have revealed that adlayers can form well-defined structures. Island formation and adsorbate segregation are phenomena that are often mainly determined by the energetics of a system. At higher temperatures entropy effects become important, but the energetics certainly remain essential for the kinetics. When we increase the temperature there is often an order-disorder phase transition. This can have a dominant effect on the experimental results as we will see in Sect. 7.4 where we discuss TPD with repulsive interactions and the butterfly in voltammetry.

Above the order-disorder phase transition temperature there is no long-range order. (For simplicity we are assuming here a system with just interactions between adsorbates at neighboring sites. The situation can become very complicated if there are also interactions at longer ranges, three-particle interactions, frustrated interactions, et cetera.) It is important to realize however that for kinetics it is not this long-range order that is most important but the short-range order, or the correlation in the occupation of neighboring sites. This has to do with the fact that there are generally reactions that involve different adsorbates at such neighboring sites, and the rate depends on both sites being occupied. One should realize that such short-range order or correlation does not vanish at the order-disorder transition. When the temperature increases the short-range order decreases, but only gradually converges to zero. This is the reason why rate equations are also not correct for catalysis at high temperatures.

And even if the temperature is so high that also the short-range order is negligible, the lateral interactions still can not be treated in the form of an effective rate constant. To see this we need a short derivation of the type we did in Sect. 4.6. Suppose we have an adsorbate A that desorbs and that has interactions with adsorbates with neighboring sites. Suppose for simplicity that the adsorption sites form a square lattice. The phenomenological equation (4.73) then says that the probability $\langle \mathrm{A} \rangle$ that a site is occupied changes according to the following equation.

$$-\frac{d\langle \mathrm{A}\rangle}{dt} = W^{(0)}_{\mathrm{des}} \left\langle \begin{matrix} & * & \\ * & \mathrm{A} & * \\ & * & \end{matrix} \right\rangle + 4W^{(1)}_{\mathrm{des}} \left\langle \begin{matrix} & * & \\ \mathrm{A} & \mathrm{A} & * \\ & * & \end{matrix} \right\rangle + 4W^{(2,1)}_{\mathrm{des}} \left\langle \begin{matrix} & \mathrm{A} & \\ \mathrm{A} & \mathrm{A} & * \\ & * & \end{matrix} \right\rangle + 2W^{(2,2)}_{\mathrm{des}} \left\langle \begin{matrix} & * & \\ \mathrm{A} & \mathrm{A} & \mathrm{A} \\ & * & \end{matrix} \right\rangle + 4W^{(3)}_{\mathrm{des}} \left\langle \begin{matrix} & \mathrm{A} & \\ \mathrm{A} & \mathrm{A} & \mathrm{A} \\ & * & \end{matrix} \right\rangle + W^{(4)}_{\mathrm{des}} \left\langle \begin{matrix} & \mathrm{A} & \\ \mathrm{A} & \mathrm{A} & \mathrm{A} \\ & \mathrm{A} & \end{matrix} \right\rangle . \tag{7.3}$$

The probabilities on the right-hand-side specify a particular occupation of a central site and its four neighbors. There are different rate constants for these occupations. The coefficients before the rate constants indicate the number of symmetry-related ways the sites can be occupied.

If we assume that the adsorbates are randomly distributed over the sites, then we can simplify this expression to

$$\begin{aligned} -\frac{d\theta_{\mathrm{A}}}{dt} = {} & W^{(0)}_{\mathrm{des}}\theta_{\mathrm{A}}(1-\theta_{\mathrm{A}})^4 + 4W^{(1)}_{\mathrm{des}}\theta^2_{\mathrm{A}}(1-\theta_{\mathrm{A}})^3 \\ & + \left[4W^{(2,1)}_{\mathrm{des}} + 2W^{(2,2)}_{\mathrm{des}}\right]\theta^3_{\mathrm{A}}(1-\theta_{\mathrm{A}})^2 \\ & + 4W^{(3)}_{\mathrm{des}}\theta^4_{\mathrm{A}}(1-\theta_{\mathrm{A}}) + W^{(4)}_{\mathrm{des}}\theta^5_{\mathrm{A}}, \end{aligned} \tag{7.4}$$

where we have used $\langle \mathrm{A} \rangle = \theta_{\mathrm{A}}$ and

$$\left\langle \begin{matrix} & \mathrm{X}_2 & \\ \mathrm{X}_1 & \mathrm{X}_0 & \mathrm{X}_3 \\ & \mathrm{X}_4 & \end{matrix} \right\rangle = \langle \mathrm{X}_0\rangle\langle \mathrm{X}_1\rangle\langle \mathrm{X}_2\rangle\langle \mathrm{X}_3\rangle\langle \mathrm{X}_4\rangle = \theta_{\mathrm{X}_0}\theta_{\mathrm{X}_1}\theta_{\mathrm{X}_2}\theta_{\mathrm{X}_3}\theta_{\mathrm{X}_4}. \tag{7.5}$$

This is as far as we can go without making any assumption besides the one about the random distribution of the adsorbates. To get a further simplification we need to make assumptions about the rate constants. If we assume that they are all equal, then the equation reduces to

$$-\frac{d\theta_{\mathrm{A}}}{dt} = W^{(0)}_{\mathrm{des}}\theta_{\mathrm{A}}\left[\theta_{\mathrm{A}} + (1-\theta_{\mathrm{A}})\right]^4 = W^{(0)}_{\mathrm{des}}\theta_{\mathrm{A}}, \tag{7.6}$$

but this assumption means that there are no lateral interactions.

For pair interactions we have

$$W_{\text{des}}^{(n,\dots)} = W_{\text{des}}^{(0)} e^{n\varphi/k_B T} \tag{7.7}$$

with $\varphi$ the interaction parameter. This simplifies the expression to

$$-\frac{d\theta_A}{dt} = W_{\text{des}}^{(0)} \theta_A \left[\theta_A e^{\varphi/k_B T} + (1-\theta_A)\right]^4. \tag{7.8}$$

We see that we retain a strongly non-linear dependence of the rate on the coverage. The system NO/Rh(111) that we will discuss here shows most of the effects caused by lateral interactions that we have mentioned above.[1] In addition it shows how lateral interactions can change the preferred adsorption site. NO on Rh(111) prefers hollow sites, but at high coverages part of the NO moves to top sites. The adsorption energy of the top site is much lower than that of the hollow sites, but by moving to top sites a strong repulsion between the adsorbates is avoided.

At temperatures below 275 K NO adsorbs molecularly [5, 6]. It prefers the threefold hcp site slightly over the threefold fcc site. The difference of the adsorption energy depends on the coverage, but it is only in the range of 3 to 8 kJ/mol [7–9]. The bridge site and certainly the top site have much smaller adsorption energies. The difference is between 20 to 24 and 61 to 64 kJ/mol, respectively.

NO starts dissociating above 275 K and forms nitrogen and oxygen atoms. Nitrogen prefers the threefold hcp site just as NO, but oxygen prefers the threefold fcc site. The difference with the other threefold site is again small: between 9 and 14 kJ/mol for nitrogen, and between 6 and 10 kJ/mol for oxygen. At low coverage all of the NO dissociates, but as the coverage increases less and less NO dissociates and the temperature at which dissociation starts increases.

An explanation why less NO dissociates at higher coverages is that there are simply no sites available. If the system is modeled with one (threefold) site per unit cell [9, 10], then the suppression of the NO dissociation can simply be explained because at high coverage there are not enough sites available for the atoms that are formed when NO dissociates. NO occupies only one site, but the atoms after dissociation occupy two. When all NO dissociates the number of occupied sites doubles. When the initial coverage of NO is higher than 0.5 ML, all sites are occupied before all NO has dissociated.

The 1-site model above has three shortcomings. First, it does not show that the dissociation of NO becomes already suppressed below 0.5 ML, and the maximum coverage of NO is only 0.75 ML. Second, it does not explain the temperature dependence. In fact, it does not have any temperature dependence related to the initial coverage of NO. Third, DFT results indicate that at least the other threefold sites should also play a role [7–9]. In fact, LEED shows that the NO adlayer with a cov-

[1] Parts of Sect. 7.2 have been reprinted with permission from C.G.M. Hermse, F. Frechard, A.P. van Bavel, J.J. Lukkien, J.W. Niemantsverdriet, R.A. van Santen, A.P.J. Jansen, Combining density-functional calculations with kinetic models: NO/Rh(111), J. Chem. Phys. **118**, 7081 (2003). Copyright 2003, American Institute of Physics.

erage of 0.75 ML has on equal distribution of the NO molecules over both threefold sits and the top site [11, 12].

It is clear from this that lateral interactions play an important role in the system. So the 1-site model should be extended with lateral interactions. This can reproduce the dissociation behavior as far as initial coverage and temperature dependence is concerned, but not the maximum NO coverage. Moreover, the lateral interactions that are needed for the 1-site model to reproduce the experimental Temperature-Programmed Reaction (TPR) spectra are an order of magnitude lower than those obtained from DFT calculations [7–9].

A correct model is obtained by including lateral interactions and the two threefold and the top sites. NO can occupy all three sites, but is has a clear preference for the threefold sites. The atoms are only allowed to occupy threefold sites. DFT calculations give lateral interactions for sites at a distance of $\frac{1}{3}\sqrt{3}d_{\mathrm{Rh-Rh}}$, $d_{\mathrm{Rh-Rh}}$, and $\frac{2}{3}\sqrt{3}d_{\mathrm{Rh-Rh}}$ with $d_{\mathrm{Rh-Rh}}$ the distance between neighboring substrate atoms. At the shortest distance of $\frac{1}{3}\sqrt{3}d_{\mathrm{Rh-Rh}}$ the interactions are so strongly repulsive that they can be modeled as site blocking. At a distance of $d_{\mathrm{Rh-Rh}}$ there is still a strong repulsion of at least 24 kJ/mol. This repulsion is still so strong that the system avoids it, and it does not show up in the results of kinetic experiments. It does however strongly determine the structure of the adlayer (it can be regarded as site blocking for the structures as well), and only when the structure of the adlayer changes the adsorbates occasionally get as close as a distance $d_{\mathrm{Rh-Rh}}$ of each other. To avoid this repulsion at coverages above 0.5 ML NO moves to top sites. There is also such strong repulsion at $\frac{2}{3}\sqrt{3}d_{\mathrm{Rh-Rh}}$ between atoms, between atoms and NO, and between NO molecules.

The system and the model of it shows many of the phenomena that result from lateral interactions that were discussed before. At low temperatures NO forms various superstructures depending on coverage. At 0.5 ML a $c(4 \times 2)$-2NO structure is formed and at 0.75 ML a $(2 \times 2)$-3NO structure (see Fig. 7.2) [11–13]. The maximum coverage of 0.75 ML is a consequence of the fact that it is the maximum coverage that has no NO molecules at distance $d_{\mathrm{Rh-Rh}}$. The phase transitions between the structures are here not caused by temperature, but by a change in coverage. The ordered structures also show an order-disorder phase transition in the kMC simulations when the temperature increased and if we leave out the NO dissociation.

When NO starts dissociating we get an adlayer with NO, N, and O atoms that segregate (see Fig. 7.3). There is little difference in N–N, O–O, and N–O interactions. This means that the nitrogen and oxygen atoms mix. The interactions with NO are however different. In particular, the repulsion between NO and the atoms is larger than the average between the atom-atom and the NO–NO interactions, mainly because the NO–NO repulsion is much weaker. As a consequence we have the segregation as in the figure. There is a strong similarity with island formation, because the interactions are formally equivalent to one with attractive interactions between the atoms and between the NO molecules.

That lateral interactions affect the kinetics is clear, but the effects are not so clear-cut as those for the structures. The suppression of the NO dissociation when

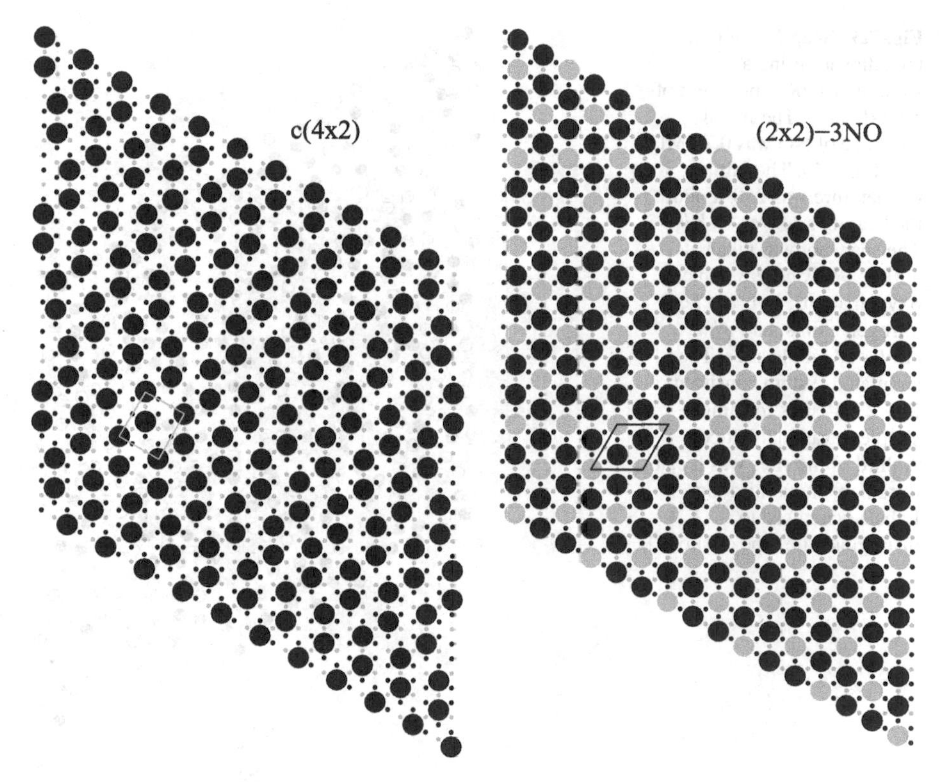


**Fig. 7.2** Ordered structures of NO/Rh(111). The coverage on the *left* is 0.5 ML and on the *right* 0.75 ML. *Large circles* depict NO molecules, *small circles* vacant sites. Top sites are *gray* and hollow sites are *black*. The structure of 0.5 ML has zigzag chains of molecules at alternating fcc and hcp sites with a vacant top site in between. The structure of 0.75 ML has a honeycomb structure of molecules at hollow sites with molecules at top sites in the centers of the hexagons [9]

the initial coverage increases is clearly due to the increase in repulsive lateral interactions. This effect becomes even noticeable at quite low coverage (see Fig. 7.4). The same thing explains the higher temperature at which the dissociation starts. At high initial coverage the dissociation is partial or even suppressed completely. In the spectra there is a second peak for dissociation at high temperature. This results from desorption of NO that frees sites and makes new dissociation reactions possible.

One consequence of the lateral interactions that is easier to explain is the temperature at which NO desorbs that does not dissociate. The adsorption energy of isolated NO has a value that suggests a much higher desorption temperature. For top sites there is a repulsion of six neighboring NO molecules at hollow sites. This is always the same, so the desorption from top sites is always at the same temperature but at a lower temperature than might be inferred from the adsorption energy for an isolated NO molecule. The desorbing NO from hollow sites that is observed in TPR spectra is surrounded by nitrogen and oxygen atoms. These NO molecules feel a strong repulsion from these atoms, and also desorb at a lower temperature.

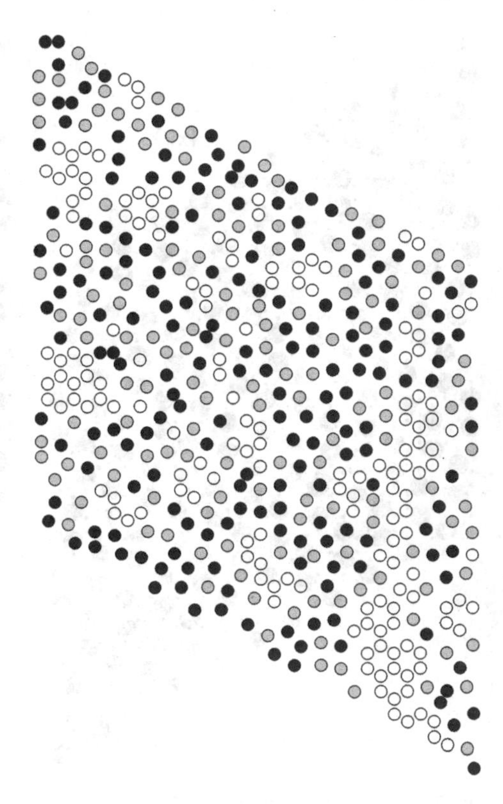

**Fig. 7.3** Snapshot of part of the adlayer during a simulated TPR experiment of NO/Rh(111). The initial coverage of NO was 0.75 ML (see Fig. 7.2). The initial temperature was 250 K and the heating rate was 10 K/s. The snapshot was taken at 490 K. *Black circles* depict oxygen atoms, *gray circles* nitrogen atoms, and *white circles* NO molecules. The simulated system consisted of 264 × 264 sites forming a simple hexagonal lattice. A third of the sites was fcc, a third hcp, and a third top. Labels were used to distinguish the type of site [9]

There are two peaks for $N_2$ desorption. There is a narrow peak at a relatively low temperature and a broad peak at high temperature. The narrow peak is associated with the desorption of NO that makes new dissociation reactions possible. This creates nitrogen atoms in the adlayer with a very high coverage and a fraction of them

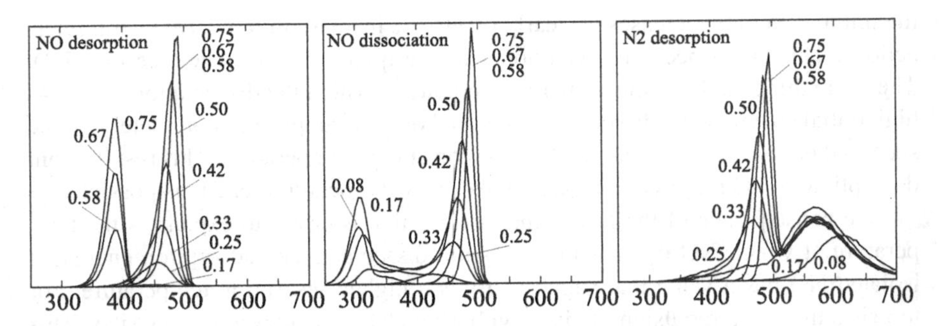


**Fig. 7.4** Temperature-Programmed Reaction spectra (rates in arbitrary units versus temperature in Kelvin) of NO/Rh(111). The heating rate is 10 K/s. There are no spectra for oxygen desorption because this takes place at higher temperatures and has not been included in the simulations. The *numbers next to the curves* are the initial coverages of NO. The spectra have been obtained by averaging over 100 kMC simulations with a simple hexagonal 264 × 264 lattice representing the top, fcc, and hcp sites [9]

is pushed of the surface. The broad peak is formed after all NO molecules have either dissociated or desorbed. It results from nitrogen surrounded by oxygen atoms.

## 7.3 $NH_3$ Induced Reconstruction of (111) Steps on Pt(111)

Another reason why macroscopic rate equations may fail is that the substrate is never homogeneous. We may have a bimetallic catalyst, but more common is that the surface has defects. As a consequence different parts of the surface have sites with different adsorption energies and reactivities. Using a model with a single coverage is then bound to fail, but even if we use different coverages for the different parts of the substrate we make an approximation that may not be justifiable.

Suppose that we have a simple adsorption-desorption equilibrium on a square lattice, but that there is a step. Suppose for simplicity that we can model this with a single row of sites that have a desorption rate constant that differs from that of other sites. The macroscopic rate equations are then

$$\frac{d\theta_n}{dt} = k_{\text{ads}}(1-\theta_n) - k_{\text{des}}^{(\text{terrace})}\theta_n + \frac{D}{a^2}[\theta_{n+1}+\theta_{n-1}-2\theta_n] \tag{7.9}$$

for $n \neq 0$ and

$$\frac{d\theta_0}{dt} = k_{\text{ads}}(1-\theta_0) - k_{\text{des}}^{(\text{step})}\theta_0 + \frac{D}{a^2}[\theta_1+\theta_{-1}-2\theta_0]. \tag{7.10}$$

Here $\theta_n$ is the probability that a site a distance $|n|a$ from the step is occupied. Sites with $n > 0$ are on one side of the step and those with $n < 0$ on the other. The parameter $a$ is the distance between the sites, $k_{\text{ads}}$ is the adsorption rate constant $k_{\text{des}}$ is the desorption rate constant, and $D/a^2$ the hopping rate constant ($D$ is the diffusion coefficient). The derivation of these rate constants is very similar to the ones in Sect. 4.6.

If the adsorbates do not hop to neighboring sites ($D = 0$), then we find

$$\theta_0 = \frac{k_{\text{ads}}}{k_{\text{ads}} + k_{\text{des}}^{(\text{step})}} \quad \text{and} \quad \theta_n = \frac{k_{\text{ads}}}{k_{\text{ads}} + k_{\text{des}}^{(\text{terrace})}} \tag{7.11}$$

for $n \neq 0$. In this case using two coverages, one for the step and one for the terraces, is justified. If the adsorbates are allowed to hop ($D \neq 0$), then this is no longer the case. We will see that we get a gradient in the coverage away from the step.

The general solution of the rate equations can be written as

$$\theta_n = \frac{k_{\text{ads}}}{k_{\text{ads}} + k_{\text{des}}^{(\text{terrace})}} + B\lambda^{-|n|}. \tag{7.12}$$

Substitution of this expression in the rate equations for the terrace sites yields

$$\lambda + \frac{1}{\lambda} = \frac{2D/a^2 + k_{\text{ads}} + k_{\text{des}}^{(\text{terrace})}}{D/a^2} \tag{7.13}$$

**Fig. 7.5** Snapshot of part of a Pt-Ru electrode during a simulation of a linear sweep voltammetry experiment in which CO is replaced by OH. The Ru part of the electrode is completely covered by CO (*gray circles*) or OH (*white circles*). The Pt part is partially covered by CO (*black circles*) but a lot of it is vacant. The formation of OH on the electrode takes place initially on the Ru side of the Pt-Ru interface. That OH reacts with CO on the Pt side of the interface which leads to the vacancies on Pt. The vacancies on Ru are filled by new OH being formed [14]

for the parameter $\lambda$. There are two solutions, but only the one with $\lambda > 1$ is physically acceptable, because the second term of the right-hand-side of equations (7.12) should go to zero for large $|n|$. The appropriate value for $\lambda$ is given by

$$\lambda = \frac{1}{2}\left[\mu + \sqrt{\mu^2 - 4}\right] \tag{7.14}$$

with $\mu$ equal to the right-hand-side of Eq. (7.13). Substitution of expression (7.12) in the rate equation for the step gives a linear equation for $B$. The solution of it is

$$B = \frac{k_{\mathrm{ads}}(k_{\mathrm{des}}^{(\mathrm{step})} + k_{\mathrm{des}}^{(\mathrm{terrace})})}{[k_{\mathrm{des}}^{(\mathrm{step})} - k_{\mathrm{ads}} + 2D(\lambda - 1)/a^2][k_{\mathrm{ads}} + k_{\mathrm{des}}^{(\mathrm{terrace})}]}. \tag{7.15}$$

The important point however is that, although all terrace sites have the same properties, their coverage is not the same because of the presence of the step.

Another example is shown in Fig. 7.5. It shows a snapshot of a model of a Pt-Ru electrode during a voltammetry experiment [14]. Initially the whole electrode is covered by CO except for a few vacancies. The initial potential is low and no reactions take place. The potential is then increased which causes water to adsorb dissociatively forming OH on the electrode. The OH reacts with CO to form $CO_2$ and a proton, which both desorb immediately. These reactions take place first at the

Pt-Ru interface. The OH forms on Ru and the CO is taken from Pt. The resulting vacancy on Ru is rapidly filled by another OH, but on Pt we get a gradient of CO molecules diffusing to the Pt-Ru interface.

When we cut a crystal we get a surface that is not necessarily stable. That surface will then change its structure. Such a change is called a reconstruction. This often leads to another form of heterogeneity of the substrate than the ones mentioned above. The structure of surfaces is also often changed by catalytic reactions [15, 16]. An example is discussed in Sect. 7.5. As a simpler model system for such a restructuring process we consider here a stepped metal surface and look at the reconstruction of a step. Due to their reduced coordination number atoms at steps often exhibit an enhanced catalytic activity [17]. We discuss here the restructuring of a stepped Pt(111) surface in the presence of adsorbed ammonia. We look at a single step and on the details of the meandering of such step [18].[2]

A Pt(443) surface in its clean state consists of monatomic (111) steps separated by seven lattice units wide (111) terrace (Fig. 7.6a and b). When this surface is exposed to ammonia at 300 K with $p_{NH_3} = 10^{-6}$ mbar first individual Pt atoms move along the step edge as shown by Figs. 7.6c and d. These atoms migrate and become again attached rigidly to the step edge. Thus holes and protrusions are formed along the step edges generating a meandering of the step edges as shown in Fig. 7.6e. This meandering takes place in about 10–15 minutes. The dynamics of the substrate changes are frozen in when after approximately 30 minutes an ordered $(2 \times 2)$ ammonia overlayer forms on the (111) terraces visible in the STM image in Fig. 7.6f.

Several aspects of the energetics of meandering are obvious. The number of Pt-Pt bonds of the step atoms decreases when the step starts to meander thus increasing the step energy of the step. Simultaneously, due to bond-order conservation the remaining bonds of the Pt step atoms will become stronger, and the $NH_3$ adsorption energy is expected to be highest on Pt step and kink atoms [19]. A first guess would therefore be that the meandering is driven by the gain in ammonia adsorption energy caused by creation of low coordinated sites. To check this explanation plane-wave DFT calculations were performed of steps with various structures with and without ammonia [20]. Ammonia adsorbs on top sites of the Pt surface. The adsorption energy at a kink site is 17 kJ/mol larger than at a straight (111) step, but it costs 41 kJ/mol to form the kink.

Ammonia adsorption on a surface with a single step was modeled allowing Pt atoms and ammonia molecules to diffuse on the upper and lower terrace but not across the step [18]. To account for the formation a $(2 \times 2)$-structure of ammonia at a 1/4 ML coverage molecules were not allowed to become nearest- or next-nearest neighbors. Adsorption of ammonia was assumed to be a simple direct adsorption.

kMC simulations without ammonia do not show any meandering at 300 K. This is not surprising as the energy changes are much higher than the thermal energy.

[2]Parts of Sect. 7.3 have been reprinted with permission from X.Q. Zhang, W.K. Offermans, R.A. van Santen, A.P.J. Jansen, A. Scheibe, U. Lins, R. Imbihl, Frozen thermal fluctuations in adsorbate-induced step restructuring, Phys. Rev. B **82**, 113401 (2010). Copyright 2010, American Physical Society.

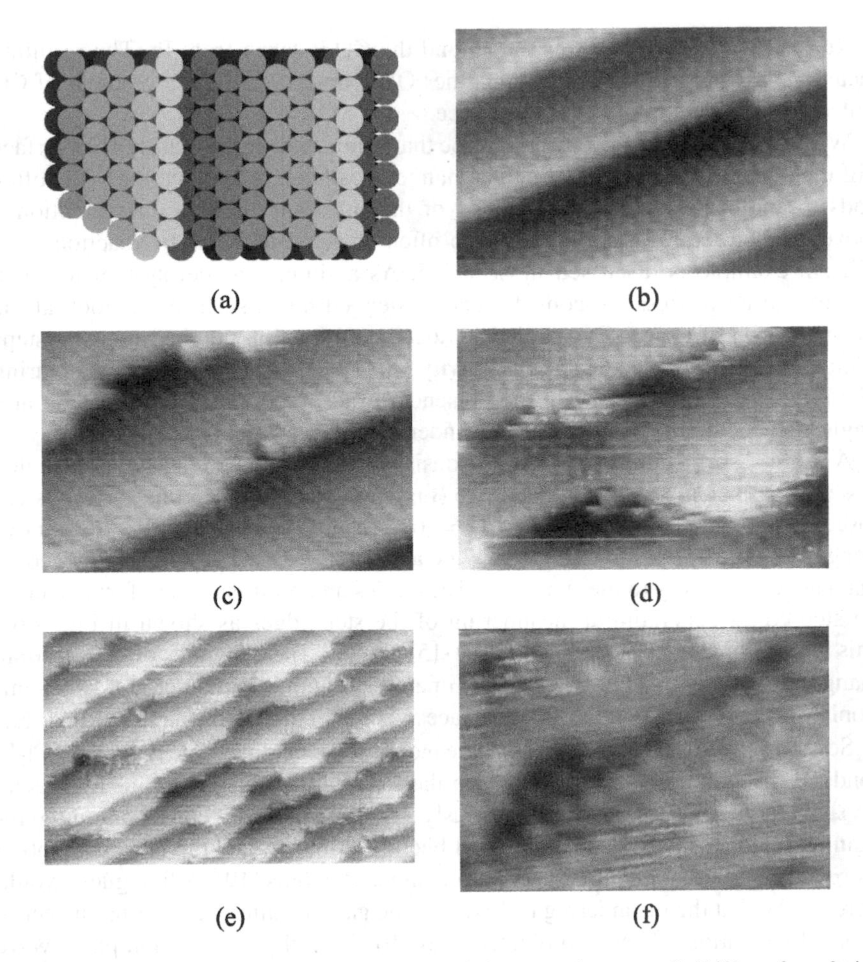


**Fig. 7.6** STM images showing the meandering of the step edges on a Pt(443) surface during exposure to ammonia at 300 K with $p_{NH_3} = 10^{-6}$ mbar. $t = 0$ refers to at the beginning of adsorption when the coverage is zero. (**a**) Structural model of Pt(443); (**b**) clean surface; (**c**) $t = 3$ min.; (**d**) $t = 15$ min.; (**e**) $t = 15$ min. The image in (**f**) at $t = 35$ min. at maximum coverage displays formations of a $2 \times 2$-$NH_3$ structure. Sizes of the imaged areas are (**b**)–(**d**) 50 Å × 30 Å, (**e**) 150 Å × 95 Å, and (**f**) 45 Å × 30 Å [18]. (Color figure online)

Only when the temperature is increased to around 600 K or higher there are deviations from the straight (111) step. Starting with a bare surface it is observed that during ammonia adsorption the initially straight (111) step starts meandering as the coverage of ammonia builds up. This is because part of the energy costs to form kinks is offset by the higher adsorption energy of ammonia at kink atoms. As a consequence there is a decrease of the stiffness of the step.

As long as there are kinks and lone Pt atoms without ammonia the shape of the step keeps changing. The meandering stops however almost completely when no more ammonia can be added to the step edge. At that point some Pt atoms at the

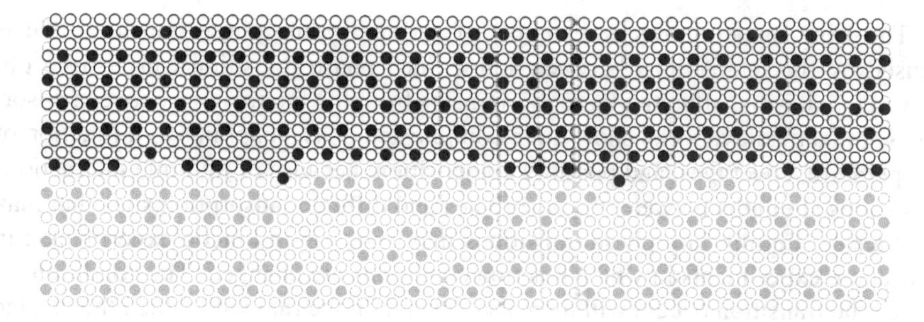

**Fig. 7.7** Snapshot of a kMC simulation showing part of a step formed with $NH_3$. *Open circles* are Pt atoms without $NH_3$, *closed circles* are Pt atoms with $NH_3$. The lower terrace is in a *lighter shade* [18]

step edge can still move, but each Pt hop is immediately followed by its reverse hop. The shape of the step is then fixed. More extensive changes to the step do not occur because they would now drive ammonia to less favorable adsorption sites, and would involve structures with energies that are substantially higher than the thermal energy. If the coverage of ammonia is increased further, then also the remaining Pt hops at the step edge become suppressed. This is because the Pt atoms at the step become blocked by ammonia at the lower terrace. It takes about 20 minutes to freeze the step, which is in good agreement with the experimental results. Also the shape of the final step is in good agreement with the STM images (see Fig. 7.7).

So the behavior of the step based on the STM experiments and the kMC simulations can be summarized as follows. Without ammonia the straight (111) step is energetically the most favorable structure, and hence the equilibrium structure. There are no fluctuations in the structure at $T = 300$ K, because structural deviations from a straight edge have energies that are substantially higher than the thermal energy. Adding ammonia still leaves the structural deviations energetically unfavorable, but the energy difference is now reduced strongly enough to allow for thermal fluctuations. This leads to meandering. When the density of ammonia at the step edge becomes high, the diffusion of Pt atoms is suppressed because any jump generates locally energetically unfavorable configurations, because ammonia would be forced to unfavorable sites. The final structure of the step can therefore be regarded as a frozen thermal fluctuation.

## 7.4 Phase Transitions and Symmetry Breaking

Phase transitions are notoriously difficult to predict accurately. The Ising model on a square lattice is a prototype model to study phase transitions. It shows an order-disorder phase transition that can be solved analytically [4]. The exact temperature of the phase transition is a factor $2\ln(1+\sqrt{2}) \approx 1.763$ lower than the temperature that is obtained from a Mean-Field Approximation similar to the one on which the macroscopic rate constants of chemical kinetics are based. Such a large discrepancy is typical, and forms a good reason for doing simulations.

The Ising model shows an example of an equilibrium phase transition that is caused by the interactions between spins. Such phase transition can also be observed in adlayers, although there it originates from the interactions between adsorbates. In fact, adlayers can show quite complicated phase diagrams as a function of temperature, coverages, and possibly composition, because of multiple interactions of various strengths between the adsorbates and different adsorption sites that may play a role. Equilibrium phase transitions may show up in kinetic experiments: in particular, experiments in which reaction conditions are changed. The conditions at which the transition occurs will then also depend on the rate with which the change of the conditions takes place.

There are also other phase transitions possible that are typical for systems with irreversible reactions. They originate from the fact that for some reaction conditions there may be more than one stable steady state, and changing reaction conditions may make the one that a system is unstable. This results then in non-equilibrium or kinetic phase transitions. When macroscopic rate equations are used they are also called bifurcations [21, 22]. In fact, much more complicated situations may arise showing oscillations, pattern formation, and chaotic behavior [23, 24].

In this section we show two examples of equilibrium phase transitions showing up in kinetic experiments. In one lateral interactions cause a phase transition, in the other the configurational entropy is responsible. We also show an example with kinetic phase transitions. All examples are kept simple so as to illustrate the phase transitions most clearly.

### *7.4.1 TPD with Strong Repulsive Interactions*

One of the clearest examples of the advantage of kMC simulations over macroscopic rate equations is formed by simple desorption with strong lateral interactions. If we take for example a square lattice, an initial situation with all sites occupied, and repulsive interactions between nearest neighbors, then the following happens. Initially, each adsorbate has the same tendency to desorb. However, if one adsorbate desorbs then its former neighbors suddenly feel less repulsion and become adsorbed more strongly. This means that desorption takes place in two stages. First about half the adsorbates desorb, because they have many neighbors. After these adsorbates have desorbed the structure of the adlayer is that of a checkerboard, with almost all adsorbates having no neighbors. Because these adsorbates feel little or no repulsion from other adsorbates they desorb at a later stages.

So we have a phase transition at half a monolayer coverage. At higher coverages the adlayer consists of patches with a checkerboard and patches with a $1 \times 1$ structure ($\theta = 1$). At lower coverages the adlayer consists again of patches with a checkerboard but with vacant areas in between. Also note that there is a symmetry breaking. Initially all sites are equivalent, but after half the adsorbates have desorbed alternate sites are occupied and vacant. Ordinary macroscopic rate equations are not able to describe this symmetry breaking, because they assume that all sites

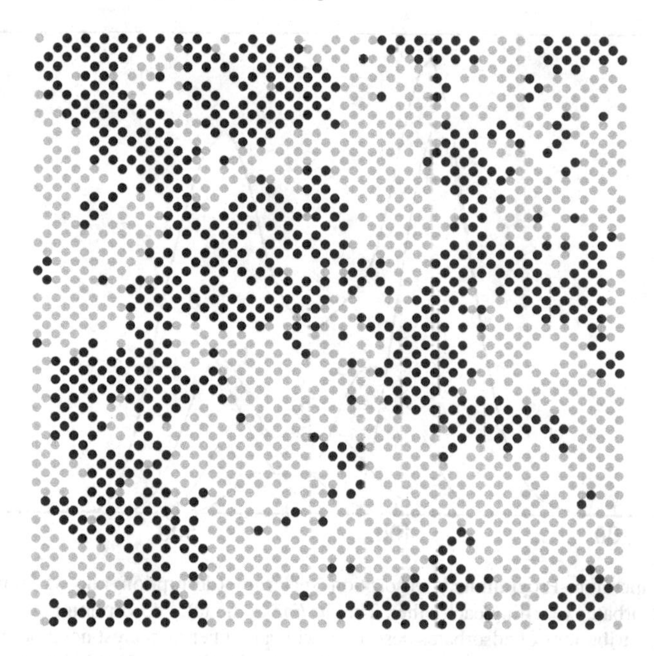

**Fig. 7.8** Domain formation during desorption with strong repulsive interactions. This is a snapshot from a simulation of a Temperature-Programmed Desorption experiment with adsorbates with a repulsion of 6.65 kJ/mol (see Fig. 7.9). The coverage is 0.29 ML. All adsorbates are equivalent but *black* and *gray* adsorbates are from different domains

are equivalent during the whole process. We can split the macroscopic rate equations in two: one for the sites with the adsorbates that desorb first, and one for the sites with adsorbates that desorb later. However, the equations for both sites are equivalent and the symmetry breaking only occurs if the initial state already has a small difference in the occupations between the sites with the early desorbers and the sites with the late desorbers. For kMC simulations such an unrealistic initial condition is not necessary. In kMC fluctuations in the times when reactions occur cause the symmetry breaking as they do in real systems. Such fluctuations are not included in macroscopic rate equations.

The fluctuations determine which adsorbates desorb first, and also affect the structure of the adlayer when the coverage has been more of less halved. A perfect checkerboard structure is only found when the adsorbates also diffuse fast and when the temperature is well below the order-disorder phase transition temperature. For small system sizes the diffusion need not be so fast as for large system sizes. Diffusion has to make sure that no domains are formed. Simulations show that even at relatively small systems with say a $128 \times 128$ lattice it is almost impossible to avoid domain formation (see Fig. 7.8).

Figure 7.9 shows a TPD spectrum for a system with repulsive interactions. We show TPD instead of isothermal desorption, because the latter shows the two stages in the desorption only in how fast the coverage and the desorption rate decreases. If the coverage is plotted logarithmically, then we get first a straight line with a

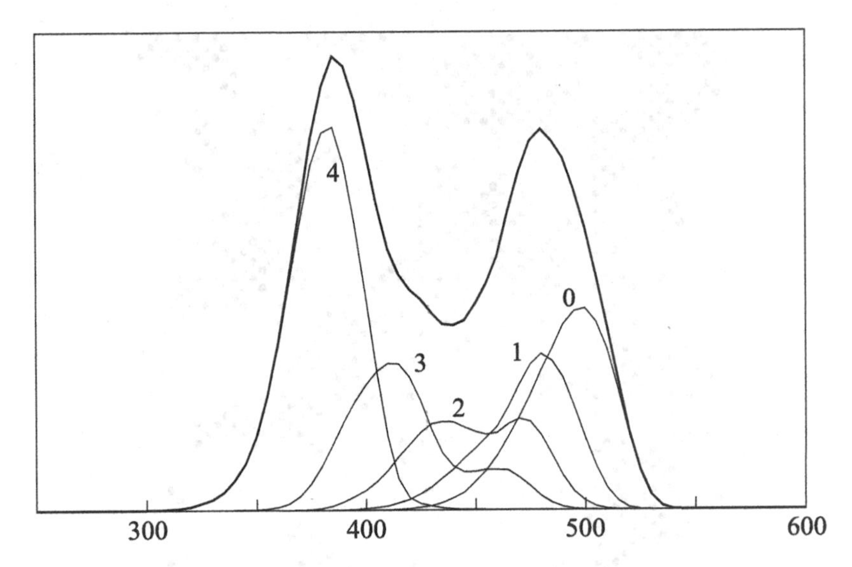


**Fig. 7.9** Temperature-Programmed Desorption spectrum (desorption rate vs. temperature in Kelvin) of adsorbates repelling each other. The *fat line* is the total desorption rate. The *thin lines* are separate contributions of adsorbates desorbing with a number of nearest neighbors given by the *number next to the curve*. Activation energy for desorption is $E_{act} = 121.3$ kJ/mol and the prefactor is $\nu = 1.435 \cdot 10^{12}$ s$^{-1}$. These numbers were taken from CO desorption from Rh(100) at low coverage. The repulsion between two adsorbates is 6.65 kJ/mol. The heating rate is 5 K/s and the initial coverage 1.0 ML

large negative slope, followed by a straight line with a smaller, in absolute sense, negative slope. In TPD the two stages are much clearer, because the desorption rate has two peaks. The figure shows that the second stage (i.e., the second peak) has also contributions from adsorbates with one, two, and even three neighbors, which is due to the fact that the first stage never forms a perfect checkerboard structure and the fact that many adsorbates first diffuse to a site with neighbors and desorb from that site. If the repulsion becomes very strong, then even more than two peaks can form.

In real systems the effects of lateral interactions are generally not so unambiguous. The lateral interactions have to be strong enough so that a well-defined structure is formed when half the adsorbates have desorbed. If this is not the case then the lateral interactions only show up by shifting or broadening a single peak when the initial coverage is increased. When lateral interactions are strong enough, then they may also push adsorbates to other adsorption sites. These sites have a lower adsorption energy, but overall the energy is lowered because it lessens the repulsion between the adsorbates. The result may even be an adlayer structure that is incommensurate with the substrate. Finally, lateral interactions need not just be between nearest neighbors. Interactions between next- and next-next-nearest neighbors are not uncommon. Longer range interactions can sometimes be excluded [25], but charged adsorbates might have long-range interactions, which may explain very broad desorption peaks [26, 27].

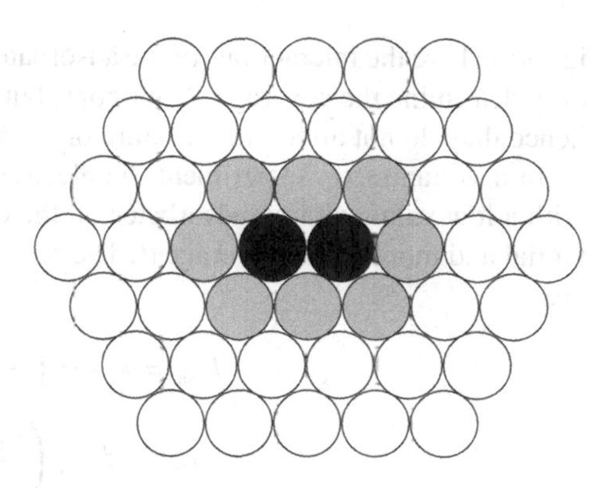

**Fig. 7.10** Lateral interaction model on a fcc(111) lattice. The adsorbed anion binds to two surface atoms in a bridged fashion (*black* atoms), making bonding to the first shell of neighboring sites (*gray*) impossible

### 7.4.2 Voltammetry and the Butterfly

The adsorption of anions on single-crystal electrode surfaces usually gives rise to the appearance of ordered adsorbate adlayers. The formation of these ordered adlayers is often accompanied by a characteristic sharply peaked current response in the cyclic voltammetry, commonly referred to as "butterfly" in the electrochemical community [28–30]. We discuss this butterfly for a model of adsorption of (bi)sulfate on fcc(111) surfaces [31].[3] The simulated voltammogram shows a broad peak associated with adsorption in a disordered phase, and a sharp one associated with a disorder-order transition in the adlayer. The disorder-order transition converts the adlayer with a coverage of 0.18 ML into a $(\sqrt{3}\times\sqrt{7})$ ordered structure with a coverage of 0.20 ML. The $(\sqrt{3}\times\sqrt{7})$ structure with 0.20 ML saturation coverage is experimentally known for the adsorption of anions from sulfuric acid solution on many fcc(111) surfaces. The anion involved is either bisulfate ($HSO_4^-$) or sulfate ($SO_4^{2-}$). The reactions are

$$A^{2-} + ** \rightleftharpoons A_{ads} + 2e^-, \tag{7.16}$$

where $**$ denotes an vacant bridge site (formed by two vacant surface atoms): each $*$ corresponds to a surface atom. Figure 7.10 shows the (111) substrate and the neighboring sites around a bridge-bonded adsorbate. There is a shell of purely hard interactions, in which the simultaneous bonding of two anions to neighboring sites is simply excluded. This model is equivalent to the elongated hard hexagon model considered by Orts et al. [32].

Because of the hard interactions the phase transition here is different from the one observed in TPD with finite repulsive interactions (see Sect. 7.4.1). The phase transition there involves a change in the energetics. This is not the case here. The system avoids the hard interactions, and the energy of the adlayer is always zero. The structure of the adlayer is only determined by the configurational entropy. (We are

---

[3]Parts of Sect. 7.4.2 have been reprinted with permission from C.G.M. Hermse, A.P. van Bavel, M.T.M. Koper, J.J. Lukkien, R.A. van Santen, A.P.J. Jansen, Modelling the butterfly: $(\sqrt{3}\times\sqrt{7})$ ordering on fcc(111) surfaces, Surf. Sci. **572**, 247 (2004). Copyright 2004, Elsevier.

ignoring here the interactions of the adsorbates with the substrate. These interactions only determine the coverage. Their contribution is constant for a give coverage, and hence they do not affect the structure of the adlayer.)

In a voltammetry experiment the electrode potential $E$ is varied: here we start with a low value without adsorbates on the electrode surface, then increase the potential and monitor the coverage $\theta$. The rate constants for adsorption and desorption are

$$k_{\text{ads}} = k^0 \exp\left(\frac{-\alpha_{\text{ads}} \gamma e E}{k_B T}\right) \tag{7.17}$$

$$k_{\text{des}} = k^0 \exp\left(\frac{\alpha_{\text{des}} \gamma e E}{k_B T}\right) \tag{7.18}$$

where $\alpha_{\text{ads}} = 1/2$ is the transfer coefficient for adsorption, $\gamma$ is the electrosorption valency (taken $-2$), and $e$ is the elementary charge. The exponent describes the potential-dependent adsorption of the anion. The definitions in Eqs. (7.17) and (7.18) imply that in our model at zero potential the adsorption rate constant is equal to the desorption rate constant: $k_{\text{ads}} = k_{\text{des}} = k^0$. The transfer coefficients are related as follows.

$$\alpha_{\text{des}} = 1 - \alpha_{\text{ads}}. \tag{7.19}$$

The diffusion steps were defined as hopping between neighboring bridge sites.

Apart from the dependence of $\theta$ on $E$, we are particularly interested in the compressibility $d\theta/dE$ of the adlayer, as this quantity is proportional to the Faradaic current measured in an electrochemical voltammetry experiment

$$j = -e\gamma \Gamma_m \nu \frac{d\theta}{dE}, \tag{7.20}$$

where $j$ is the Faradaic current in A/cm$^2$, $\Gamma_m$ is the number of surface sites per unit surface area (taken to be $1.5 \times 10^{15}$ sites/cm$^2$), and $\nu$ is the sweep rate (typically 50 mV/s).

Figure 7.11 shows the voltammogram obtained from kMC simulations. The disorder-order transition at 0.11 V is particularly sensitive to the level of equilibration: an insensitivity of this peak to reducing the sweep rate indicates that the surface is well equilibrated. The simulations were therefore done by choosing the rates of adsorption, desorption and diffusion such that upon reducing the sweep rate from 50 to 5 mV/s the disorder-order transition peak is shifted by less than 5 mV. This is also what is expected experimentally. The values fulfilling this requirement are $k^0 = 10^3$ s$^{-1}$ and $k^0_{\text{diff}} = 10^5$ s$^{-1}$. The sweep rate used was 50 mV/s.

Going from more negative to more positive potential, the anion adsorbs between $-0.1$ and 0.1 V in a disordered phase (see snapshots in Fig. 7.12a, b). This results in a broad adsorption peak in the voltammogram. There is a disorder-order transition at 0.11 V, indicated by the sharp peak in the voltammogram. At this voltage the anion coverage rapidly increases from 0.18 to 0.20 ML, which is the saturation coverage. This result was previously obtained by Orts et al. using an elongated hard hexagon type model [32]. The onset of the disorder-order transition is shown in Fig. 7.12c.

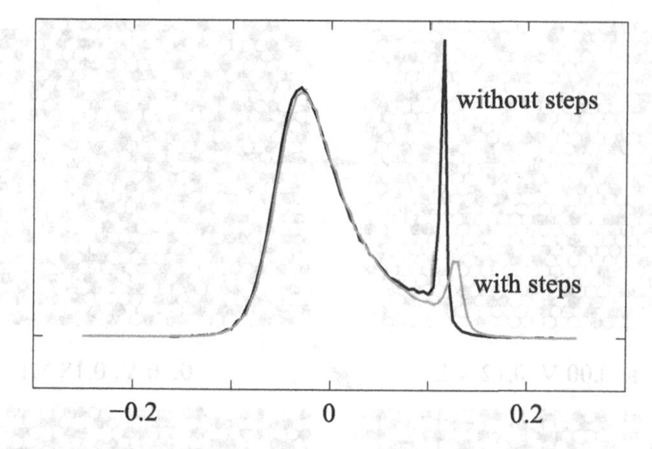


**Fig. 7.11** Voltammograms (current in arbitrary units versus electrode potential in V) averaged over 500 kMC simulations with a hexagonal 128 × 128 lattice, temperature $T = 300$ K, and sweep rate of 50 mV/s. The width of the terraces in the surface with the steps is 16 rows

$(\sqrt{3} \times \sqrt{7})$ Islands are forming, and, as indicated in the figure, these can have three different orientations. After the disorder-order transition (at potentials of 0.12 V and above) only large $(\sqrt{3} \times \sqrt{7})$ domains are present on the surface. These typically consist of more than one thousand adsorbates. At 0.10 V, the surface is occupied by anions which have three different orientations. Packing of equal amounts of these anions of different orientations is only possible up to about 0.18 ML. This causes a decrease in the adsorption current close to 0.10 V. It is only with the alignment of the adsorbates in the same direction that more anions can adsorb, since the ordered $(\sqrt{3} \times \sqrt{7})$ structure allows for a tighter packing of anions.

The presence of defects in the surface prevents long-range ordering of adsorbates. The voltammogram is therefore expected to change when steps are introduced. We have investigated this effect by introducing step sites at regular intervals in our model. Bonding to these step sites is excluded, and the adsorbates are also not allowed to diffuse across the steps. Our results indicate that for different terrace sizes only the disorder-order transition peak changes: the broad peak associated with adsorption in a disordered phase hardly changes. The disorder-order transition peak decreases in intensity with decreasing terrace size (Fig. 7.11). The saturation coverage decreases as well. For terrace sizes of 64 and larger, the orientation of the $(\sqrt{3} \times \sqrt{7})$ domains is independent of the step orientation: i.e., islands of all the three different orientations of the $(\sqrt{3} \times \sqrt{7})$ structure are found. The intensity of the disorder-order transition peak is comparable to the one for the case without steps. For terrace sizes of 16 and 32, the $(\sqrt{3} \times \sqrt{7})$ domains align to the steps, as shown in Fig. 7.13. For an even smaller terrace size of 8, no ordering is observed, even if the sweep rate is reduced to 5 mV/s.

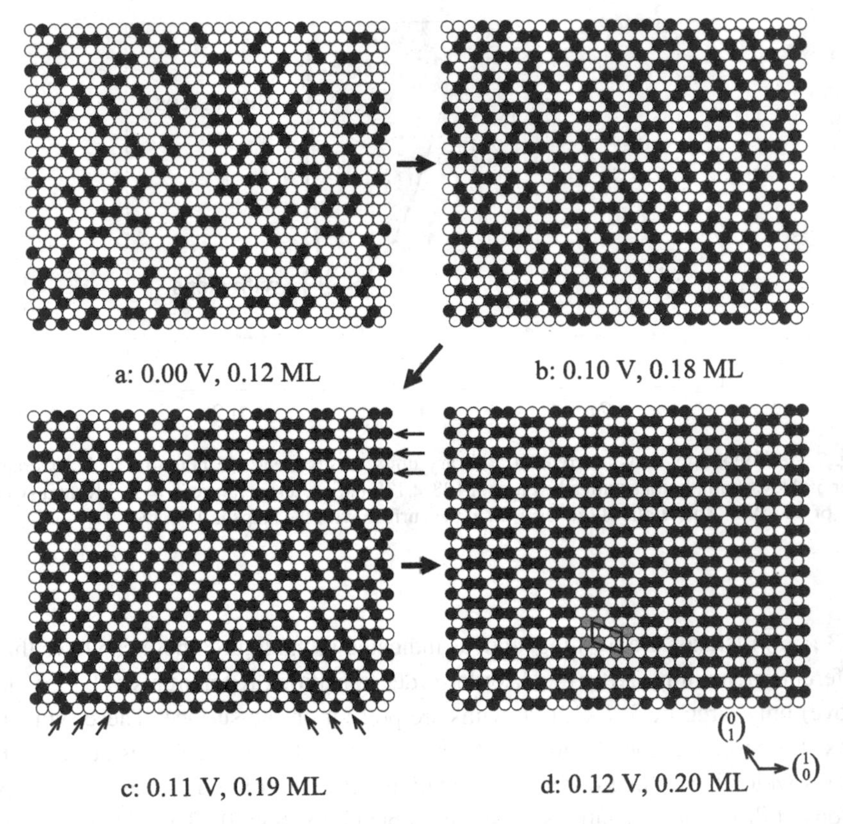


**Fig. 7.12** Snapshots of the surface during anion adsorption. Before the disorder-order transition (panel (**a**) and (**b**)) there is no ordering; during the disorder-order transition $(\sqrt{3} \times \sqrt{7})$ islands grow ((**c**); the three different domain orientations are indicated by the *small arrows*); after the disorder-order transition large islands dominate (**d**). The $(\sqrt{3} \times \sqrt{7})$ unit cell is indicated in (**d**)

### *7.4.3 The Ziff–Gulari–Barshad Model*

Although the work by Ziff, Gulari, and Barshad does not represent the first application of Monte Carlo to model surface reactions it is probably the most influential work of its type [33]. There are several reasons for that. It deals with CO oxidation which was and still is a very important process in catalysis and surface science. It is a very simple model, which makes it generic. Its simplicity also makes it possible to analyze in detail the relation between the microscopic reactions and the macroscopic properties. It has shown the shortcomings of the macroscopic rate equations and what the origin of these shortcomings were. It has shown that kMC simulations can describe kinetic or non-equilibrium phase transitions. In fact, apart from the first-order phase transition also known from macroscopic rate equations, it has shown that there is a continuous phase transition as well.

We present here the model, which we will call the Ziff–Gulari–Barshad (ZGB) model, in its generic form. There are two adsorbates: A and B. If one wants to use

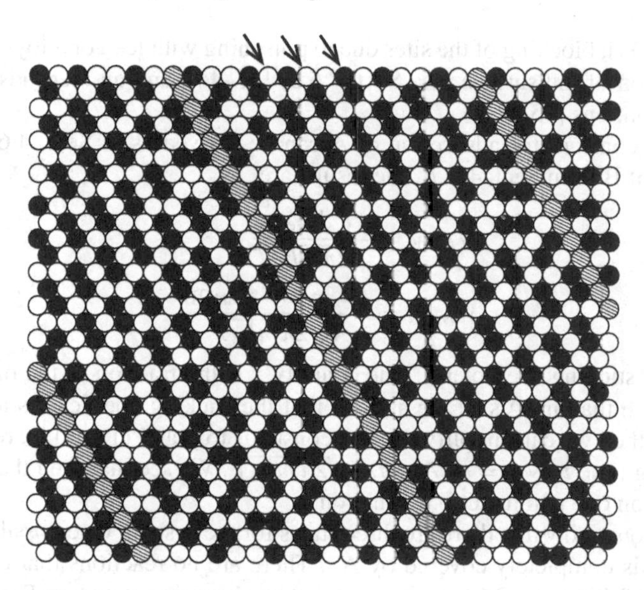

**Fig. 7.13** Snapshot of the surface at 0.25 V after anion adsorption for the model with first neighbor shell exclusion and a terrace width of 16. Bonding to the steps (*shaded circles*) is forbidden. The *arrows* show the alignment of the adsorbate islands (in *black*) to the steps (*shaded*), the coverage equals 0.19 ML

the ZGB-model for CO oxidation, then A stands for CO and B for atomic oxygen. There are three reactions in the model. Adsorbate A can adsorb at single vacant sites. Adsorbate B can adsorb, but as it forms diatomic molecules in the gas phase, two neighboring vacant sites are needed. An A will react with a B if they are nearest neighbors. This reactions is infinitely fast, so this takes place immediately after an adsorption. We can write the reaction as

$$\begin{aligned} \mathrm{A(gas)} + * &\rightarrow \mathrm{A(ads)} \\ \mathrm{B_2(gas)} + 2* &\rightarrow 2\mathrm{B(ads)} \\ \mathrm{A(ads)} + \mathrm{B(ads)} &\rightarrow \mathrm{AB(gas)} + 2* \end{aligned} \tag{7.21}$$

where $*$ is a vacant site, "ads" stands for an adsorbed species, and "gas" for a species in the gas phase. The reactions involving two sites can only take place on neighboring sites. Focusing on the sites only we have

$$\begin{aligned} * &\rightarrow \mathrm{A} \\ 2* &\rightarrow 2\mathrm{B} \\ \mathrm{A} + \mathrm{B} &\rightarrow 2* \end{aligned} \tag{7.22}$$

The adsorbates do not diffuse in the original model, and the lattice is a square one. There are many extensions to this model dealing, among others, with desorption of CO and oxygen [34], diffusion of the adsorbates [34], an Eley–Rideal mechanism for the oxidation step, physisorption of the reactants, lateral interactions between the adsorbates [34], blocking of the sites due to poisoning with lead or alloying [35, 36], reconstruction of the surface (see Sect. 7.5.2) [37–44], and an inert adsorbate that causes oscillations [45, 46].

The rate constant for adsorption of A can be derived as in Sect. 4.6.3, and for adsorption of B as in Sect. 4.6.7. The results are

$$W_{\mathrm{A,ads}} = \frac{yPA_{\mathrm{site}}\sigma_{\mathrm{A}}}{\sqrt{2\pi mk_{\mathrm{B}}T}} \tag{7.23}$$

$$W_{\mathrm{B,ads}} = \frac{2(1-y)PA_{\mathrm{site}}\sigma_{\mathrm{B}}}{4\sqrt{2\pi mk_{\mathrm{B}}T}}. \tag{7.24}$$

The $\sigma$'s are sticking coefficients. The quantity $y$ is the fraction of the molecules in the gas phase that are A's. If we assume that the sticking coefficients are equal to each other, then we can simplify the rate constants to $y$ and $(1-y)/2$, respectively, by replacing time $t$ by $\tau = tPA_{\mathrm{site}}\sigma_{\mathrm{A}}/\sqrt{2\pi mk_{\mathrm{B}}T}$. We see that then the model depends only on one parameter: i.e., only on $y$.

Simulations show that there are three states for the system. One possibility is that the surface is completely covered by A's. There are no reactions that can take the system out of this state. Such a state is called on absorbing state (see Sect. 2.2.1). In catalysis one talks in such a case of A poisoning, because it leads to the undesirable situation that the reactivity is zero. There is another absorbing state, but then with B poisoning. If the parameter $y$ is below a critical value $y_1$, then the system will always evolve into the B poisoning state. If the parameter $y$ is above another critical value $y_2$, then the system will always evolve into the A poisoning state. For $y < y_1$ there are so many $B_2$ molecules in the gas phase that B adsorption will outcompete the A's for the vacant sites that are formed by the reaction between the A's and B's. To same thing happens for $y > y_2$ except in this situations the A adsorption wins. At $y_1$ and at $y_2$ there is a kinetic or non-equilibrium phase transition.

For $y_1 < y < y_2$ there is a third state with A's and B's on the surface and a non-zero reactivity. Figure 7.14 shows how the reactivity and the coverage dependence on $y$. Note that all quantities change discontinuously at $y_2$. The phase transition at that value of $y$ is therefore called a first-order transition. At $y_1$ the quantities change continuously, so we have a continuous (or second-order) phase transition. Macroscopic rate equations also predict the first-order phase transition, but not the continuous one. Moreover, the first-order phase transition is predicted by the macroscopic rate equations to be at $y_1 = 2/3$, whereas the best estimates from kMC simulations are $y_2 = 0.52560 \pm 0.00001$ [47, 48]. The continuous phase transition is estimated to be at $y_1 = 0.39065 \pm 0.00010$ [49].

Figure 7.15 shows the reason for the discrepancy between the kMC results and those of macroscopic rate equations. The adlayer is definitely not a random mixture of adsorbates. The reason is the fast reaction between the A's and B's. This reaction causes segregation of the adsorbates. Isolated A's will not last long on the surface, because a B may adsorb on one of the vacant neighboring sites that will immediately react with the A which will remove the A from the surface. For a similar reason isolated B's will be rare. Only islands of the same kind of adsorbate can last, because the particles in the center of an island have no neighboring sites onto which other

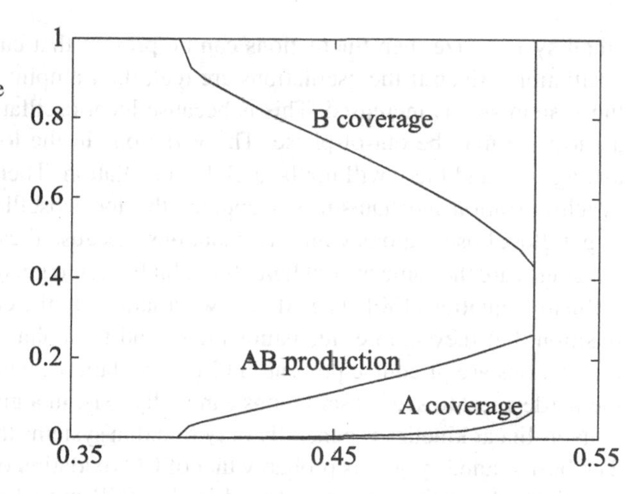


**Fig. 7.14** Phase diagram of the Ziff–Gulari–Barshad model. The coverages and the AB formation per unit time per site are shown as a function of the $y$ parameter

particles can adsorb with which they will react. These islands are formed randomly. Islands of B are larger. They need be because A's need only one vacant site for adsorption and can relatively easily break them up. Islands of A can be smaller, B's need two neighboring vacant sites for adsorption and have more difficulty to remove A's.

## 7.5 Non-linear Kinetics

The ZGB-model of Sect. 7.4.3 has shown that also the reactions themselves can result in a structured adlayer. There have been some claims that this was because there was no diffusion in the ZGB-model but, as the A + B model has shown, also with diffusion the adlayer is not necessarily homogeneous.

Here we will look at other systems for which reactions and other processes can lead to interesting structured adlayers but also to oscillations and other forms of non-linear kinetics. These systems showing non-linear kinetics have mainly been studied with rate equations or with reaction-diffusion equations [23, 24, 50]. Such systems are difficult for kMC simulations. If one does a kMC simulation with a

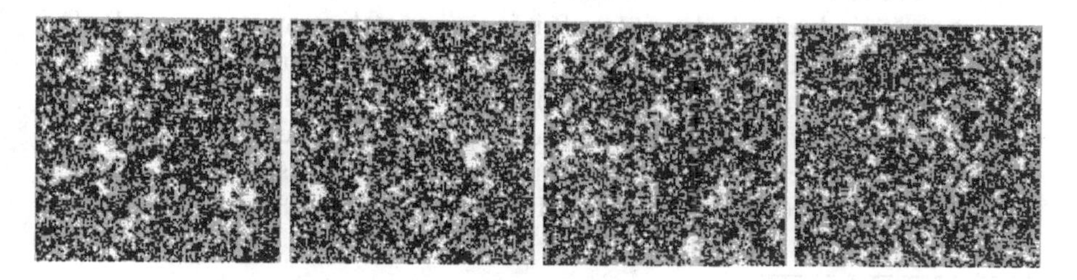

**Fig. 7.15** Snapshots of the adlayer in the Ziff–Gulari–Barshad model for $y = 0.5255$ a value just below the first-order transition point. The *black* islands are formed by B's. The A's form the *white* islands. Vacant sites are *gray*

small system size then fluctuations can be present that can easily be mistaken for oscillations. Even if the oscillations are real, their amplitude often decreases when the system size is increased. This is because local oscillations at different parts on a substrate may be out-of-phase. The variations in the local reactivities will then average out and there will not be a global oscillation. There needs to be a so-called synchronization mechanism that couples the local oscillations. It is much easier to get global oscillations with rate equations, because they assume a priori that the coverages are the same everywhere. It is a bit harder to get oscillations with reaction-diffusion equations [50]. They do allow variations in the coverages as a function of position, but they ignore fluctuations that tend to dephase local oscillations. Such fluctuations are of course present in kMC simulations, which means that they form the hardest test to see if oscillations can really exist in a given system.

Non-linear kinetics is generally only found in systems that are quite complicated. The best studied system is probably that of CO oxidation on Pt surfaces [23, 24, 37, 51]. Apart from the processes found in the ZGB-model one needs also diffusion, but most importantly a reconstruction of the substrate. There are only a few simple models that show oscillations [23, 45, 52]. The models that are therefore used to study non-linear kinetics are not nearly as realistic as the others discussed in this chapter.

### 7.5.1 The Lotka Model

We first discuss what is probably the simplest model that shows oscillations: the Lotka model [52–54].[4] This model is not only interesting because of the oscillations, but also because it shows behavior that is qualitatively different from what is predicted by macroscopic rate equations. The model has been studied in by Mai et al. [53, 54], who used it in testing the applicability of different approximations. They showed that even quite complicated approximations seem to fail in the description of this reaction system. They concluded that there exists a critical value $\bar{\zeta} \approx 0.11$ for the only parameter in the model below which the coverages oscillate in time.

The model consists of two kinds of particles, denoted A and B, with the following reactions. A's adsorb with rate constant $\zeta$, B's desorb with rate constant $1-\zeta$ and when there is an A adsorbed next to a B it is immediately converted into a B. The sites form a simple square lattice. We note that the whole system depends only on a single parameter $\zeta$. In effect this means that we have scaled time so that the sum of the rate constants for A adsorption and B desorption equals one.

This model has several potential experimental realizations, simulating a class of possible realistic feedback mechanisms. For example, A and B could be the same

[4]Parts of Sect. 7.5.1 have been reprinted with permission from J.P. Hovi, A.P.J. Jansen, R.M. Nieminen, Oscillating temporal behavior in an autocatalytic surface reaction model, Phys. Rev. E **55**, 4170 (1997). Copyright 1997, American Physical Society.

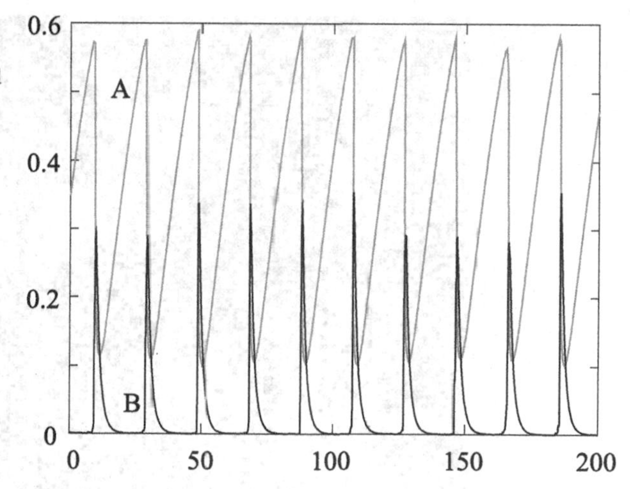


**Fig. 7.16** Oscillation in the coverage of A and B of the Lotka model for $\zeta = 0.05$ and a lattice size of 2048 × 2048 [52]

molecule, but adsorbed differently. Then A may denote a chemisorbed and B a physisorbed molecule, as neighboring B weakens the bonding of A and turns it to B. Alternatively, B may be adsorbed on a reconstructed site, and such a molecule induces the same reconstruction on a neighboring unreconstructed occupied site. On the other hand, A could be B plus an additional ligand. Such a ligand could desorb immediately if A and B come into contact, leaving two B's on the surface.

Because each site must be either vacant or occupied by an A or a B, and from the fact that averaged over time the number of adsorbing A's must equal the number of desorbing B's, we have [52]

$$\bar{\theta}_{\mathrm{B}} = \zeta(1 - \bar{\theta}_{\mathrm{A}}) \tag{7.25}$$

with $\bar{\theta}_{\mathrm{X}}$ the coverage of X averaged over time. The macroscopic rate equations do not predict any oscillations. Instead they predict a steady state with $\theta_{\mathrm{A}} = 0$ and $\theta_{\mathrm{B}} = \zeta$. To obtain this result one must use a finite rate constant of the conversion $\mathrm{A} + \mathrm{B} \to 2\mathrm{B}$ and then take the limit to infinity.

Figure 7.16 shows the temporal evolution of the coverages for $\zeta = 0.05$. It can be seen that there are very well-defined oscillations with a large amplitude. An analysis of these oscillations showed that near $\zeta = 0$ they are related to synchronized avalanches, which occur with a well defined frequency and come in all possible sizes: i.e., they exhibit power-law scaling [52]. Although no complete theory could be presented for the emergence of the oscillations, possibly the simplest heuristic explanation for this synchronization of the avalanches is the following. The typical cycle consists of a sudden increase of coverage of B followed by a slower decrease. In the end of this decrease period, as most B's have desorbed and the coverage is very small. A clusters are then free to grow. As there are very few conversions of A's to B's, A clusters grow until there is a nonzero probability that an arbitrary cluster is separated from a B molecule only by a single vacant perimeter site. This happens when $\theta_{\mathrm{A}}$ is close to the critical value. Figure 7.17 shows a snapshot taken just after an increase in the coverage of B. The avalanches are recognizable in the form of large connected clusters of B's.

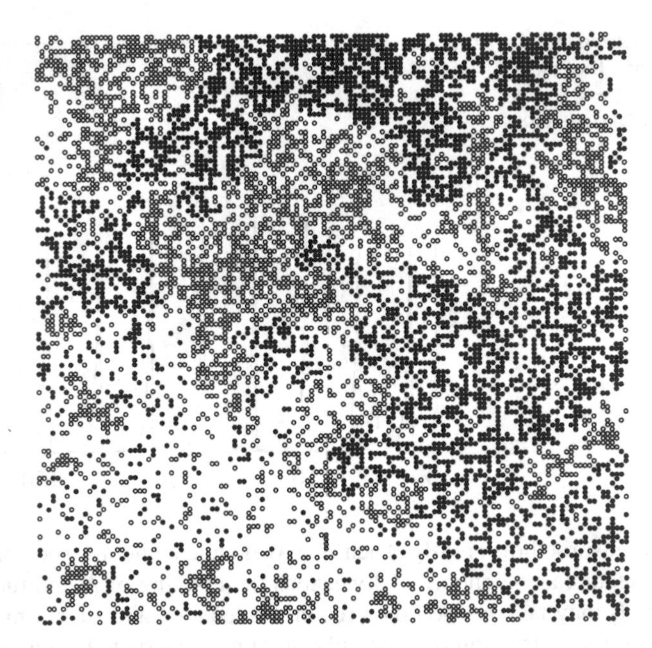

**Fig. 7.17** Snapshot of part of the adlayer in the Lotka model for $\zeta = 0.05$ and a lattice size of $2048 \times 2048$. The A's are depicted by *white circles*. They form large clusters. The B's are depicted by *black circles*. They form the same kind of clusters but with many B's removed because of the fast desorption. The *lower left* shows an open area where such a cluster of B's has almost completely been removed

## 7.5.2 Oscillations of CO Oxidation on Pt Surfaces

One problem for which extensive kMC simulations have been done by various groups is the problem of CO oscillations on Pt(100) and Pt(110). A crucial role in these oscillations is played by the reconstruction of the surface, and the effect of this reconstruction on the adsorption of oxygen. The explanation of the oscillations is as follows. A bare Pt surface reconstructs into a structure with a low sticking coefficient for oxygen. This means that predominantly CO adsorbs on bare Pt. However, CO lifts the reconstruction. The normal structure has a high sticking coefficient for oxygen. So after CO has adsorbed in a sufficient amount to lift the reconstruction oxygen can also adsorb. The CO and the oxygen react, and form $CO_2$. This $CO_2$ rapidly desorbs leaving bare Pt which reconstructs again. An important aspect of this process, and also other oscillatory reactions on surfaces, is the problem of synchronization. The cycle described above can easily take place on the whole surface, but oscillations on different parts on the surface are not necessarily in phase, and the overall reactivity of a surface is then constant. To get the whole surface oscillating in phase there has to be a synchronization mechanism.

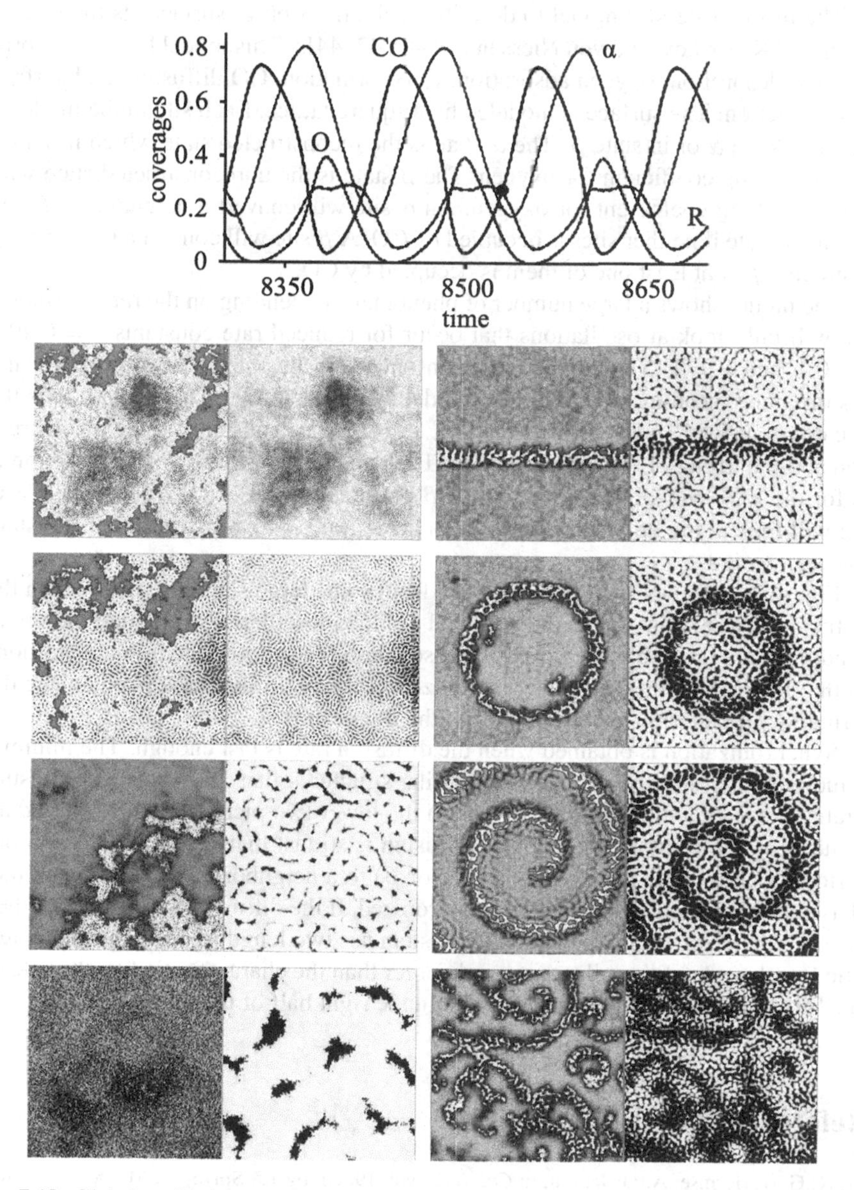


**Fig. 7.18** Global oscillations and pattern formation with $D = 250$. The *top* shows temporal variations of the coverages, the fraction of the substrate in the $\alpha$ phase, and the $CO_2$ production $R$. Each picture has two parts. In the *left part* we plot the chemical species: CO particles are *gray* and O particles are *white*, and vacant sites are *black*. The *right part* shows the structure of the surface: $\alpha$ phase sites are *black*, and $\beta$ phase sites are *white*. Sections of the *upper-left* corner with $L = 8192$, 4096, 1024, and 256 are shown on the *left half* of the figure. The sections correspond to the *dot* in the temporal plot at the *top*. On the *right half* of the figure we have a wave front, a target, a spiral, and turbulence ($L = 2048$), which can be obtained with different initial conditions [51]

The most successful model to describe oscillations on Pt surfaces is the one by Kortlüke, Kuzovkov, and von Niessen [38–40, 42–44].[5] This model has CO adsorption and desorption, oxygen adsorption, $CO_2$ formation, CO diffusion, and surface reconstruction. The surface is modeled by a square lattice. Each site in the model is either in state $\alpha$ or in state $\beta$. The $\alpha$ state is the reconstructed state which has a reduced sticking coefficient for oxygen. The $\beta$ state is the unreconstructed state with a high sticking coefficient for oxygen. An $\alpha$ site will convert a neighboring $\beta$ site into an $\alpha$ state if neither sites is occupied by CO. A $\beta$ site will convert a neighboring $\alpha$ site into $\beta$ if at least one of them is occupied by CO.

The model shows a large number of phenomena depending on the rate constants. We will only look at oscillations that occur for reduced rate constants $y = 0.494$, $k = 0.1$, and $V = 1$ [51]. The first rate constant, $y$, is the one for CO adsorption and has the same meaning as in the ZGB-model (see Sect. 7.4.3). The second, $k$, is the rate constant for CO desorption. The last, $V$, is the rate constant for the reconstruction and the lifting of the reconstruction. The rate constant for oxygen adsorption is as for the ZGB-model $(1 - y)/2$ on the $\beta$ phase, and $s_\alpha(1 - y)/2$ on the $\alpha$ phase. We will look at the Pt(110) surface which has $s_\alpha = 0.5$. The diffusion rate constant has been varied.

Figure 7.18 shows snapshots obtained from some large simulations in which the diffusion is just about fast enough to lead to global oscillations provided the initial conditions are favorable. However, it is also possible to choose the initial conditions so that the oscillations are not synchronized properly. In that case one can see the formation of patterns as the right half of the figure shows.

Synchronization is obtained when the diffusion rate is fast enough. The minimal value is related to the so-called Turing-like structures that are formed in the substrate. These structure can best be seen in the lower two pictures on the left and all pictures on the right of Fig. 7.18. If diffusion is so fast that within one oscillatory period CO can move from one phase ($\alpha$ or $\beta$) to a neighboring island of the other phase, then the oscillation are well synchronized. If the diffusion rate is slower, then we get pattern formation. Note that the system has two length scales. The characteristic length scale of the adlayer is much larger than the characteristic length scale of the Turing-like structures as can be seen in the right half of the figure.

## References


1. C.G.M. Hermse, A.P.J. Jansen, in *Catalysis*, vol. 19, ed. by J.J. Spivey, K.M. Dooley (Royal Society of Chemistry, London, 2006)
2. A.P.J. Jansen, Phys. Rev. B **69**, 035414 (2004)
3. A.P.J. Jansen, C. Popa, Phys. Rev. B **78**, 085404 (2008)
4. R.J. Baxter, *Exactly Solved Models in Statistical Mechanics* (Academic Press, London, 1982)


[5]Parts of Sect. 7.5.2 have been reprinted with permission from R. Salazar, A.P.J. Jansen, V.N. Kuzovkov, Synchronization of surface reactions via turing-like structures, Phys. Rev. E **69**, 031604 (2004). Copyright 2004, American Physical Society.

5. H.J. Borg, J.F.C.J.M. Reijerse, R.A. van Santen, J.W. Niemantsverdriet, J. Chem. Phys. **101**, 10052 (1994)
6. R.M. van Hardeveld, M.J.P. Hopstaken, J.J. Lukkien, P.A.J. Hilbers, A.P.J. Jansen, R.A. van Santen, J.W. Niemantsverdriet, Chem. Phys. Lett. **302**, 98 (1999)
7. D. Loffreda, D. Simon, P. Sautet, J. Chem. Phys. **108**, 6447 (1998)
8. D. Loffreda, D. Simon, P. Sautet, J. Catal. **213**, 211 (2003)
9. C.G.M. Hermse, F. Frechard, A.P. van Bavel, J.J. Lukkien, J.W. Niemantsverdriet, R.A. van Santen, A.P.J. Jansen, J. Chem. Phys. **118**, 7081 (2003)
10. A.P.J. Jansen, C.G.M. Hermse, F. Fréchard, J.J. Lukkien, in *Computational Science—ICCS 2001*, ed. by V. Alexandrov, J. Dongarra, B. Juliano, R. Renner, C. Tan (Springer, Berlin, 2001), pp. 531–540
11. C.T. Kao, G.S. Blackman, M.A.V. Hove, G.A. Somorjai, C.M. Chan, Surf. Sci. **224**, 77 (1989)
12. I. Zasada, M.A.V. Hove, G.A. Somorjai, Surf. Sci. **418**, L89 (1998)
13. K.B. Rider, K.S. Hwang, M. Salmeron, G.A. Somorjai, Phys. Rev. Lett. **86**, 4330 (2001)
14. M.T.M. Koper, J.J. Lukkien, A.P.J. Jansen, R.A. van Santen, J. Phys. Chem. B **103**, 5522 (1999)
15. M. Flytzani-Stephanopoulos, L.D. Schmidt, Prog. Surf. Sci. **9**, 83 (1979)
16. T.C. Wei, J. Phillips, Adv. Catal. **41**, 359 (1996)
17. D.W. Blakely, G.A. Somorjai, Surf. Sci. **65**, 419 (1977)
18. X.Q. Zhang, W.K. Offermans, R.A. van Santen, A.P.J. Jansen, A. Scheibe, U. Lins, R. Imbihl, Phys. Rev. B **82**, 113401 (2010)
19. E. Shustorovich, Surf. Sci. Rep. **6**, 1 (1986)
20. W.K. Offermans, A.P.J. Jansen, R.A. van Santen, G. Novell-Leruth, J.M. Ricart, J. Pérez-Ramirez, J. Phys. Chem. C **111**, 17551 (2007)
21. S. Wiggins, *Introduction to Applied Nonlinear Dynamical Systems and Chaos* (Springer, New York, 1990)
22. J.K. Hale, H. Koçak, *Dynamics and Bifurcations* (Springer, New York, 1991)
23. M.M. Slin'ko, N.I. Jaeger, *Oscillating Heterogeneous Catalytic Systems*. Studies in Surface Science and Catalysis, vol. 86 (Elsevier, Amsterdam, 1994)
24. R. Imbihl, G. Ertl, Chem. Rev. **95**, 697 (1995)
25. Q. Ge, R. Kose, D.A. King, Adv. Catal. **45**, 207 (2000)
26. M.T.M. Koper, J. Electroanal. Chem. **450**, 189 (1998)
27. M.T.M. Koper, A.P.J. Jansen, J.J. Lukkien, Electrochim. Acta **45**, 645 (1999)
28. J. Clavilier, in *Interfacial Electrochemistry, Theory, Experiment and Applications*, ed. by A. Wieckowski (Dekker, New York, 1999), pp. 231–248
29. B.M. Ocko, J.X. Wang, T. Wandlowski, Phys. Rev. Lett. **79**, 1511 (1997)
30. A.M. Funtikov, U. Stimming, R. Vogel, J. Electroanal. Chem. **428**, 147 (1997)
31. C.G.M. Hermse, A.P. van Bavel, M.T.M. Koper, J.J. Lukkien, R.A. van Santen, A.P.J. Jansen, Surf. Sci. **572**, 247 (2004)
32. J.M. Orts, L. Blum, D. Huckaby, J.M. Feliu, A. Aldaz, Abstr. Pap. Am. Chem. Soc. **220**, 91 (2000)
33. R.M. Ziff, E. Gulari, Y. Barshad, Phys. Rev. Lett. **56**, 2553 (1986)
34. H.P. Kaukonen, R.M. Nieminen, J. Chem. Phys. **91**, 4380 (1989)
35. J.P. Hovi, J. Vaari, H.P. Kaukonen, R.M. Nieminen, Comput. Mater. Sci. **1**, 33 (1992)
36. J. Mai, A. Casties, W. von Niessen, V.N. Kuzovkov, J. Chem. Phys. **102**, 5037 (1995)
37. R.J. Gelten, A.P.J. Jansen, R.A. van Santen, J.J. Lukkien, J.P.L. Segers, P.A.J. Hilbers, J. Chem. Phys. **108**, 5921 (1998)
38. V.N. Kuzovkov, O. Kortlüke, W. von Niessen, Phys. Rev. Lett. **83**, 1636 (1999)
39. O. Kortlüke, V.N. Kuzovkov, W. von Niessen, Phys. Rev. Lett. **83**, 3089 (1999)
40. V.N. Kuzovkov, O. Kortlüke, W. von Niessen, Phys. Rev. E **66**, 011603 (2002)
41. R.J. Gelten, R.A. van Santen, A.P.J. Jansen, in *Molecular Dynamics: From Classical to Quantum Methods*, ed. by P.B. Balbuena, J.M. Seminario (Elsevier, Amsterdam, 1999), pp. 737–784
42. O. Kortlüke, V.N. Kuzovkov, W. von Niessen, Phys. Rev. Lett. **81**, 2164 (1998)

43. V.N. Kuzovkov, O. Kortlüke, W. von Niessen, J. Chem. Phys. **108**, 5571 (1998)
44. O. Kortlüke, V.N. Kuzovkov, W. von Niessen, J. Chem. Phys. **110**, 11523 (1999)
45. A.P.J. Jansen, R.M. Nieminen, J. Chem. Phys. **106**, 2038 (1997)
46. A.P.J. Jansen, J. Mol. Catal. A, Chem. **119**, 125 (1997)
47. R.M. Ziff, B.J. Brosilow, Phys. Rev. A **46**, 4630 (1992)
48. B.J. Brosilow, E. Gulari, R.M. Ziff, J. Chem. Phys. **98**, 674 (1993)
49. I. Jensen, H. Fogedby, R. Dickman, Phys. Rev. A **41**, 3411 (1990)
50. P. Grindrod, *The Theory and Applications of Reaction-Diffusion Equations: Patterns and Waves* (Clarendon, Oxford, 1996)
51. R. Salazar, A.P.J. Jansen, V.N. Kuzovkov, Phys. Rev. E **69**, 031604 (2004)
52. J.P. Hovi, A.P.J. Jansen, R.M. Nieminen, Phys. Rev. E **55**, 4170 (1997)
53. J. Mai, V.N. Kuzovkov, W. von Niessen, A Lotka type model for oscillations in surface reactions (1995). Private communication
54. J. Mai, V.N. Kuzovkov, W. von Niessen, Oscillations beyond the mean-field description: The Lotka model (1995). Private communication

# Chapter 8
# New Developments

**Abstract** Kinetic Monte Carlo (kMC) is a very versatile and powerful method to study the kinetics of surface reactions, but there are nevertheless some systems and phenomena for which one would like it to be more efficient or one would like to extend it. We discuss what kind of fast processes may pose a problem, when they become problematic, and what can be done about it so that kMC simulations can be done with longer time scales. Coarse-graining is presented as a method to do simulations with larger length scales. Mass transport and heat transfer are discussed in the context of continuum models that are coupled to kMC simulations. Finally, off-lattice kMC simulations are introduced to show that one can also do kMC simulations of systems that can not be described with a lattice-gas model.

## 8.1 Longer Time Scales and Fast Processes

In many simulations the period that one has to simulate a system is determined by how long it takes the system to reach the state of interest (e.g., a steady state), and the time one needs to simulate the system to get good statistics for the results. Neither times are generally very long. There are exceptions however. Relaxation times sometimes are long, or one is interested in processes with intrinsically long time scales: e.g., time-dependent experiments like Temperature-Programmed Desorption or oscillations. It is not always possible to get to the time scale of interest with a conventional kinetic Monte Carlo (kMC) simulation.

Even if the real time that one wants to simulate a system is not long, a simulation may still take a lot of computer time because some of the processes are very fast. The computer time is determined by the number of processes that actually take place. As this number also determines if one has good statistics, one might assume that with faster processes one need not simulate a system so long. That is correct, except when there are both fast and slow processes in the system. To get good statistics on the slow processes one then needs to simulate an inordinate large number of fast processes.

A.P.J. Jansen, *An Introduction to Kinetic Monte Carlo Simulations of Surface Reactions*, Lecture Notes in Physics 856,
DOI 10.1007/978-3-642-29488-4_8, © Springer-Verlag Berlin Heidelberg 2012

### 8.1.1 When Are Fast Processes a Problem?

The theory of kMC does not pose any restrictions on the rate constants of the processes being simulated. So there is no fundamental problem when there are processes in a reaction system with very dissimilar rate constants. However, fast processes can make a simulations slow and inefficient.

First, there is the question of which algorithm to use. This is discussed in Chap. 3. The problem is oversampling, which can become very inefficient with rate constants that differ a lot. The First Reaction Method (FRM) has no problems at all with fast processes because there is no oversampling, but the Random Selection Method will only rarely work well. The Variable Step Size Method (VSSM) works well if no oversampling is used, but otherwise may be inefficient. The procedure in Sect. 3.9 can be used to show rapidly which algorithm works best.

A different problem, one not so easily solved, is when almost all processes that actual take place in a system are fast processes but one is really interested in the few slow ones. This is for example quite common when one is interested in the chemical reactions on a catalyst's surface, but most processes are diffusional hops of the adsorbates. A kMC simulation will then spend almost all time on diffusion whatever the algorithm one uses.

It is important to understand the precise nature of the problem. A fast process in itself is not a problem For example, suppose that a molecule has a high rate constant for adsorption. The consequence of this is that a molecule will adsorb as soon as there is a vacancy. The number of such adsorption process will remain small, because adsorption will stop when all vacancies are occupied. It only becomes a problem if the molecules also have a high rate constant for desorption. In that case new vacancies will be formed with a high rate, and a simulation ends up in spending all time simulating an adsorption-desorption equilibrium. There is also no problem if only the rate constant for desorption is high, because then the number of processes is limited by the slow adsorption. Of course, there will not be many molecules on the surface, but that is as it should be, as it is a consequence of the nature of the adsorption-desorption equilibrium.

So there will only be problems with fast processes if a fast process can be followed by infinite many other fast processes. Adsorption-desorption is one example. Another would be a fast reaction that also has a fast reverse reaction. The most common case is however fast diffusion.

### 8.1.2 A Simple Solution

The simplest approach is to reduce the rate constants of all fast processes involved by some constant factor. This is the same as increasing the energies of the transition states of these processes by the same amount. The idea is that what all these processes do is to equilibrate a subset of all configurations. This equilibration can however also be accomplished with much smaller rate constants. The pragmatic approach is then to start with a very large reduction of the rate constants: so large, that the fast processes no longer dominate the time it takes to do a simulation. Then one should increase the rate constants and look at how the results of the simulation change. At a certain point the results will converge as a function of these rate constants. At this point the simulation yields the same results as a simulation with the real rate constants, but with minimal computer time. We have noticed that in the case of diffusion this approach can easily speed up a simulation by three or more orders of magnitude. Note that reducing the rate constants of the fast processes does not slow down the evolution of the system because this is determined by the slow processes. The same behavior is seen but with fewer fast processes.

The idea of reducing the rate constants has been worked out and automated by Chatterjee and Voter in their Accelerated Superbasin kMC [1]. The approach keeps track of how often configurations are revisited. If this occurs often, then this is indicative of the system being trapped in a set of configurations connected by fast processes. This is the superbasin of the approach. The energies of the transition states of these fast processes is then increased. Depending on how much that these energies are changes, the procedure may have to be repeated a number of times until the fast processes are slowed down so much that an escape from the superbasin occurs. Chatterjee and Voter give criteria to determine when a revisit is indeed pointing to the existence of a superbasin.

Changing the rate constants as described here introduces no errors in the probabilities of the configurations if the system is in equilibrium. The energies of transition states do not determine or affect equilibrium properties in any way. Many situations exist however for surface reactions where there is an irreversible process: e.g., the desorption of a molecule that has been created on the surface. In these situations the system is not in equilibrium, but at steady state. Forward and reverse reactions have different rates in such a system, and the difference is proportional to the overall reactivity. Changing the energies of the transition states changes initially the rates of forward and reverse reactions by the same factor, and that changes their difference as well. As a consequence the coverages and the reactivity will subsequently change as well, and the behavior of the system is no longer the same. In general, this effect is small if only transition states of fast processes are changed, and the approach then is a good approximation.

### *8.1.3 Reduced Master Equations*

A more sophisticated approach using the same equilibration idea is to remove the fast processes completely. Let $C(\alpha)$ be the set of all configurations that can be reached from configuration $\alpha$ by some fast processes only. We can partition all configurations in such sets. Let $P_{C(\alpha)}$ be the probability that the system is in one of the configurations of $C(\alpha)$: i.e.,

$$P_{C(\alpha)} = \sum_{\alpha' \in C(\alpha)} P_{\alpha'}. \tag{8.1}$$

We then have

$$\frac{dP_{C(\alpha)}}{dt} = \sum_{C(\beta)} [\Omega_{C(\alpha)C(\beta)} P_{C(\beta)} - \Omega_{C(\beta)C(\alpha)} P_{C(\alpha)}] \tag{8.2}$$

with the new rate constants

$$\Omega_{C(\alpha)C(\beta)} = \sum_{\alpha' \in C(\alpha)} \sum_{\beta' \in C(\beta)} W_{\alpha'\beta'} \frac{P_{\beta'}}{P_{C(\beta)}}. \tag{8.3}$$

This equation is again a master equation provided that all $P_{\beta'}/P_{C(\beta)}$ are constant, which they are for infinitely fast processes. The reason is that such fast processes bring all $\beta' \in C(\beta)$ to equilibrium instantaneously and the ratio is nothing but a conditional probability that the system is in $\beta'$ given that it is in one of the configurations of $C(\beta)$. This ratio is determined by the fast processes only.

The applicability of Eq. (8.2) depends a how easily $P_{\beta'}/P_{C(\beta)}$ can be determined. For diffusion in a one-dimensional zeolite this proved to be easy, and instead of a master equation with configurations a master equations with the numbers of molecules could be used [2]. For higher dimensional system things might not be so easy.

To get $P_{\beta'}/P_{C(\beta)}$ in the general case, one has to solve

$$\sum_{\beta'' \in C(\beta)} [W_{\beta'\beta''} P_{\beta''} - W_{\beta''\beta'} P_{\beta'}] = 0 \tag{8.4}$$

with $\beta' \in C(\beta)$ and subject to the restriction

$$\sum_{\beta' \in C(\beta)} P_{\beta'} = P_{C(\beta)}. \tag{8.5}$$

We can remove the need to know $P_{C(\beta)}$ by rewriting this as

$$\sum_{\beta'' \in C(\beta)} \left[ W_{\beta'\beta''} \frac{P_{\beta''}}{P_{C(\beta)}} - W_{\beta''\beta'} \frac{P_{\beta'}}{P_{C(\beta)}} \right] = 0 \tag{8.6}$$

with the restriction

$$\sum_{\beta' \in C(\beta)} \frac{P_{\beta'}}{P_{C(\beta)}} = 1. \tag{8.7}$$

This does not yield the $P_{\beta'}$'s, but only the $P_{\beta'}/P_{C(\beta)}$'s, but that is also really all we need to know.

Equations (8.6) and (8.7) form a set of linear equations that need to be solved. In the case of the one-dimensional zeolite mentioned above this could be done analytically [2], but if this is not possible one has to hope that the number of equations, which equals the number of configurations in $C(\beta)$, is not too large.

Mastny et al. have used this reduced master equation (8.3) for a simple model of CO oxidation [3]. They have defined the sets $C(\alpha)$ as all configurations with the same number of adsorbates of each type just as was done for the one-dimensional zeolite. They have tried two approaches to solve Eq. (8.6). In the first they used very small lattices ($5 \times 5$ or $10 \times 10$). With these lattices there are only several hundreds to several thousands equations (8.3) and the rate constants for all of them were determined by doing normal but short kMC simulations with these small lattices. Only diffusion of the adsorbates was included in these simulations. The probability that a certain configuration $\beta$ was found in such a simulation was then equal to $P_\beta / P_{C(\beta)}$, and that information was then used to compute $\Omega_{C(\alpha)C(\beta)}$ via Eq. (8.3).

Their second approach consisted of a derivation of the exact rate equations for the coverages (see Sect. 4.6). The coverage dependence of the rates was then determined using kMC simulations with large lattices, but again also with only diffusion. Instead of investigating all possible combinations of the number of adsorbates, which would not have been possible, they looked at a discrete sampling of coverages and then interpolated.

Both approaches were shown to give good results. An advantage of both approaches is that it is much easier to study how the behavior of a system depends on the rate constants of the processes than with kMC. The reason is that the effect of the diffusion can be determined independently from that of the other processes: i.e., the kMC simulations with only diffusion need to be done only once and the results can be used in all subsequent simulations or calculations.

### *8.1.4 Dealing with Slightly Slower Reactions*

A more sophisticated approach was introduced by Mason et al. in their work on flicker processes [4, 5]. It has the advantage that it can also be used when there is not such a clearcut distinction between fast and slow processes, although the extra computational costs of the more complicated algorithm may not be worthwhile if the "fast" processes become too slow. A flicker process is change $\alpha \to \beta$ of configurations immediately followed by the reverse change $\beta \to \alpha$. In their work Mason et al. grouped all changes $\alpha \to \beta \to \alpha$ for a fixed $\alpha$ and all possible $\beta$'s, and computed a time when the chain of flicker processes is broken. They also determined which configuration the system then goes to. Consequently, the flicker processes need no longer be simulated explicitly which can speed up a simulation substantially.

#### 8.1.4.1 The General Case

We will first discuss a more general situation. Suppose at time $t = 0$ we are at configuration $\alpha_0$ and there are a number of processes that we consider fast, that we do not want to simulate explicitly, and that can bring the system to a set of configurations $C(\alpha_0)$. We also define the complement $K(\alpha_0)$ of $C(\alpha_0)$, which contains all configurations not in $C(\alpha_0)$. The situation is sketched in Fig. 8.1. We want to know first what the first configuration in $K(\alpha_0)$ will be when the system moves outside $C(\alpha_0)$. We want to know second at what time this will occur. And third, we want an algorithm that can generate this information.

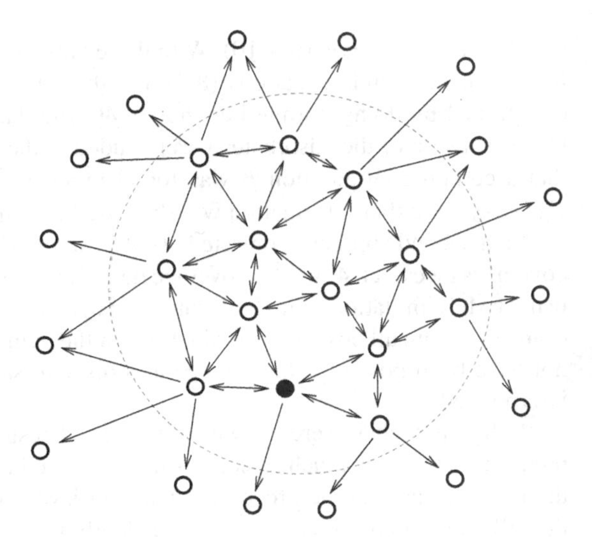

**Fig. 8.1** At time $t = 0$ the system is in configuration $\alpha_0$ which is depicted by the *black dot*. There are fast processes between the configurations in $C(\alpha_0)$, which is depicted by the *open dots* enclosed by the *dashed curve*. Slower processes bring the system to configurations not in $C(\alpha_0)$: i.e., in the complement $K(\alpha_0)$. *Single-headed arrows* indicate irreversible processes, *double-headed* ones reversible processes. (Color figure online)

The problem described above is a first-passage problem [6]. We can simplify the situation by removing from $K(\alpha_0)$ all configurations that can not be reached directly from $C(\alpha_0)$: i.e., we include only those configurations $\gamma$ for which there is a process $\beta \to \gamma$ with $\beta \in C(\alpha_0)$ and $\gamma \in K(\alpha_0)$. Moreover, we remove all processes from $K(\alpha_0)$ to $C(\alpha_0)$. The remaining configurations in $K(\alpha_0)$ are then absorbing states (see Sect. 2.2.1). That is the situation that is sketched in Fig. 8.1. We want an algorithm that combines several processes in one multistep process. This can be accomplished by separating the processes that take place from the determination of the times. That this is possible is most easily seen in the VSSM algorithm (see Sect. 3.2). The time determined by Eq. (3.11) does not depend on the process that will take place, nor does the time affect the determination of the process. It is possible to determine first which processes will take place without determining any times, and to determine times after that [7–9].

Let's call a series of subsequent processes $\alpha_0 \to \alpha_1 \to \alpha_2 \to \ldots \to \alpha_N$ a path. A process $\alpha_n \to \alpha_{n+1}$ will take place with probability

$$P_{\alpha_{n+1}\alpha_n} = \frac{W_{\alpha_{n+1}\alpha_n}}{R_{\alpha_n}} \tag{8.8}$$

with

$$R_\alpha = \sum_{\beta \in C(\alpha_0) \cup K(\alpha_0)} W_{\beta\alpha}. \tag{8.9}$$

The path $\alpha_0 \to \alpha_1 \to \alpha_2 \to \ldots \to \alpha_N$ will be taken by the system with probability

$$\prod_{n=1}^{N} P_{\alpha_n\alpha_{n-1}}. \tag{8.10}$$

If a system starts at configuration $\alpha_0$ then the probability that the system ends up in the absorbing state $\beta \in K(\alpha_0)$ equals the sum of the probabilities of all paths starting at $\alpha_0$ and ending at $\beta$.

The evaluation of this sum can most easily be computed with a transition matrix **P** defined by

$$\mathbf{P}_{\alpha_j\alpha_i} = P_{\alpha_j\alpha_i} \tag{8.11}$$

with $\alpha_j, \alpha_i \in C(\alpha_0)$. The matrix elements of **P** can be interpreted as the probabilities of paths that consist of only a single process. From

$$\left(\mathbf{P}^2\right)_{\alpha_j\alpha_i} = \sum_{\alpha_k} P_{\alpha_j\alpha_k} P_{\alpha_k\alpha_i} \tag{8.12}$$

we see that $\mathbf{P}^2$ has the probabilities of paths consisting of two processes starting and ending at configurations of $C(\alpha_0)$. Similarly $\mathbf{P}^n$ has the probabilities of paths consisting of $n$ processes starting and ending at configurations of $C(\alpha_0)$. Now the probability that the system starts at $\alpha_0$ and ends up in $\beta \in K(\alpha_0)$ after $n$ processes in $C(\alpha_0)$ equals

$$\sum_{\alpha\in C(\alpha_0)} P_{\beta\alpha}(\mathbf{P}^n)_{\alpha\alpha_0}. \tag{8.13}$$

The probabilities that the system exits $C(\alpha_0)$ after $n$ processes is obtained by summing over all $\beta \in K(\alpha_0)$. These probabilities allow us to determine the number of processes that take place in $C(\alpha_0)$ before the system exits $C(\alpha_0)$, and the probabilities for the configurations in $K(\alpha_0)$ the system then goes to.

Now suppose that we are also able to determine the precise path that the system takes in $C(\alpha_0)$, then the time when the system exits $C(\alpha_0)$ can be determined as follows. Let the path be $\alpha_0 \to \alpha_1 \to \alpha_2 \to \ldots \to \alpha_n \to \beta$. The time it takes for all processes in this path to occur is then

$$\sum_{m=0}^{n} \Delta t_m \tag{8.14}$$

with $\Delta t_m$ a random number taken from the probability distribution

$$R_{\alpha_m} \exp[-R_{\alpha_m} t]. \tag{8.15}$$

The algorithm is then as follows

1. Determine the number of processes in $C(\alpha_0)$ before the system exits $C(\alpha_0)$ by summing over the configurations $\beta$ in Eq. (8.13).
2. Determine the configuration $\beta \in K(\alpha_0)$ that the system moves to after exiting $C(\alpha_0)$. Use again Eq. (8.13).
3. Determine the path from $\alpha_0$ to $\beta$ with the correct number of processes.
4. Determine the time that it takes to follow this path. Use Eq. (8.14).

Steps 1 and 2 are selections that can be done with one of the procedures discussed in Sect. 3.3.1. See also Sect. 8.1.4.5. Step 3 has not been discussed yet. In general it is a very complicated problem. So we will discuss only the most common cases but with all steps in detail.

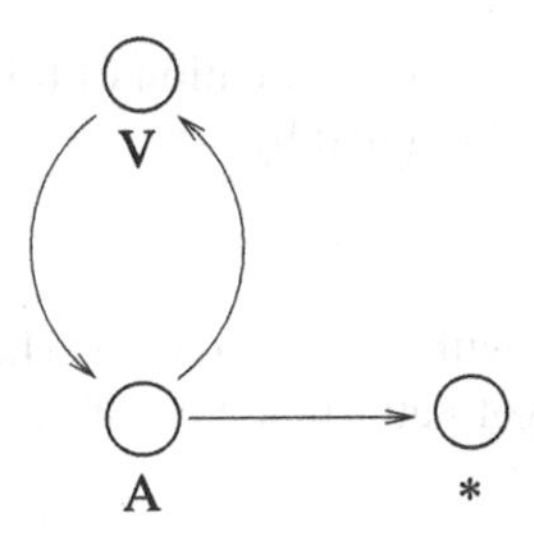


**Fig. 8.2** The three configurations involved in fast adsorption-desorption with a slower reaction of the adsorbed species. *V* stands for a vacant adsorption site, *A* for the adsorbed species, and $*$ for the absorbing state with the species that has reacted

#### 8.1.4.2 Adsorption-Desorption

Suppose we have a single adsorption site and an adsorbate with high adsorption and desorption rate constants. Figure 8.2 shows the three configurations involved. We have an adsorption rate constant $W_{\mathrm{ads}}$ changing V to A, a desorption rate constant $W_{\mathrm{des}}$ changing A to V, and a reaction rate constant $W_{\mathrm{rx}}$ for the reaction A $\to *$. Our approach is useful if $W_{\mathrm{ads}}, W_{\mathrm{des}} \gg W_{\mathrm{rx}}$, but it is also correct if this inequality does not hold. From the rate constants we can derive the probabilities of the processes

$$P_{\mathrm{ads}} = 1, \tag{8.16}$$

$$P_{\mathrm{des}} = \frac{W_{\mathrm{des}}}{W_{\mathrm{des}} + W_{\mathrm{rx}}}, \tag{8.17}$$

and

$$P_{\mathrm{rx}} = \frac{W_{\mathrm{rx}}}{W_{\mathrm{des}} + W_{\mathrm{rx}}}. \tag{8.18}$$

The transition matrix **P** then becomes

$$\mathbf{P} = \begin{pmatrix} 0 & P_{\mathrm{des}} \\ 1 & 0 \end{pmatrix} \tag{8.19}$$

with the states ordered V, A. It is easy then to derive

$$\mathbf{P}^{2n} = P_{\mathrm{des}}^{n} \begin{pmatrix} 1 & 0 \\ 0 & 1 \end{pmatrix} \tag{8.20}$$

and

$$\mathbf{P}^{2n+1} = P_{\mathrm{des}}^{n} \begin{pmatrix} 0 & P_{\mathrm{des}} \\ 1 & 0 \end{pmatrix}. \tag{8.21}$$

We start in configuration V. The probability that the reaction to $*$ takes place after adsorption V $\to$ A, and $n$ adsorption-desorption cycles A $\to$ V $\to$ A, is then

$$\left(\mathbf{P}^{2n+1}\right)_{\mathrm{AV}} P_{\mathrm{rx}} = P_{\mathrm{des}}^{n} P_{\mathrm{rx}}. \tag{8.22}$$

We note that these probabilities are properly normalized as

$$\sum_{n=0}^{\infty} P_{\mathrm{des}}^{n} P_{\mathrm{rx}} = \frac{P_{\mathrm{rx}}}{1 - P_{\mathrm{des}}} = 1. \tag{8.23}$$

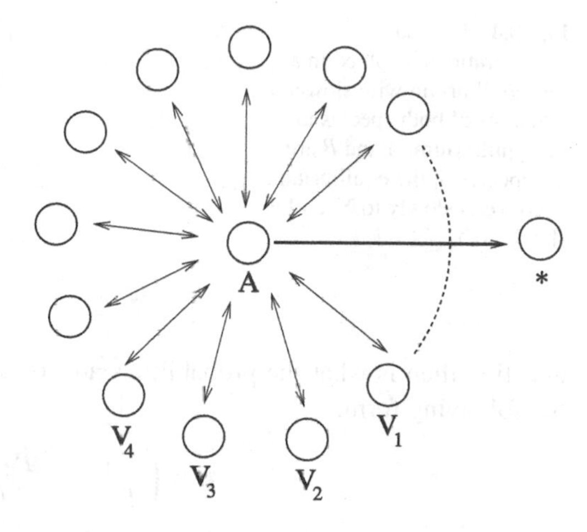


**Fig. 8.3** Fast reactions from A to one of the $V_n$'s and back occur repeatedly before the system converts to $*$. All processes $A \rightarrow V_n$ have the same rate constant $W_{A \rightarrow V}$. All processes $V_n \rightarrow A$ have the rate constant $W_{V \rightarrow A}$. The conversion to $*$ has rate constant $W_{A \rightarrow *}$

We can use these probabilities to generate the number of cycles $A \rightarrow V \rightarrow A$ that take place before the reaction occurs with one the selection methods in Sect. 3.3.1.

To determine when the reaction takes place is easy in this case, because given the number $n$ of cycles $A \rightarrow V \rightarrow A$ it is clear what path the system takes. With $n$ such cycles there are $n+1$ adsorptions, $n$ desorptions, and one reaction. So we need to generate $n+1$ intervals $\Delta t$ from the probability distribution $W_{\text{ads}} \exp[-W_{\text{ads}} \Delta t]$ and also $n+1$ intervals from $(W_{\text{des}} + W_{\text{rx}}) \exp[-(W_{\text{des}} + W_{\text{rx}}) \Delta t]$. If we add up all these intervals, we get the time when the reaction takes place if we start in V at time $t = 0$. Note that the intervals for the desorption and the reaction are taken from the same distribution, because they occur in the same configuration A.

If we don't have one adsorption site but many, then we can regard the adsorption, the desorption-adsorption cycles, and the reaction at each site as one multistep process, determine a time for each of them, and use FRM to determine which multistep process occurs first and to do the simulation (see Sect. 3.5). This avoids a much larger matrix $\mathbf{P}$ describing all possible occupations of the various sites.

The adsorption-desorption process above can easily be extended to the flicker processes studied by Mason et al. [4]. Figure 8.3 shows multiple configurations to which the system can convert from and to A. (We use a notation similar to the one of the adsorption-desorption case, but change the meaning.) This is essentially the same process as adsorption-desorption except that the rate constant $W_{\text{ads}}$ should be replaced by $N W_{V \rightarrow A}$ and $W_{\text{des}}$ by $N W_{A \rightarrow V}$.

#### 8.1.4.3 Fast Equilibrium

Suppose we have a fast conversion reaction $A \rightarrow B$ and a fast reverse $B \rightarrow A$ on a single site. Also suppose that there are two slower reactions: $A \rightarrow X$ and $B \rightarrow Y$ (see Fig. 8.4). The important difference with the cases above is that there are two ways in which the cycles of fast reactions $A \rightarrow B \rightarrow A$ can end. The additional

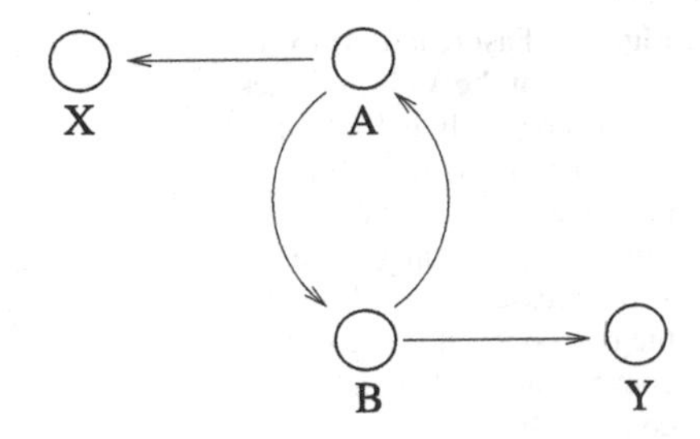


**Fig. 8.4** The four configurations involved in a fast equilibrium with slower reactions of both species in the equilibrium. *A* and *B* are the species of the equilibrium. A converts slowly to X, and B slowly to Y

question then is what the probabilities are for ending at X or Y. The **P** matrix has the following form.

$$\mathbf{P} = \begin{pmatrix} 0 & P_{\mathrm{B}\to\mathrm{A}} \\ P_{\mathrm{A}\to\mathrm{B}} & 0 \end{pmatrix} \tag{8.24}$$

with

$$P_{\mathrm{A}\to\mathrm{B}} = \frac{W_{\mathrm{A}\to\mathrm{B}}}{W_{\mathrm{A}\to\mathrm{B}} + W_{\mathrm{A}\to\mathrm{X}}} \tag{8.25}$$

and

$$P_{\mathrm{B}\to\mathrm{A}} = \frac{W_{\mathrm{B}\to\mathrm{A}}}{W_{\mathrm{B}\to\mathrm{A}} + W_{\mathrm{B}\to\mathrm{Y}}}. \tag{8.26}$$

We have

$$\mathbf{P}^{2n} = (P_{\mathrm{A}\to\mathrm{B}} P_{\mathrm{B}\to\mathrm{A}})^n \begin{pmatrix} 1 & 0 \\ 0 & 1 \end{pmatrix} \tag{8.27}$$

and

$$\mathbf{P}^{2n+1} = (P_{\mathrm{A}\to\mathrm{B}} P_{\mathrm{B}\to\mathrm{A}})^n \begin{pmatrix} 0 & P_{\mathrm{B}\to\mathrm{A}} \\ P_{\mathrm{A}\to\mathrm{B}} & 0 \end{pmatrix}. \tag{8.28}$$

If we start at A, then the probability to end at X after $n$ cycles A $\to$ B $\to$ A equals

$$(P_{\mathrm{A}\to\mathrm{B}} P_{\mathrm{B}\to\mathrm{A}})^n P_{\mathrm{A}\to\mathrm{X}} \tag{8.29}$$

with

$$P_{\mathrm{A}\to\mathrm{X}} = \frac{W_{\mathrm{A}\to\mathrm{X}}}{W_{\mathrm{A}\to\mathrm{B}} + W_{\mathrm{A}\to\mathrm{X}}}. \tag{8.30}$$

The probability to end at Y after A $\to$ B and $n$ cycles B $\to$ A $\to$ B equals

$$P_{\mathrm{A}\to\mathrm{B}} (P_{\mathrm{B}\to\mathrm{A}} P_{\mathrm{A}\to\mathrm{B}})^n P_{\mathrm{B}\to\mathrm{Y}} \tag{8.31}$$

with

$$P_{\mathrm{B}\to\mathrm{Y}} = \frac{W_{\mathrm{B}\to\mathrm{Y}}}{W_{\mathrm{B}\to\mathrm{A}} + W_{\mathrm{B}\to\mathrm{Y}}}. \tag{8.32}$$

We have

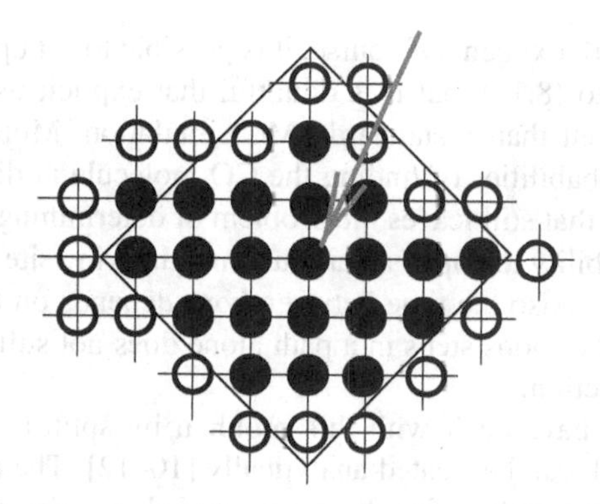

**Fig. 8.5** A collection of adsorption sites for a CO molecule. The molecule can hop between neighboring sites. The *open dots* depict sites at which the molecule can react with an oxygen atom to form $CO_2$. At time $t = 0$ the CO molecule is assumed to have adsorbed at the site indicated by the *arrow*. The *large square* centered at this site is an example of a protective domain. (Color figure online)

$$\begin{aligned}\sum_{n=0}^{\infty}(P_{A\to B}P_{B\to A})^n P_{A\to X} + \sum_{n=0}^{\infty} P_{A\to B}(P_{B\to A}P_{A\to B})^n P_{B\to Y} &= \frac{P_{A\to X}}{1-P_{A\to B}P_{B\to A}} + \frac{P_{A\to B}P_{B\to Y}}{1-P_{B\to A}P_{A\to B}} \\ &= \frac{(1-P_{A\to B})+(1-P_{B\to A})P_{A\to B}}{1-P_{A\to B}P_{B\to A}} = 1 \qquad (8.33)\end{aligned}$$

as it should be. We also see that the probability to end up at X equals $P_{A\to X}/(1 - P_{A\to B}P_{B\to A})$ and the probability to end up at Y equals $P_{A\to B}P_{B\to Y}/(1 - P_{B\to A}P_{A\to B})$.

The times it takes to end at X or Y after a number of cycles are sums of intervals taken from the probability distributions $f_A(\Delta t) = (W_{A\to B} + W_{A\to X}) \times \exp[-(W_{A\to B} + W_{A\to X})\Delta t]$ and $f_B(\Delta t) = (W_{B\to A} + W_{B\to Y})\exp[-(W_{B\to A} + W_{B\to Y})\Delta t]$. To end up at X after $n$ cycles A → B → A takes $n + 1$ intervals from $f_A$ and $n$ intervals from $f_B$. To end up at Y after A → B and $n$ cycles B → A → B takes $n + 1$ intervals from $f_A$ and $n + 1$ intervals from $f_B$.

#### 8.1.4.4 Fast Diffusion

The analytical approaches for adsorption-desorption and fast equilibria are not always possible for diffusion. To see the problem let's look at a simple model. Suppose we have a single CO molecule that can hop between a number of sites. Some of these sites are next to sites occupied with oxygen atoms. If the CO molecule is at such a site, it can react with on oxygen atom to form $CO_2$. The situation is sketched in Fig. 8.5 for a square lattice of adsorption sites.

The problem is that, because of the irregular shape of the set of sites that CO can hop to, it will be very hard to derive expressions for finding CO at the edge sites where it can react with oxygen. Of course, it is possible to set up a transition matrix $\mathbf{P}$ as in Eqs. (8.11) to (8.13), but it is doubtful that explicit use of such a matrix is really more efficient than a standard kMC simulation. Moreover, the matrices $\mathbf{P}^n$ may give the probabilities of finding the CO molecule at different sites after a number of steps, but that still leaves the problem of determining the path. Note that each site has a probability to hop to a particular neighboring site that depends on the number of neighbors. Also the time between hops depends on that number. Hence knowing the number of hops/steps in a path alone does not suffice for determining the time until the reaction.

Oppelstrup et al. have dealt with this problem by splitting the whole path in pieces, each of which can be treated analytically [10–12]. The idea is to define so-called protective domains (see Fig. 8.5). Inside such a domain the particle freely moves and diffusion is the only process possible. The shape of the domain is such that the diffusion can be treated analytically. This is used to determine the time that the CO molecule reaches the boundary of the domain for the first time. (Hence the name for their method: first-passage kMC.) The system is then advanced to that time. Then it is checked if the CO will react at its new site. If not, a new protective domain is defined and the procedure is repeated. Otherwise, CO will react and be removed from the surface.

The example of CO diffusion and $CO_2$ formation is really just a simple application of the method of Oppelstrup et al. [10–12]. They have extended their approach to many diffusing adsorbates that can react with each other. An important aspect of their approach is that in such situation it is not necessary to update the position and protective domains of all adsorbates when one of them reaches the boundary of its protective domain. Only the positions of that adsorbate and neighboring adsorbates need to be updated. This asynchronous algorithm makes the method size-independent and efficient.

The method works best if the protective domains are large, because then many diffusional hops can be treated in one fell swoop. This is the case at low coverages. At high coverages and low reaction rate constants the method is not useful.

#### 8.1.4.5 Algorithmic Aspects

Determining the number of reactions that take place in $C(\alpha_0)$ before a configuration in $K(\alpha_0)$ is reached is a selection problem (see Sect. 3.3.1). Suppose that the probability that $n$ steps take place in $C(\alpha_0)$ equals $S_n$. We define the cumulative probability

$$C_n = \sum_{m=0}^{n} S_m \tag{8.34}$$

and a uniform deviate $r \in [0, 1)$. The number of steps $n$ is given by

$$C_{n-1} < r \le C_n \tag{8.35}$$

with $C_{-1} = 0$ by definition.

It may be possible to do this selection easily by using the fact that there is a simple expression for $C_n$. For example, for the adsorption-desorption equilibrium case (Sect. 8.1.4.2) the probability of the number of steps is given by $P_{\text{rx}} P_{\text{des}}^n$ with $n$ the number of A $\to$ V $\to$ A cycles. So the number of steps is $2n + 1$ and $S_{2n+1} = P_{\text{rx}} P_{\text{des}}^n$. We also have $S_{2n} = 0$, because after $2n$ steps the system is in configuration V from which no reaction is possible. Consequently

$$C_{2n} = C_{2n+1} = P_{\text{rx}} \frac{1 - P_{\text{des}}^{n+1}}{1 - P_{\text{des}}}. \tag{8.36}$$

If we now look for $C_{n-1} < r \le C_n$ with $r \in [0, 1)$ a uniform deviate, then we can easily determine $n$ by solving

$$P_{\text{rx}} \frac{1 - P_{\text{des}}^{\mu+1}}{1 - P_{\text{des}}} = r \tag{8.37}$$

for $\mu \in \mathcal{R}$. The number of cycles is simply the largest integer smaller or equal to $\mu$.

A similar solution can be found for the fast equilibrium (Sect. 8.1.4.3). Instead of one summation, we have two: one for the paths that end up in X and one for those that end up in Y. We first do a simple selection for whether we end up in X or in Y, and then use a summation as for the adsorption-desorption case. For the situation of the fast diffusion the situation is not so easy however, because there is no simple expression for the probability of the number of processes in $C(\alpha_0)$, and we need to work with Eq. (8.35).

The determination of the time when the system exits $C(\alpha_0)$ has been presented so far as a summation of the time intervals of all the individual processes. When the number of these processes becomes large, this is a costly procedure. In fact, one may wonder if treating the problem as a single multistep process is really so much more efficient than a normal kMC simulation. It is therefore useful to be able to determine the time when the system exits $C(\alpha_0)$ using a (hopefully small) number of random number generations that does not depend on the number of processes in the path in $C(\alpha_0)$.

The examples above show that many of the time intervals are taken from the same probability distribution. We therefore will look here at the probability distribution of the sum

$$\sum_{m=0}^{n} \Delta t_m \tag{8.38}$$

with all $\Delta t_m$ taken from the same probability distribution

$$P_1(\Delta t) = W e^{-W \Delta t}. \tag{8.39}$$

The subscript of $P_1$ indicates that this is a probability distribution for a single process. For two subsequent processes we have

$$P_2(\Delta t) = \int_0^{\Delta t} d\tau \, P_1(\Delta t - \tau) P_1(\tau) = \frac{W^2}{2} \Delta t e^{-W \Delta t}. \tag{8.40}$$

Using

$$P_{n+1}(\Delta t) = \int_0^{\Delta t} d\tau P_1(\Delta t - \tau) P_n(\tau) \tag{8.41}$$

it can be shown that

$$P_n(\Delta t) = \frac{W^n}{n!} \Delta t^{n-1} e^{-W\Delta t}. \tag{8.42}$$

To determine the time $t_x$ that the system exits $C(\alpha_0)$ after $n$ processes we have to solve

$$\int_0^{t_x} d\tau P_n(\tau) = r \tag{8.43}$$

with $r$ a uniform deviate on the unit interval.

For large $n$ the integral in Eq. (8.43) does not yield a convenient expression. Instead it seems better to use tabulation. We write

$$\int_0^t d\tau P_n(\tau) = \frac{W^n}{n!} \int_0^t d\tau \tau^{n-1} e^{-W\tau} = \frac{1}{n!} \int_0^{Wt} dx x^{n-1} e^{-x} = \pi_n(Wt) \tag{8.44}$$

with

$$\pi_n(\tau) = \frac{1}{n!} \int_0^{\tau} dx x^{n-1} e^{-x}. \tag{8.45}$$

So we compute and tabulate $\pi_n$ for many values of $n$ and then use that tabulation to solve Eq. (8.43).

### *8.1.5 Two Other Approaches*

The time step in all of the algorithms that we have discussed so far, except the one in Sect. 3.7.1, is not under our control. This is sometimes a disadvantage: e.g., when one wants to determine the temporal dependence of a probability distribution. In such a situation one would like to generate data at fixed points in time. Of course, one can use the fixed time step method of Sect. 3.7.1, but that is very inefficient. Moreover, the time step of that method is in general much smaller than the difference in time between two subsequent data sampling moments.

Kantorovich has investigated the possibility of an algorithm with a fixed large time step [9]. Because the average times between the occurrence of two subsequent processes is generally smaller than the desired time step, this algorithm simulates also multistep processes. The difference between the approach in Sect. 8.1.4 is the way that the path of subsequent processes ends. In Sect. 8.1.4 this was determined by the occurrence of a particular slow process. Here it is determined by the system reaching a certain time.

The approach is based on the integral formulation of the master equation.

$$\mathbf{P}(t) = \left[ \mathbf{Q}(t,0) + \int_0^t dt' \mathbf{Q}(t,t')\mathbf{W}(t')\mathbf{Q}(t',0) \right.$$
$$\left. + \int_0^t dt' \int_0^{t'} dt'' \mathbf{Q}(t,t')\mathbf{W}(t')\mathbf{Q}(t',t'')\mathbf{W}(t'')\mathbf{Q}(t'',0) + \ldots \right]$$
$$\times \mathbf{P}(0) \tag{8.46}$$

with

$$\mathbf{Q}(t',t) = \exp\left[ - \int_t^{t'} dt'' \mathbf{R}(t'') \right], \tag{8.47}$$

$$\mathbf{W}_{\alpha\beta} = W_{\alpha\beta}, \tag{8.48}$$

and

$$\mathbf{R}_{\alpha\beta} = \begin{cases} 0, & \text{if } \alpha \neq \beta, \\ \sum_\gamma W_{\gamma\beta}, & \text{if } \alpha = \beta \end{cases} \tag{8.49}$$

(see Sects. 3.2.1 and 3.6). Note that this is the form for time-dependent rate constants. If the rate constants are time independent, then the argument of the matrix $\mathbf{W}$ can be left out and $\mathbf{Q}(t,t')$ depends only on the difference $t - t'$ (see Sect. 3.2.1). The different terms on the right-hand-side of Eq. (8.46) stand for no processes taking place between times 0 and $t$, one process taking place, two processes taking place, et cetera.

Kantorovich proceeds by computing all of the contributions to $\mathbf{P}(t)$ analytically up to a preset maximum number of processes. This definitely yields the most detailed information on the evolution of the system, but it is very hard to do in general. By taking the time step not too large, the maximum number of processes can be chosen small, and the whole method may become feasible. However, it remains to be seen how useful the method is from a practical point of view.

The $\tau$-leap method also simulates many processes as one [13–15]. It was developed for Dynamic Monte Carlo (DMC) simulations of rate equations, where only information on the total number of chemical species is used. The processes that are taken together are assumed to be independent of each other: i.e., the occurrence of one does not affect the others. In general, this is an approximation. It has been shown however that the errors thus introduced are negligible provided the relative change in the number of chemical species at each step is small [13]. This is the so-called leap condition.

In normal lattice-gas kMC it seems impossible to fulfill this leap condition. A site is either vacant or occupied and relative changes are always as large as they can be. Vlachos has tried to developed a method that leads to a much less restrictive leap condition [16]. He groups all processes with the same rate constant together. The $\tau$-leap method is then applied to the processes in one group. The leap condition then becomes that the number of processes occurring in one time step must be small compared to the total number of processes in a group, but this number may still be large.

The method has so far only been used for one system [16]. There it showed substantial savings in computer time. It was also noted however that the spatial information in kMC, but absent in DMC, causes new problems for the $\tau$-leap method.

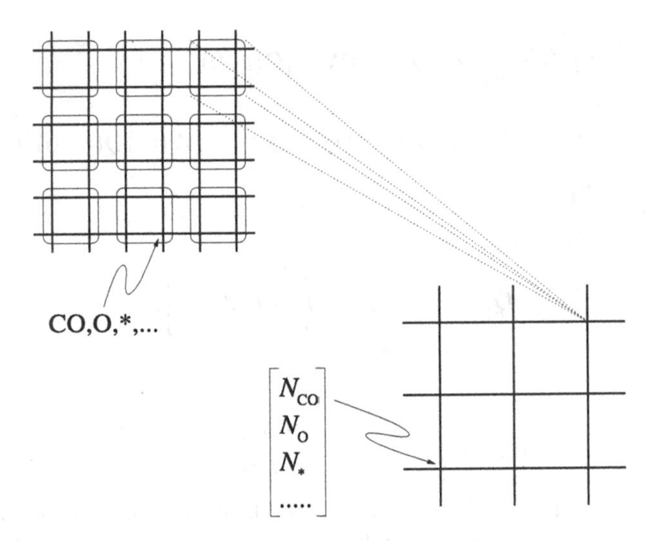


**Fig. 8.6** Sketch of the coarse-graining of a lattice. The *top-left* part shows a square lattice with each lattice point representing an adsorption site. This lattice is partitioned in $2 \times 2$ blocks, and each of these blocks is represented by a single lattice point of the new lattice on the *bottom-right*. Each lattice point of the original microscopic lattice has a label indicating the occupation of the adsorption site. Each lattice point of the new coarse-grained lattice has information on the number and type of adsorbates in the block. (The example here is of CO oxidation with sites being either vacant $*$, occupied by CO, or occupied by atomic oxygen)

For example, it may occur that during a step two diffusional hops are chosen that bring two different adsorbates to the same site, which should not be possible. Strong spatial correlation in general (e.g., when one has pattern formation or an ordered adlayer) might pose problems for the approach. In individual causes a solution to these problems may be found, but it is not clear yet how easy it is to use the method for an arbitrary system.

## 8.2 Larger Length Scales

Very often the characteristic length scale of a system is small: e.g., the correlation length of the occupation of sites is generally only a few times the distance between neighboring sites. If the substrate has however a long characteristic length scale or when there is pattern formation one may want to simulate a larger system than is possible with conventional kMC.

A possible approach is to use coarse-graining. The idea is to replace the lattice of sites by a lattice with a larger lattice spacing. Instead of a lattice point representing a single site, points of the new lattice represent a block of sites. Such a point does not specify an occupation of a single site, but for each type of adsorbate the number of such adsorbates in the block of sites is given (see Fig. 8.6).

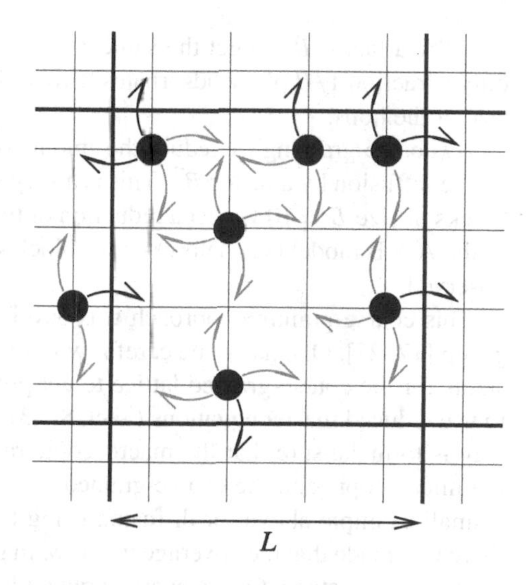


**Fig. 8.7** The *arrows* show some diffusional hops and what happens with them when we do coarse-graining with a $4 \times 4$ block as shown. The *gray arrows* do not take an adsorbate outside the block and can be ignored, because they do not change the number of adsorbates in the block. The *black arrows* do take them to another block and do need to be simulated

An immediate advantage of blocking the lattice is that much less time needs to be spent on simulating diffusion. There are two reasons for this. The first, and most obvious, is that diffusional hops within the blocks need not be simulated. They do not change the numbers of the adsorbates within a block (i.e., they don't change the configuration of the new coarse-grained lattice), and can therefore be ignored. We can estimate how much time this reduces the amount of time spent on simulating diffusion. Suppose we have a square lattice of adsorption sites that we partition in square blocks of $B \times B$ lattice points. We only need to simulate diffusion that takes an adsorbate from one block to another. So hops originating from the $(B-2)^2$ lattice points in the interior of the block can be ignored. Only hops from the $4(B-1)$ sites on the edges are important. However, also hops from these sites that remain in the same block can be ignored. This means that from a site at the corner of a block hops in two directions need to be taken into account, and from the other sites on the edges only hops in one direction (see Fig. 8.7). There are four corner sites which account for 8 hops and $4(B-2)$ other sites at the edges accounting for as many hops. So in total $4B$ hops need to be simulated. Normally each site has four possible hops, so that a block has $4B^2$ hops. We see that we get a reduction of a factor $B$ in the number of hops that we need to simulate. This number is a consequence of the different ways in which the number of sites in a block and the number of sites on the edges scale with the size of the block. Different partitions have the same scaling and yield therefore the same reduction.

There is yet another reduction factor which is equal to $B$, and which is a consequence of the fact that the lattice points of the coarse-grained lattice are a factor $B$ farther apart than those in the microscopic lattice. The hops are therefore larger. An alternative way to see this is that the diffusion constant should be the same for both lattices. As has been shown in Sect. 4.6.5 we have $D = W_{\text{hop}}a^2$ in the limit of low coverages with $D$ the diffusion constant, $W_{\text{hop}}$ the hopping rate constant, and $a$ the distance between the sites. Increasing the latter by a factor $B$ means decreasing

$W_{\text{hop}}$ by a factor $B^2$ to get the same $D$. One factor $B$ originates from the fact that only a fraction $1/B$ of the adsorbates leaves a block and the other from the increased size of the hops.

By coarse-graining we reduce the amount of computer time that we have to spend on the diffusion by a factor $B^2$. This is a very substantial reduction. Even with small blocks of size $B = 10$ we get a reduction of two orders of magnitude. In simulations of the A + B model (see below) with a block size of $B = 32$ a speed up of 1000 was observed.

This coarse-graining approach was developed most extensively by the Vlachos group [17–27]. One has to be careful with the definition and values of the rate constants for the coarse-grained lattice to get proper simulations. It would be possible to use reduced master equations (Sect. 8.1.3), but an alternative and probably easier way is to make sure that the macroscopic rate equations that can be derived from the microscopic and the coarse-grained lattice are the same [25]. We already gave a small example above for diffusion using the diffusion constant. The assumption there was made that the coverage was low. In general the coverage can have an effect on the rate constants for the coarse-grained lattice however.

Let's look at simple desorption. We assume that there is only one type of adsorbate and that desorption involves only one site, no lateral interactions, and that there is only one site in the unit cell. We can show that

$$\frac{dN}{dt} = -W_{\text{des}} N \tag{8.50}$$

with $N$ the number of adsorbates in a block of sites, and $W_{\text{des}}$ the rate constant for desorption for the microscopic lattice (see Sect. 4.6.2). It looks as if the rate constant for the coarse-grained lattice is the same as the one for the microscopic lattice. Indeed, we reached a similar conclusion in Sect. 4.6.2 with respect to the rate constant for macroscopic rate equations. However, we want to do here kMC simulations with the coarse-grained lattice, and in these simulations elementary desorption events reduce the number of adsorbates by one, although the rate of desorption is proportional to the number of adsorbates as the expression above shows. This means that the rate constant to be used in a kMC simulation with a coarse-grained lattice should be $W_{\text{des}} N$. Similar changes to the rate constants can be derived for other processes. For an extensive discussion see reference [25].

For processes that involve only one site (simple desorption, simple adsorption, isomerization) the macroscopic rate equations have a closed form. For processes that involve two or more sites that is no longer the case. This is related to an important weakness of the approach. The problem is that the kinetics within the blocks is not completely determined only by the number of adsorbates. The way adsorbates are distributed within blocks is important as well. The problem is the same as the one we encountered for the derivation of the macroscopic equations for bimolecular reactions (Sect. 4.6.6). One therefore has to introduce an approximation. The most widely used approximation is Mean Field (MF): i.e., one assumes that the adsorbates are randomly distributed over the sites in a block. It is not clear if this is good enough, as can be shown for the simple A + B model [28]. In this model there

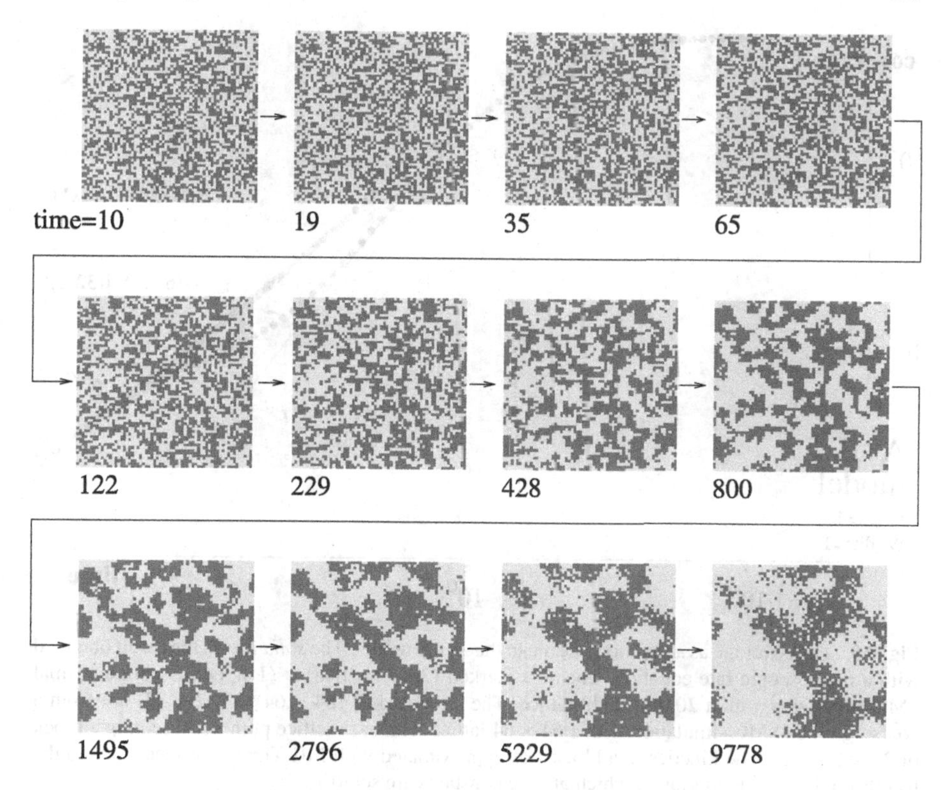


**Fig. 8.8** Snapshots taken at different times for the A + B model. The snapshots show A- and B-areas. A site belongs to an A-area when it is occupied by an A or when the nearest adsorbate is an A. A similar definition holds for B-areas. It can be seen that the adsorbates segregate in spite of diffusion. As a consequence reactions takes place only where an A-area meets a B-area. Time is in seconds, and the snapshots are from a simulation in which all rate constants are set to 1 $s^{-1}$. (Color figure online)

is only one reaction: an A and a B next to each other can react to form an AB that immediately desorbs. The A's and B's also diffuse by hopping to neighboring sites. The model shows anomalous kinetics. Because the adsorbates can only react to form a product that desorbs, the coverages decrease in time. If we start with equal numbers of A and B, macroscopic rate equations predict that at large times $t$ the coverage decreases as $t^{-1}$. Scaling arguments show however that it should be as $t^{-1/2}$ [28]. The reason is that the adsorbates segregate and the reaction only occurs where areas with only A's and those with only B's meet (see Fig. 8.8). kMC simulations confirm this behavior.

One needs large lattices to study the low coverages in the A+B model, and therefore one might want to do coarse-graining. If one uses a Mean Field Approximation (MFA) for the blocks, then one gets the following results (see Fig. 8.9). Initially, the A's and B's are randomly distributed and the rate equations and both simulations (normal and coarse-grained kMC) give identical results. When the areas with only A's or only B's are larger than the size of the blocks, then the coarse-grained

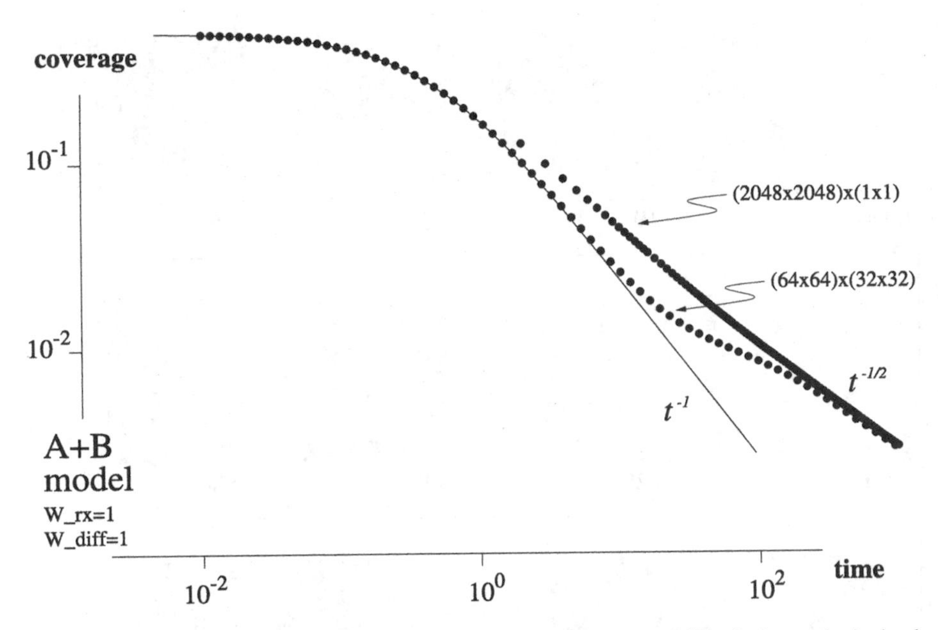


**Fig. 8.9** The coverage as a function of time in the A + B model. The *solid line* is the result obtained with a macroscopic rate equation. The *dots* marked (2048 × 2048) × (1 × 1) are from a normal kMC simulation with a 2048 × 2048 lattice. The *dots* marked (64 × 64) × (32 × 32) are from a coarse-grained kMC simulation with a 64 × 64 lattice with each lattice point representing a block of 32 × 32 sites. The kinetics in a block was approximated with MF. Time is in seconds, and the results are from a simulation in which all rate constants are set to 1 $s^{-1}$

simulation also gives the same results as the normal kMC simulations, because the kinetics is determined by the boundaries of the areas, which are well approximated by the blocks. Between these limits (very early or very late in the simulations) normal and coarse-grained kMC give different results. After the segregation has started, but when the areas are still smaller than the blocks, the coarse-grained simulation follows the results of the macroscopic rate equation instead of the normal kMC however. The reason is that the adsorbates in a block are not randomly distributed and MF does not hold.

A possible solution to this problem is to use a two-step procedure. First, normal kMC simulations are done for the blocks. These simulations are small, so this does not add too much computational effort to the coarse-grained simulation. These kMC simulations are used to determine the kinetics within the blocks. Second, the kinetics thus obtained are used in the coarse-grained simulations. This two-step procedure works well for the A + B model: the result is the same as for a normal kMC simulations with the full lattice. This approach is very similar to the one Mastny et al. used in their approach with reduced master equations (see Sect. 8.1.3) [3], and Chatterjee et al. for kMC simulations of adsorption and desorption [22]. For more complex reaction systems this has not been used yet.

Figure 8.10 shows the result of a large number of kMC simulations have been done for the A + B model with different initial numbers of A's and B's. The lattice

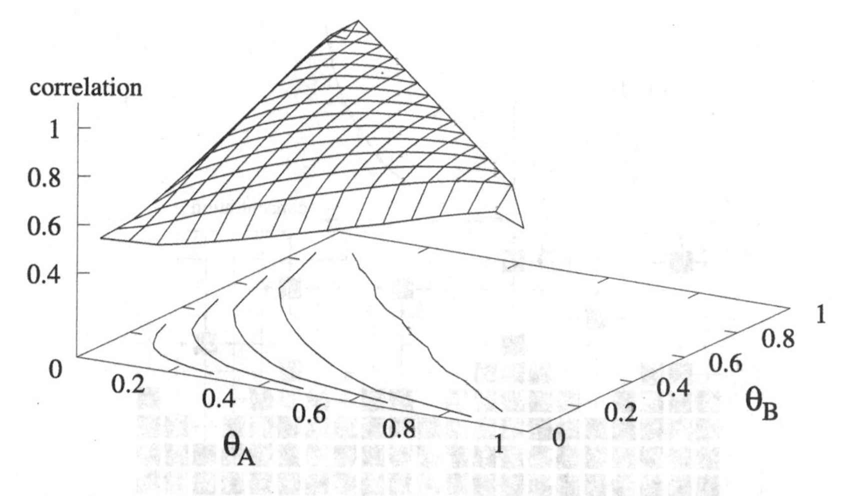


**Fig. 8.10** Correlation $F_{AB}$ in the occupation of neighboring sites for the A + B model. The figure shows the probability that one site of a pair of neighboring sites is occupied by an A and the other by a B. This correlation is related via $W_{rx}\theta_A\theta_B F_{AB}$ to the rate of the reaction, where $\theta_X$ is the coverage of X and $W_{rx}$ is the rate constant of the reaction. The correlation was determined from simulations in which all rate constants were set to 1 $s^{-1}$

size was 128 × 128. These simulations were used to construct a table with the rate of the A + B reaction for any number of A's and B's in a 32 × 32 block. This information was then used in the coarse-grained kMC simulation. This approach is not easy to use for more complicated systems. The A + B model has only two reacting species, so the table is only two-dimensional and therefore small. Moreover, it was also possible to do kMC simulations for the blocks with any possible combination of numbers of A's and B's.

For more complicated systems it may well be necessary to only study those coverages that are really relevant. It is also necessary to have another representation of the rates as a function of the coverages. Such a representation must be flexible so that it can incorporate incomplete and noisy data of kMC simulations. It must also be able to inter- and extrapolate this data to be used in coarse-graining kMC simulations. Neural networks might be useful, but this is very much unexplored territory [29, 30].

## 8.3 Embedding kMC in Larger Simulations

The kMC simulations that we have discussed focus on the processes taking place on the surface of a catalyst. In a real catalytic system there are also other processes: in particular, processes dealing with transfer of matter and heat. It is quite possible that they affect the performance of a catalyst as much as the surface processes. Some studies have been done to extend the kMC simulations to include these other processes.

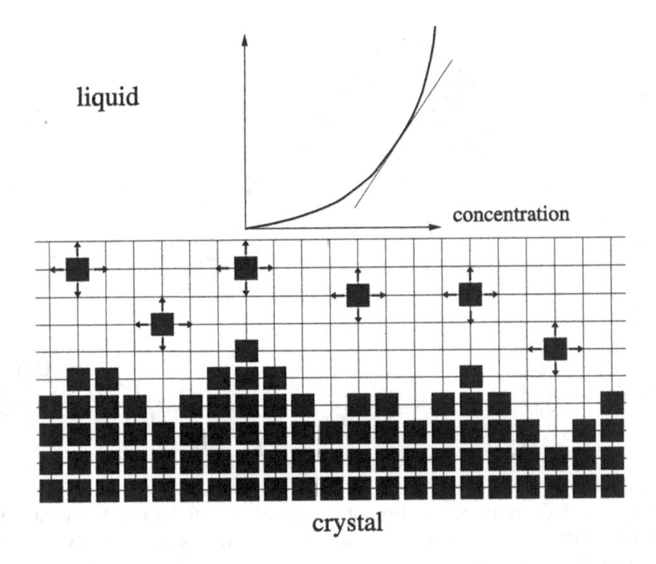


**Fig. 8.11** A lattice-gas model of the surface of a growing copper crystal coupled to a continuum model representing the liquid [31–35]. The *bottom* part shows a two-dimensional cross section of a three-dimensional lattice. The *black squares* represent the copper atoms. The isolated copper atoms depict atoms still in solution. The *arrows* indicate hops that model the diffusion of these atoms. The flux of the atoms through the interface between the lattice gas and the continuum model can be computed from the difference in frequency of upward and downward hops. The *top* part shows the continuum model that specifies the concentration of the copper atoms in solution as a function of distance to the surface. The flux through the interface can be computed in this model from the gradient of the concentration. Both fluxes should be equal

The main focus has been on the mass transport from the liquid or gas phase to the surface. This has received most attention not in catalysis, but in crystal growth. In a series of papers from the University of Illinois the electrodeposition of copper was investigated [31–35]. The diffusion of the copper in the liquid was modeled with a continuum diffusion equation. This is a typical approach in modeling the system except for the surface processes. The diffusion model was one-dimensional: only the direction perpendicular to the surface was modeled and the concentration of copper was assumed to be independent on the other directions. The diffusion equation was coupled to a three-dimensional kMC simulation. The third dimension was used in the kMC simulation to allow for the growth of the crystal.

A major concern in simulations where kMC is combined with a continuum model is the coupling between the methods. For the copper deposition this was done as follows (see Fig. 8.11). The bottom part of the third dimension of the lattice in the kMC simulation was used for the growing copper crystal, but the top part was used to model the part of the liquid in contact with the surface. The diffusion of the copper in that part of the liquid was modeled as hops on the three-dimensional lattice. The coupling was accomplished by equating the flux of the copper in the liquid through the interface between the continuum part and the kMC lattice. In the continuum part this flux can be derived from the gradient of the concentration perpendicular to the surface. In the lattice part this flux can be computed by averaging over the number of hops perpendicular to the surface.

Besides mass transport, heat transfer can also be an issue. This was investigated for pure heat transfer and a model of desorption by Castonguay and Wang [36]. Two approaches were tested. The first incorporated heat into the kMC simulations by modeling it as so-called thermal bits that could be exchanged between lattice points. The second used a Poisson heat equation that was solved between occurrences of regular processes. Both approaches were shown to give the same accurate results, but the approach with the Poisson heat equation was faster.

Matera and Reuter took this approach to a new level by including both mass and heat transport [37, 38]. These were modeled by the transient Navier–Stokes equation and conservation of energy and atomic species. They applied their method to a ab-initio model of CO oxidation on $RuO_2(110)$. This process had been studied before and a comparison between the simulations with and without modeling the mass and heat transfer showed substantial differences [39, 40]. The surface kinetics was sometimes hidden in the overall reactivity by transport limitations, and the overall kinetics became more complicated because of various steady states that arose solely from the coupling between the gas transport and surface processes.

The approaches described above all modeled only a relatively small patch of a catalyst's surface compared to what is found in a reactor. A much more ambitious approach was developed by Majumder and Broadbelt [41]. They modeled a whole flow reactor. (Actually, they modeled only a pore in the support with the catalyst deposited on the inner surface of the pore. The model however can just as well be used for a whole flow reactor. The main difference is the size of the system.) This reactor was envisioned as a pipe (see Fig. 8.12). The flow was described by

$$u\frac{\partial C_i}{\partial z} = D_i\frac{\partial^2 C_i}{\partial y^2} \tag{8.51}$$

with $C_i$ the concentration of component $i$, $D_i$ its diffusion constant, $u$ the local velocity of the fluid phase, $z$ the coordinate along the pipe, and $y$ the coordinate perpendicular to the surface. Boundary conditions were $C_i = C_i^{(0)}$ for $y > 0$ and $z = 0$ with $C_i^{(0)}$ a constant, $\partial C_i/\partial y = 0$ at the center of the pipe, and

$$D_i\frac{\partial C_i}{\partial y} = R_i^{(\text{ads})} - R_i^{(\text{des})} \tag{8.52}$$

with $R_i^{(\text{ads})}$ the adsorption rate and $R_i^{(\text{des})}$ the desorption rate of component $i$. This last boundary condition couples the kMC simulation to the continuum model for the liquid.

In a flow reactor the processes on the catalyst change the concentrations in the liquid. This however takes place on a much larger length scale than can be modeled by a single kMC simulation. The reactor was therefore spatially discretized in a grid which was used the solve the continuum model. At the boundary where the catalyst's surface was found, the last boundary condition (8.52) was handled in one of two ways. For some lattice points, periodically spaced, a kMC simulation was done to determine the adsorption and desorption rates. For the other grid points an

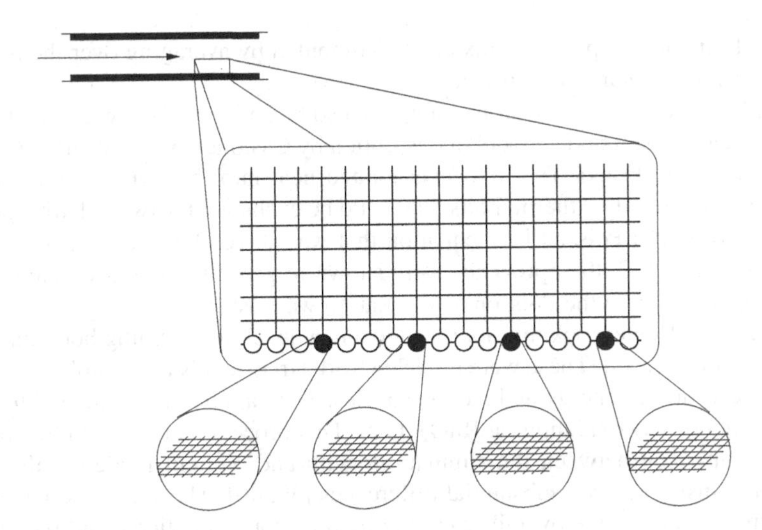

**Fig. 8.12** Modeling a flow reactor [41]. The reactor is modeled as a pipe as shown in the *top-left*. The reactants stream in from the left. The *black strips* depict the catalyst. The *middle* shows an enlargement of part of the reactor including same of the catalyst, and the lattice that is used to solve the continuum equation for the flow. The *points* at the bottom of the lattice represent the catalyst. The *closed spheres* indicate the lattice points that are coupled to a kMC simulations shown at the bottom. The reactivity of the catalyst at the *open spheres* are derived from interpolating kMC results

interpolation scheme was used based on the kMC results of nearest lattice points for which simulations were done.

The approach was used to study two simple models. The first had adsorption of A's on a single site, conversion of the A's into B's, and then desorption of B's. This model was mainly used to make sure that the whole approach gave correct results, as it could also be solved analytically. The other model consisted of adsorption of A's and B's, which together reacted to form C's which immediately desorbed. This model was used to highlight the difference with the conventional approach with rate equations.

## 8.4 Off-lattice kMC

Although lattice-gas models are very versatile, there are situations in which it is simply too difficult or cumbersome to model a system with a lattice. Sometimes it is possible to extend the lattice-gas concept while retaining its advantages. For example, Bos et al. have used multiple lattices to study massive phase transformations metals of fcc to bcc [42, 43]. The two lattices (fcc and bcc) cannot be modeled as one lattice that includes all the fcc and bcc sites. So the atoms are allowed to choose between the sites of an fcc and of a bcc lattice. In addition some random sites are included to make the transition from one to another lattice easier. These sites are dynamically added and removed at positions where they prove to be most needed [43]. This approach has still all the advantages of a model with a single lattice, but it is more flexible. We will look here however at not using a lattice at all.

In the derivation of the master equation

$$\frac{dP_\alpha}{dt} = \sum_\beta [W_{\alpha\beta} P_\beta - W_{\beta\alpha} P_\alpha] \tag{8.53}$$

the subscripts $\alpha$ and $\beta$ refer to catchment regions or basins of attraction of the corresponding minimum of the potential-energy surface (PES) of a system (see Sect. 2.2.2) [44]. This means that it is possible to do kMC simulations in which the processes are hops on the PES from on basin of attraction of the PES to a neighboring one. This can in principle be done for any PES, which would extend the applicability of kMC compared to its use in combination with lattice-gas models enormously. Viewed like this kMC is competitive with a number of methods based on Molecular Dynamics for simulating atomic systems over long time scales: see the reviews of Henkelman et al. and Voter et al. [45, 46].

The way kMC works for hopping on a PES is as follows. Suppose that at a certain time $t_n$ the system is near minimum $\alpha_n$ of the PES. Then a list of hops $\alpha_n \to \beta$ need to be made, with $\beta$ a neighboring minimum: i.e., a minimum that can be reached from the current one along a continuous curve via a single saddle point of the PES. This means determining all saddle points or transition states of the PES surrounding the minimum $\alpha_n$. The rate constant $W_{\beta\alpha_n}$ needs to be computed for each transition state that is found. The transition to another minimum $\alpha_{n+1}$ is then made with a probability proportional to $W_{\alpha_{n+1}\alpha_n}$ and time is advanced to

$$t_{n+1} = t_n - \frac{\ln r}{\sum_\beta W_{\beta\alpha_n}} \tag{8.54}$$

with $r$ a uniform deviate on the unit interval. Alternatively, instead of the VSSM algorithm described here, one of the other algorithms of Chap. 3 can be used just as well.

The important difference with lattice-gas kMC is that the processes are no longer known before a kMC simulation is started, but they have to be determined on-the-fly. This makes this general kMC method much more costly. Determining a transition state for a process is costly in itself, but the task here is especially daunting because in principle all transition states around a minimum have to be determined. Nevertheless, this approach has been used, even in combination with Density-Functional Theory calculations, in the pioneering work by Henkelman and Jónsson [47, 48]. They applied it to diffusion of clusters of Al atoms on an Al surface. This work showed a very important advantage of the method. Because the processes are not precomputed, surprising processes may show up and may turn out to be highly relevant. In Henkelman and Jónsson's work diffusion of the clusters took place with concerted motions that included the atoms of the surface as well. Early application were also done on the formation of defects during crystallization [49], and the segregation of atoms [50].

Although there are methods that can insure finding all transition states around a minimum [51–53], these are extremely costly. Sometimes it is possible, similarly to lattice-gas kMC, to say in advance what the processes are. For example, in a study on diffusion in binary alloys a vacancy could move in only twelve possible directions [54]. Consequently, the transition states were at least approximately known and needed only to be refined. The same approach could be used in a study on H diffusion in iron [55], and in a similar study of vacancy diffusion in crystalline silicon a list could be made during a simulation with topological characteristics of diffusion mechanisms to facilitate finding transition states [56]. In general, however, this is not possible. In their seminal work Henkelman and Jónsson started each search for a new transition state with a random displacement of all atoms and then did 25 to 50 searches from each minimum [47]. In a later study more selective initial displacements were tested [57]. These studies naturally lead to the question how one knows that all transition states have been found.

This question was given a statistical answer by Xu and Henkelman [57]. They argued that if there are $T$ transition states, one has already found $F$ different ones, and each transition state has the same probability of being found, then the probability of having found all transition states equals $F/T$. This equation is not useful as one does not know $T$. However, if one finds only one new transition state per $S$ searches, then $F/T = 1 - 1/S$. So the latter expression can be used as a criterion to stop looking for new transition states. One simply requires $1 - 1/S$ to be larger than some preset value.

Some Bayesian statistics can put this criterion on a firmer footing. The probability that the next transition state search, number $n$, will find a new transition state ($X_n = 1$) is given by

$$P(X_n = 1|F_{n-1}T) = \frac{T - F_{n-1}}{T}, \tag{8.55}$$

and the probability of finding one that has been found before ($X_n = 0$) equals

$$P(X_n = 0|F_{n-1}T) = \frac{F_{n-1}}{T}. \tag{8.56}$$

Here $F_{n-1}$ is the number of transition states found after $n - 1$ searches. From Bayes's theorem

$$P(X_n|F_{n-1}T)P(T|F_{n-1}) = P(X_nT|F_{n-1}) = P(T|F_{n-1}X_n)P(X_n|F_{n-1}) \tag{8.57}$$

we get

$$P(T|F_{n-1}X_n) \propto P(X_n|F_{n-1}T)P(T|F_{n-1}). \tag{8.58}$$

So if we have the probabilities $P(T|F_{n-1})$ that there are $T$ transition states, then a new transition state search will change these probabilities. If it leads to a new transition state, then the probabilities should be multiplied by $(T - F_{n-1})/T$, if not by $F_{n-1}/T$. (The probabilities of course also need to be normalized.) Subsequent transition state searches will make $P(T|F_n)$ sharper peaked. One should stop if the maximum $P(T|F_n)$ becomes larger than some preset value.

Initially we have $F_0 = 0$ and let's assume $P(T|F_0) = 1/M$ with $M$ a number that is larger than the number of transition states. Suppose we do $S$ searches. We can then write

$$P(T|F_S) = \frac{A(T, F_S)}{T^S} \tag{8.59}$$

From Eqs. (8.55) to (8.58) we can see that $A(T, F_S)$ is a product of factors $T - F_m$, $F_m$, and a normalization constant. The $F_m$'s here are the number of transition states found so far during the searches. Because we started with $F_0 = 0$ these factors yield $T(T-1)(T-2)\ldots(T-F_S+1)$. There is one factor for every time that a new transition state was found. For searches that did not yield a new transition state we get some factor $F_m$. Because such a factor does not affect the $T$-dependence of $P(T|F_S)$ we can include these factors in the normalization constant. So we have

$$P(T|F_S) = \begin{cases} N\frac{T!}{(T-F_S)!T^S}, & \text{if } T \geq F_S, \\ 0, & \text{if } T < F_S, \end{cases} \tag{8.60}$$

with $N$ a normalization constant.

The probability $P(T|F_S)$ has a single maximum. If we want to know if we have found all transition states, then we want that maximum to be at $T = F_S$. From

$$\frac{P(T = F_S + 1|F_S)}{P(T = F_S|F_S)} = \frac{F_S^S}{(F_S+1)^{S-1}} < 1 \tag{8.61}$$

we then get

$$S > \frac{\ln(F_S+1)}{\ln(F_S+1) - \ln F_S}. \tag{8.62}$$

This is however only a lower bound for the number of searches that we need to do. It insures that the number of transition states that we have found is the most likely number of transition states, but in general we want the probability that we have found all transition states to be higher: i.e., we will require $P(T = F_S|F_S) > P_{\text{found}}$ with $P_{\text{found}}$ to be some number close to 1. Figure 8.13 shows how many searches one needs to do to achieve such higher probabilities. We see that the number of searches is a factor of two to three higher than the lower bound depending on the actual number of transition states and how certain one wants to be that one has found all transition states. We also see that when there are more transition states that we also need to do more searches per transition state.

An even better criterion can be given if one realizes that one does not really need to known all transition states. One only needs to know that one that corresponds to the process that will take place first. Instead of the VSSM expression (8.54) one can also use FRM and compute a time for each transition state (see Sect. 3.5). Then the problem becomes finding the transition state with the smallest value for that time: i.e., the one corresponding the first process to occur. We will not determine all transition states to compare all times, but we will give a statistical criterion that gives us a probability that we have found the first process to take place.

We define two new stochasts. Stochast $Y_n = 1$ if at transition state search $n$ we find a transition state of an earlier process than any we have found so far, and $Y_n = 0$

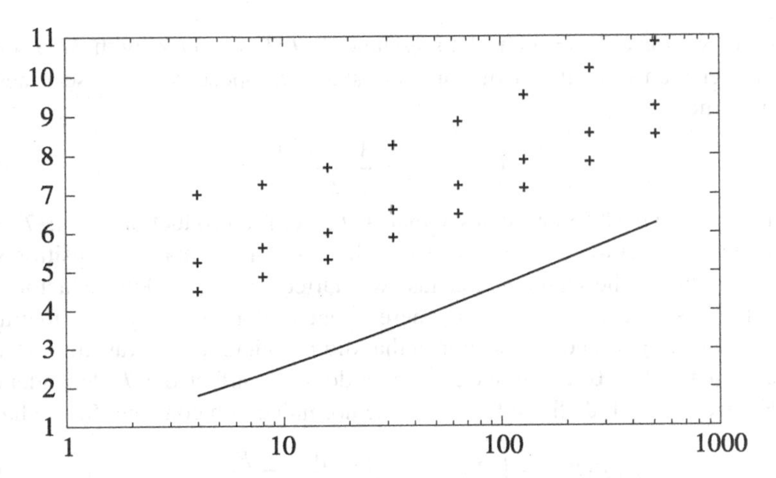


**Fig. 8.13** The number of transition states searches per transition state that are needed to achieve a probability of 0.99, 0.95, or 0.9 (*markers* from top to bottom) that all transition states have been found as a function of the actual number of transition states. The *curve* indicates the situation that the number of transition states that have been found is the most likely number of transition states

if that is not the case. Stochast $R_n$ is the rank number of the transition state that has been found after $n$ searches corresponding to the earliest process: i.e., the first of all processes has rank $R_n = 1$, the second $R_n = 2$, et cetera. Of course, we do not know the value of $R_n$, but we can make some statistical inferences from finding new transition states corresponding to earlier processes.

We have

$$P(X_n Y_n | T F_{n-1} R_{n-1}) = \begin{cases} \frac{R_{n-1}-1}{T}, & \text{if } X_n = Y_n = 1, \\ \frac{F_{n-1}}{T}, & \text{if } X_n = Y_n = 0, \\ \frac{T-(R_{n-1}-1)-F_{n-1}}{T}, & \text{if } X_n = 1 \text{ and } Y_n = 0. \end{cases} \tag{8.63}$$

Note that $X_n = 0$ and $Y_n = 1$ is not possible: if we find a transition state of an earlier process it must also be a new transition state. Bayes's theorem then gives

$$P(T R_{n-1} | F_{n-1} X_n Y_n) \propto P(X_n Y_n | T F_{n-1} R_{n-1}) P(T R_{n-1} | F_{n-1}). \tag{8.64}$$

This is however not quite what we want. We do not want the probabilities for $R_{n-1}$ but for $R_n$. If $X_n = 0$ or $Y_n = 0$, then there is no new transition state of an earlier process, and $P(T R_n | F_{n-1} X_n Y_n) = P(T R_{n-1} | F_{n-1} X_n Y_n)$. If $Y_n = 1$, and hence $X_n = 1$, then the new transition state corresponds to an earlier process: i.e., it has a rank number smaller than $R_{n-1}$. Of course, we do not know the value of $R_n$, but we can use the same assumption as before that all transition states with $R_n < R_{n-1}$ are equally likely. This means that the probabilities $P(T R_{n-1} | F_{n-1} X_n Y_n)$ should be equally divided into the probabilities $P(T R_n | F_{n-1} X_n Y_n)$ with $R_n < R_{n-1}$, or

$$P(T R_n | F_{n-1} X_n Y_n) = \sum_{R_{n-1}=R_n+1}^{T} \frac{1}{R_{n-1}-1} P(T R_{n-1} | F_{n-1} X_n Y_n). \tag{8.65}$$

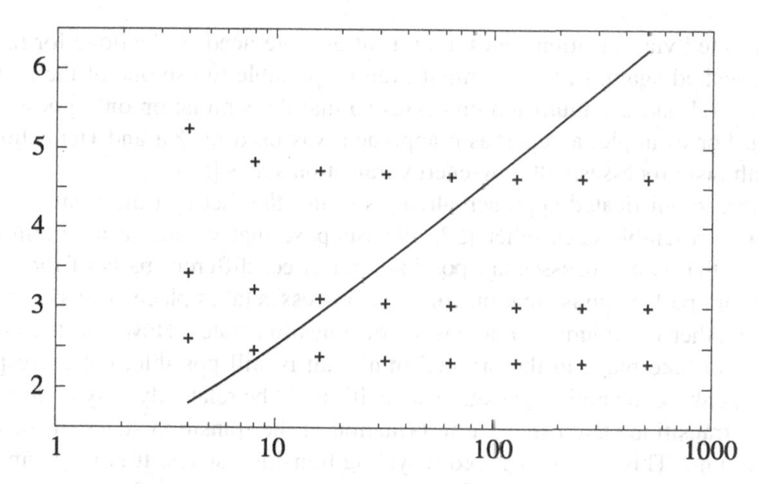


**Fig. 8.14** The number of transition states searches per transition state that are needed to achieve a probability of 0.99, 0.95, or 0.9 (*markers* from top to bottom) that the first process to occur has been found as a function of the actual number of transition states (see Fig. 8.13). The *curve* indicates the situation that the number of transition states that have actually been found is also the most likely number of transition states

This expression will give a strong concentration of probabilities for lower values of $R_n$. This is advantageous, because we really want to know the probability that we have found the first process of all: i.e., we want $P(T, R_n = 1|F_{n-1}X_nY_n)$.

We start as before with assuming that $P(T|F_0) = \sum_{R_0=1}^{T} P(TR_0|F_0) = c/M$ with $c$ a normalization constant. We also assume that all values for $R_0$ are equally likely, so $P(TR_0|F_0) = c/MT$. Figure 8.14 shows the number of transition search that needs to be done to find the first process with different probabilities. We notice two differences with respect to trying to find all transition states. First the number of searches is significant smaller. The curve in Fig. 8.14 is the same one as in Fig. 8.13 and indicates a lower bound for the number of searches to find all transition states. If there are many transition states we need not even do so many searches if we only want to find the first process to occur. This makes sense, because it is of course harder to find all transition states than to find one particular transition state. The second differences is that the number of searches per transition state does not increase with the actual number of transition states, but seems to be constant for many transition states and even decreases if there are few transition states. This is a more favorable behavior than was seen for finding all transition states.

Xu and Henkelman also looked at the case when not all transition states are equally likely to be found, which is more realistic for the methods that are generally used in the transition state searches. They have a higher probability of finding low-energy transition states.

Some other techniques have been used to deal with the problem of finding transition states. An obvious one is to save transition states that have already been found [57, 58]. For example, when a process has just occurred, the transition state for the reverse process is then of course known. By constructing the graph of minima that are connected via transition states, no transition state needs to be done for minima that are visited again. In fact, it might even be possible to use one of the methods in Sect. 8.1.4 and use multistep processes so that the simulation only goes to new minima. For example, a superbasin approach was used by Xu and Henkelman to deal with fast processes with low-energy transition states [57].

A more sophisticated approach attempts to use the fact that different transition states may resemble each other [57, 59]. Suppose that we are at a certain minimum and that two processes are possible that affect different parts of the system that are far apart. Suppose that one of these processes takes place. The system then gets to another minimum with a new set of transition states. However, the process that did not take place in the original minimum is still possible, but corresponds strictly speaking to another transition state. It should be relatively easy to determine this new transition state using the information of the transition states of the original minimum. This has been called recycling transition states. It is very similar to what is done in the algorithms of lattice-gas kMC when lists of processes are not recomputed but only updated.

There have been few studies yet using off-lattice kMC so far. It remains to be seen how big the problem is of finding transition states on-the-fly. Increasing the system size seems to increase the number of transition states exponentially [60]. Recycling may change this unfavorable scaling just as it does for lattice-gas kMC. Much more work needs definitely to be done.

For some systems off-lattice kMC needs not be so much more time consuming than lattice-gas kMC. For reactions in solvents it is possible to use lattice-gas models [61–65]. The drawback is that a lot of computer time is spent on diffusion of particles that need to find each other first before they can react. It has recently been shown that this need not be necessary [66]. It can be assumed that diffusion in the solvent is a Brownian motion. This can be treated analytically, which leads to effective rate constants that depend on the distance between the reacting particles when they are formed and their diffusion constants. Moreover, there is also a factor that is an intrinsic rate constant, which can be computed before doing the kMC simulation. This makes the simulation more efficient than simulations in which rate constants are computed on-the-fly [67, 68].

## References

1. A. Chatterjee, F. Voter, J. Chem. Phys. **132**, 194101 (2010)
2. S.V. Nedea, A.P.J. Jansen, J.J. Lukkien, P.A.J. Hilbers, Phys. Rev. E **65**, 066701 (2002)
3. E.A. Mastny, E.L. Haseltine, J.B. Rawlings, J. Chem. Phys. **125**, 194715 (2006)
4. D.R. Mason, R.E. Rudd, A.P. Sutton, Comput. Phys. Commun. **160**, 140 (2004)
5. R.E. Rudd, D.R. Mason, A.P. Sutton, Prog. Mater. Sci. **52**, 319 (2007)
6. N.G. van Kampen, *Stochastic Processes in Physics and Chemistry* (North-Holland, Amsterdam, 1981)
7. S.X. Sun, Phys. Rev. Lett. **96**, 210602 (2006)
8. S.A. Trygubenko, D.J. Wales, Graph transformation method for calculating waiting times in Markov Chains. http://arXiv.org/abs/cond-mat/0603830 (2006)

9. L. Kantorovich, Phys. Rev. B **75**, 064305 (2007)
10. T. Oppelstrup, V.V. Bulatov, G.H. Gilmer, M.H. Kalos, B. Sadigh, Phys. Rev. Lett. **97**, 230602 (2006)
11. T. Oppelstrup, V.V. Bulatov, A. Donev, M.H. Kalos, G.H. Gilmer, B. Sadigh, Phys. Rev. E **80**, 066701 (2009)
12. A. Donev, V.V. Bulatov, T. Oppelstrup, C.H. Gilmer, B. Sadigh, M.H. Kalos, J. Comput. Phys. **229**, 3214 (2010)
13. D.T. Gillespie, J. Chem. Phys. **115**, 1716 (2001)
14. D.T. Gillespie, L.R. Petzold, J. Chem. Phys. **119**, 8229 (2003)
15. Y. Cao, L. Petzold, M. Rathinam, D.T. Gillespie, J. Chem. Phys. **121**, 12169 (2004)
16. D.G. Vlachos, Phys. Rev. E **78**, 046713 (2008)
17. M.A. Katsoulakis, A.J. Majda, D.G. Vlachos, Proc. Natl. Acad. Sci. USA **100**, 782 (2003)
18. M.A. Katsoulakis, A.J. Majda, D.G. Vlachos, J. Comput. Phys. **186**, 250 (2003)
19. M.A. Katsoulakis, D.G. Vlachos, J. Chem. Phys. **119**, 9412 (2003)
20. A. Chatterjee, D.G. Vlachos, J. Chem. Phys. **121**, 11420 (2004)
21. M.A. Snyder, A. Chatterjee, D.G. Vlachos, Comput. Chem. Eng. **29**, 701 (2005)
22. A. Chatterjee, D.G. Vlachos, J. Chem. Phys. **124**, 064110 (2006)
23. A. Chatterjee, D.G. Vlachos, J. Comput. Phys. **211**, 596 (2006)
24. A. Chatterjee, D.G. Vlachos, J. Comput.-Aided Mater. Des. **14**, 253 (2007)
25. S.D. Collins, A. Chatterjee, D.G. Vlachos, J. Chem. Phys. **129**, 184101 (2008)
26. J. Dai, D. Seider, T. Sinno, J. Chem. Phys. **128**, 194705 (2008)
27. S. Are, M.A. Katsoulakis, A. Szepessy, Chin. Ann. Math., Ser. B **30**, 653 (2009)
28. S. Redner, in *Nonequilibrium Statistical Mechanics in One Dimension*, ed. by V. Privman (Cambridge University Press, Cambridge, 1996), pp. 3–27
29. K. Gurney, *An Introduction to Neural Networks* (University College London Press, London, 1997)
30. J. Hertz, A. Krogh, R.G. Palmer, *Introduction to the Theory of Neural Computation* (Addison-Wesley, Redwood City, 1991)
31. T.J. Pricer, M.J. Kushner, R.C. Alkire, J. Electrochem. Soc. **149**, C396 (2002)
32. T.J. Pricer, M.J. Kushner, R.C. Alkire, J. Electrochem. Soc. **149**, C406 (2002)
33. T.O. Drews, E.G. Webb, D.L. Ma, J. Alameda, R.D. Braatz, R.C. Alkire, AIChE J. **50**, 226 (2004)
34. E. Rusli, T.O. Drews, R.D. Braatz, Chem. Eng. Sci. **59**, 5607 (2004)
35. Z. Zheng, R.M. Stephens, R.D. Braatz, R.C. Alkire, L.R. Petzold, J. Comput. Phys. **227**, 5184 (2008)
36. T.C. Castonguay, F. Wang, J. Chem. Phys. **128**, 124706 (2008)
37. S. Matera, K. Reuter, Catal. Lett. **133**, 156 (2009)
38. S. Matera, K. Reuter, Phys. Rev. B **82**, 085446 (2010)
39. K. Reuter, M. Scheffler, Phys. Rev. B **68**, 045407 (2003)
40. K. Reuter, M. Scheffler, Phys. Rev. B **73**, 045433 (2006)
41. D. Majumder, L.J. Broadbelt, AIChE J. **52**, 4214 (2006)
42. C. Bos, F. Sommer, E.J. Mittemeijer, Acta Mater. **52**, 3545 (2004)
43. C. Bos, F. Sommer, E.J. Mittemeijer, Acta Mater. **53**, 5333 (2004)
44. P.G. Mezey, *Potential Energy Hypersurfaces* (Elsevier, Amsterdam, 1987)
45. G. Henkelman, G. Jóhannesson, H. Jónsson, in *Progress in Theoretical Chemistry and Physics*, ed. by S.D. Schwarts (Kluwer Academic, London, 2000)
46. A.F. Voter, F. Montalenti, T.C. Germann, Annu. Rev. Mater. Res. **32**, 321 (2002)
47. G. Henkelman, H. Jónsson, J. Chem. Phys. **115**, 9657 (2001)
48. G. Henkelman, H. Jónsson, Phys. Rev. Lett. **90**, 116101 (2003)
49. F. Much, M. Ahr, M. Biehl, W. Kinzel, Comput. Phys. Commun. **147**, 226 (2002)
50. T.F. Middleton, D.J. Wales, J. Chem. Phys. **120**, 8134 (2004)
51. K.M. Westerberg, C.A. Floudas, J. Chem. Phys. **110**, 9259 (1999)
52. Y. Lin, M.A. Stadtherr, J. Chem. Phys. **121**, 10159 (2004)
53. Y. Lin, M.A. Stadtherr, J. Comput. Chem. **26**, 1413 (2005)

54. E.A. Bleda, X. Gao, M.S. Daw, Comput. Mater. Sci. **43**, 608 (2008)
55. A. Ramasubramaniam, M. Itakura, M. Ortiz, E.A. Carter, J. Mater. Res. **23**, 2757 (2008)
56. F. El-Mellouhi, N. Mousseau, L.J. Lewis, The kinetic activation-relaxation technique: A powerful off-lattice on-the-fly kinetic Monte Carlo algorithm. http://arXiv.org/abs/0805.2158v1 (2008)
57. L. Xu, G. Henkelman, J. Chem. Phys. **129**, 114104 (2008)
58. D.R. Mason, T.S. Hudson, A.P. Sutton, Comput. Phys. Commun. **165**, 37 (2005)
59. A. Kara, O. Trushin, H. Yildirim, T.S. Rahman, J. Phys., Condens. Matter **21**, 084213 (2009)
60. R.A. Olsen, G.J. Kroes, G. Henkelman, A. Arnaldsson, H. Jónsson, J. Chem. Phys. **121**, 9776 (2004)
61. M. Jorge, S.M. Auerbach, P.A. Monson, J. Am. Chem. Soc. **127**, 14388 (2005)
62. M.H. Ford, S.M. Auerbach, P.A. Monson, J. Chem. Phys. **126**, 144701 (2007)
63. L. Jin, S.M. Auerbach, P.A. Monson, J. Phys. Chem. C **114**, 14393 (2010)
64. A. Malani, S.M. Auerbach, P.A. Monson, J. Phys. Chem. Lett. **1**, 3219 (2010)
65. A. Malani, S.M. Auerbach, P.A. Monson, J. Phys. Chem. C **115**, 15988 (2011)
66. X.Q. Zhang, A.P.J. Jansen, Phys. Rev. E **82**, 046704 (2010)
67. X.Q. Zhang, T.T. Trinh, R.A. van Santen, A.P.J. Jansen, J. Am. Chem. Soc. **133**, 6613 (2011)
68. X.Q. Zhang, T.T. Trinh, R.A. van Santen, A.P.J. Jansen, J. Phys. Chem. C **115**, 9561 (2011)

# Glossary

**absorbing state** A state of a master equation from which there is no transition possible to another state (Sect. 2.2.1). If the master equation describes the evolution of an adlayer, then a state is a configuration and an absorbing state corresponds to poisoning.

**activation barrier** When one or more molecules react the potential energy increases from a minimum value, reaches a maximum, and then decreases to another minimum value. The activation barrier is the difference between the maximum and the initial minimum value. It is an energy threshold that needs to be overcome for a reaction to occur (Sect. 4.2). Sometimes zero-point energy is added to the initial minimum and the maximum to get a zero-point corrected activation barrier (Sect. 4.3.2). The concept is closely related to the activation energy.

**activation energy** A parameter in the Arrhenius expression for rate constants that describes how a rate constant changes with temperature. The activation energy is closely related to the activation barrier, as the latter is the dominant contribution to the activation energy, but they are not quite the same (Sect. 4.2.2).

**adlayer** The layer on top of the substrate where the adsorbates are found.

**adsorption** Process in which an atom or a molecule from the gas phase or from a solution attaches itself to a surface. It is the reverse of a desorption.

**adsorption site** A preferred position for an atom, a molecule, or a molecular fragment on a surface. This position corresponds to a minimum of the potential-energy surface (Sect. 2.1).

**Arrhenius form** An expression that describes the rate constant as depending exponentially on the reciprocal temperature. It has two parameters: the activation energy and the prefactor (Sect. 4.2.2).

**bookkeeping site** A lattice point in a lattice-gas model that does not correspond to an adsorption site, but that is used to store information that simplifies running a simulation (Sect. 5.5.4).

**cluster expansion** A mathematical expression that can be used to express the lateral interactions in an adlayer. It is a sum of terms that correspond to interactions between adsorbates and substrate, between pairs of adsorbates, between triplets of adsorbates, between quadruplets of adsorbates, et cetera (Sect. 4.5.1).

A.P.J. Jansen, *An Introduction to Kinetic Monte Carlo Simulations of Surface Reactions*, Lecture Notes in Physics 856,
DOI 10.1007/978-3-642-29488-4, 

**coarse-graining** A technique in which an atomic scale description is replaced by one with a longer length scale. In kinetic Monte Carlo coarse-graining is used when the (catchment regions of the) minima of a potential-energy surface are used instead of the precise values of the coordinates and velocities of the atoms (Sect. 2.2.2), and when lattice is used with each lattice point corresponding to a block of sites instead of a single site (Sect. 8.2).

**configuration** The lattice points in lattice-gas models that are used in kinetic Monte Carlo all have a label. A particular assignment of values to all the labels in a lattice-gas model is called a configuration. If each lattice point corresponds to an adsorption site (as is often the case), then a configuration is also a particular way to distribute adsorbates over the adsorption sites (Sect. 2.1.3).

**configuration space** If we take all coordinates of all $N_{\text{atom}}$ atoms of a molecular system and put them together to form one vector with $3N_{\text{atom}}$ components, then the vector space formed by such vectors is the configuration space of the molecular system. Potential-energy surfaces are defined on the configuration space, and the space is also important in statistical physics.

**coordination number** The number of neighbors of a adsorption site or an atom (Sects. 4.6.6, 4.6.7, and 6.7).

**coverage** The density of the adsorbates. It is most often defined as the number of adsorbates per number of unit cell of the substrate, but also as the number of adsorbates per number of adsorption sites or per unit area.

**defect** A change in structure that destroys periodicity or translational symmetry (Sect. 5.5.3).

**desorption** A process in which an atom or a molecule detaches itself from a surface and goes into the gas phase or a solution. It is the reverse of an adsorption.

**diffusion** A random motion of an atom, a molecule, or a molecular fragment that is a consequence of collisions with other particles or the interaction with the thermal motion of other particles. It corresponds to random hops from one adsorption site to a neighboring one in a lattice-gas model.

**diffusion coefficient** A parameter describing the rate of diffusion. Apart from a constant is a proportionality factor between the square of a displacement and time (Sect. 4.6.5).

**disabled process** A process that is no longer possible because of changes in the system by other processes (Sects. 3.2.3 and 3.3.2). It is also called a null event.

**dissociation** A process in which a molecule is split into two or more parts. It is the reverse of a formation.

**dividing surface** In the derivation of the master equation phase space is partitioned into regions that correspond to the minima of the potential-energy surface or configurations of an adlayer. The boundaries separating the regions in phase space are called dividing surfaces (see Sect. 2.2.2). The term originates from a derivation of Variational Transition-State Theory in which phase space is split into a part corresponding to reactants and a part corresponding to products separated by a dividing surface.

**dynamic correction factor** Sometimes a process occurs but it is immediately followed by the reverse of the process. A calculation of the rate constants for the processes will give higher values when we regard forward and reverse process actually having taken place, than when we assume that effectively nothing has happened. The latter is regarded the correct approach. The factor that corrects for overestimated rate constants that are obtained from the former approach is called the dynamic correction factor (Sect. 4.4.3).

**Eley–Rideal reaction** Chemical reaction with reactants both on a surface and in the gas phase or in a solution (Sect. 4.4.4).

**enabled process** A process that is possible (Sect. 3.2.3). The term is mainly used to distinguish it from a disabled process.

**event** The actual occurrence of a process (Sects. 2.1 and 3.2.2).

**event list** A list of all possible processes that can occur at a certain moment in a simulation (Sect. 3.2.3). This list includes besides the processes that actual will occur (i.e., the events) also other processes. These other processes never occur, because, before they can occur, the system changes in such a way that they are no longer possible. We prefer the term "list of all processes" in this book. This is less succinct but also less confusing.

**First Reaction Method** An algorithm of kinetic Monte Carlo in which a time is computed for each possible process. The process that actually takes place is then the one with the smallest value for the time (Sect. 3.5).

**flicker process** A process that is very rapidly followed by its reverse process (Sect. 8.1.4).

**floppy molecule** A floppy molecule is a molecule with a potential-energy surface that has many minima that are separated by barriers that can easily overcome by thermal excitation or tunneling.

**fluctuation** Random variations of some property around an average value (Sect. 5.2).

**formation** A process in which atoms or molecular fragments get together to form a molecule. It is the reverse of a dissociation.

**immediate process** A process with an infinitely large rate constant (Sect. 5.3). The time in a simulation does not change when an immediate process takes place. Immediate processes are mainly used for changes in the configuration that are too complicated to be described by a single process. They are often used together with bookkeeping sites.

**interaction model** A particular way to describe the lateral interactions between adsorbates (Sect. 4.5.1).

**inverted list** A list that specifies the processes that are possible for each adsorbate (Sect. 3.3.2). This list can be regarded as the inverse (hence its name) of the event list.

**label** Each lattice point in a lattice-gas model has a label. This label specifies the status of the lattice point (Sect. 2.1.3). We represent labels in this book by strings because of their flexibility (Sect. 5.3). The most common use of labels is to indicate the occupation of adsorption sites, but they have many other uses in particular in combination with bookkeeping sites and immediate processes (Chaps. 5 and 6).

**Langmuir–Hinshelwood reaction** Chemical reaction with all reactants on a surface (Sect. 4.4.1).

**lateral interaction** Interaction between two or more adsorbates (Sect. 4.5). Lateral interactions include direct interactions, but also interactions that are a consequence of adsorbates changing the electronic structure of the substrate or displacing the atoms of the substrate and thereby inducing stress.

**lattice** A collection of discrete points that is invariant under certain shifts or translations (Sect. 2.1). For a simple lattice the vectors connecting the points can be written as a linear combination of basis vectors with integer coefficients. These basis vectors are called primitive vectors. A composite lattice is a collection of two or more interpenetrating simple lattices with the same primitive vectors.

**lattice gas** A lattice with a label for each lattice point (Sect. 2.1).

**long-range order** Strong correlation in the structure of a system over a much longer distance than the range of the interactions in the system (Sects. 7.2 and 7.4.1).

**macroscopic equation** An equation that describes how a statistical average of some property changes with time (Sect. 4.6.1). It is also called the phenomenological equation. It is derived from the master equation. The name indicates that one generally uses it for macroscopic properties, although it also holds for other properties. If the property is a coverage, then it is often also called macroscopic rate equation or just rate equation. Note however that the macroscopic equation is exact, whereas a rate equation often implies that a Mean Field Approximation has been used.

**Markov chain** A sequence of Markov processes.

**Markov process** A process that depends only on the current state of a system and not on the system's history (Sect. 2.2.1).

**master equation** A linear equation that describes how the probabilities of various states of a system change in time (Sect. 2.2.1). The equation conserves the total probability of all states. For kinetic Monte Carlo the states are configurations of a lattice-gas model.

**Mean Field Approximation** Name for various similar approximations. For lattice-gas models it assumes a random distribution of the adsorbates over the adsorption sites.

**noise** Random variations of some property around an average value (Sect. 5.2). Another term for noise is fluctuation.

**null event** A process that is no longer possible because of changes in the system by other processes. We prefer the term "disabled process" in this book (Sect. 3.3.2).

**occupation** A specification of the adsorbate on a site (Sect. 2.1).

**overfitting** A situation in a fitting process where the fit not only describes the underlying data but also the errors in that data (Sect. 4.5.1).

**oversampling** A two-step procedure. In the first step numbers are generated or items chosen with (transition) probabilities that are too high. In the second step there is a correction for the incorrect (transition) probability (Sects. 3.3.1, 3.3.4, and 3.4).

**partition function** A function of the thermodynamic variables that define a statistical ensemble and from which other thermodynamic variables can be calculated (Sect. 4.3).

**phase space** If we take all coordinates and all conjugate momenta of all $N_{\text{atom}}$ atoms of a molecular system and put them together to form one vector with $6N_{\text{atom}}$ components, then the vector space formed by such vectors is the phase space of the molecular system.

**phase transition** Discontinuity in the properties of a system when external conditions are changed (Sect. 7.4).

**phenomenological equation** An equation that describes how a statistical average of some property changes with time (Sect. 4.6.1). It is also called the macroscopic equation. It can be derived from the master equation.

**poisoning** Situation in which no chemical reactions on a surface are possible anymore because the presence of some chemical species prevents the adsorption of one or more reactants (Sect. 7.4.3).

**potential-energy surface** A function that gives the potential energy of a molecular system as a function of the position of the atoms. If there are $N_{\text{dof}}$ degrees of freedom, then the function can be represented as a surface in a space of dimension $N_{\text{dof}} + 1$.

**pre-exponential factor** It is the factor in the Arrhenius form of a rate constant besides the exponential factor that describes the temperature dependence of a rate constant. The rate constant becomes equal to the prefactor at infinitely high temperature (Sect. 4.2.2).

**prefactor** Short for pre-exponential factor.

**primitive vector** A vector of a minimal set of vectors that generate a simple lattice. The lattice is obtained by taking all linear combination of the primitive vectors with integer coefficients (Sect. 2.1.1). A primitive vector is also called a primitive translation.

**process** A description of how labels of a lattice-gas model can change or the change itself. The description consists of a list of lattice points, the labels that are to be changed, and the labels into which they are changed (Sect. 2.1.3). Processes are often actual physical processes or chemical reactions. We use in this book also processes that are only defined to facilitate a simulation and that often use book-keeping sites and labels that do not correspond to adsorbates.

**Random Selection Method** An algorithm of kinetic Monte Carlo in which it is assumed that all processes have the same rate constant and that each process can take place anywhere on the surface. Oversampling is then used to correct for this obviously erroneous assumption (Sect. 3.4).

**rate** The magnitude of the change of some property with time. Mathematically it is the first derivative of that property with respect to time. In kinetics the property is generally a concentration, a density, or a coverage.

**rate constant** A proportionality constant between a rate and some other property of a system. In kinetics this property is generally of the same type as the property of the rate: i.e., we write the rate with which a coverage changes as a rate constant times some expression of the same or other coverages. Note that we make a strict distinction between a rate and a rate constant. There are two kinds of rate constants in this book. One is found in the master equation and is also called the transition probability per unit time (Sect. 2.2). The other kind is found in macroscopic rate equations (Sect. 4.6).

**rate equation** An equation that describes how a property changes with time (Sect. 4.6.1). It is also called the macroscopic rate equation and the property is usually a concentration or a coverage (Sect. 4.6).

**reaction-diffusion equation** An extension of the macroscopic rate equation for the coverage or concentration in which the coverage or concentration is allowed to vary with position. In addition to the terms in the macroscopic rate equation there is also a term for describing diffusion.

**short-range order** Strong correlation in the structure of a system over a distance comparable to the range of the interactions in the system (Sect. 7.2).

**site** Short for adsorption site.

**sticking coefficient** The fraction of the total number of atoms and molecules in the gas phase or solution that hit a surface and that also adsorb (Sect. 4.4.3).

**sublattice** A lattice can either be simple or composite. A composite lattice consists of two or more identical simple lattices that are shifted with respect to each other. Such a simple lattice is called a sublattice (Sect. 2.1).

**superbasin** A collection of catchment regions (or basins of attraction) of minima of a potential-energy surface. The catchment regions should be neighbors and the superbasin is a connected region (Sect. 8.1.2).

**supertype** A collection of different types of processes that are given the same rate constant. A supertype is used together with oversampling to correct for the fact that some types of processes forming the supertype have a smaller rate constant than is given to the supertype (Sect. 3.3.4).

**surface reconstruction** A structure of a surface that is different from the one that is obtained by cutting a crystal, or the change that leads to that structure (Sects. 5.5.3, 7.3, and 7.5).

**symmetry breaking** Phenomenon where a system changes in such a way that the symmetry group of the system is reduced to that of a subgroup (Sect. 7.4).

**Temperature-Programmed Desorption** Experimental technique in which an adlayer is heated up and the desorption of atoms and molecules is monitored. If during the experiment there are also reactions on the surface, then one also talks about Temperature-Programmed Reaction (Sects. 3.6 and 7.2).

**transition matrix** A matrix that gives the probabilities with which a system changes from one configuration into another (Sect. 8.1.4). The transition matrix is closely related to the matrix of transition probabilities of the master equation.

**transition probability** This is short for transition probability per unit time. It is a parameter in the master equation (Sect. 2.2.1). Because of the similarity and because the term is better known, we mostly use the term “rate constant” instead of transition probability in this book.

**transition state** When one or more molecules react their potential energy increases from a minimum value, reaches a maximum, and then decreases to another minimum value. The structure of the molecule(s) corresponding to the maximum energy is called the transition state (Sect. 4.2).

**Transition-State Theory** An approximation to calculate rate constants. There are many assumptions that can be made that all lead to the same expression. In relation to the derivation of the master equation in this book the assumption is that recrossings of the dividing surface is ignored (Sects. 4.2 and 4.4.3).

**unit cell** A representative part of a periodic system or system with translational symmetry. The whole system can be obtained by making copies the unit cell and shifting them over all symmetry translations (Sect. 2.1).

**Variable Step Size Method** An algorithm of kinetic Monte Carlo in which a single time is computed for each configuration that is encountered during a simulation and where processes are chosen proportional to their rate constants (Sects. 3.2 and 3.3). The method is also often called the n-fold way.

**voltammetry** Experimental technique in which the potential of an electrode is changed and the current is measured that is caused by reactions on the electrode's surface (Sects. 3.6 and 7.4.2).

**zero-point energy** The difference in energy between the energy of the minimum of a potential-energy surface and the ground state energy (Sect. 4.3.2).

**Ziff–Gulari–Barshad model** A simple model for CO oxidation that has only three processes: adsorption of CO, dissociative adsorption of molecular oxygen, and formation and immediate desorption of $CO_2$ (see Sect. 7.4.3).

# Index

**A**
Absorbing state, 24, 202, 216
Activation barrier, 11, 75, 78
Activation energy, 76, 78
    Brønsted–Polanyi relation, 103
    definition of, 76, 77
Adlayer, 22, 28, 75, 105, 158
    island, 178
    ordered, 2, 94, 159, 178, 179, 226
    structure, 2, 3, 11, 21, 59, 95
    superstructure, 178
Adsorption, 87–92, 147, 212, 218, 219
    dissociative, 114, 115, 165, 174
    modeling, 128, 129, 136, 149, 156, 165, 167–170, 174, 176
    precursor-mediated, 167–170
    simple, 108–110, 128, 129, 156, 170, 174, 176
    sticking coefficient, 89–91
Adsorption site, 14, 126
    label, 18, 126, 171
    lattice point, 14, 126
    modeling, 133–141, 171
    occupation, 14, 18
Algorithm
    First Reaction Method, 53
    parallel, 61–64
    Random Selection Method, 52
    rejection-free, 48
    Variable Step Size Method, 40–50
Arrhenius form, 76, 77

**B**
Bookkeeping site, 144, 146, 160, 161
    label, 141

**C**
Cluster expansion, 94–96
    truncation, 95
Coarse-graining, 226–231
Configuration, 18, 22
    initial, 177–179
Continuum model, coupling to, 231–234

**D**
Defect, 9, 140
Desorption, 86, 87, 89, 92, 122, 156, 169, 170, 212, 218, 219
    associative, 130, 131, 165
    modeling, 128–131, 156, 165, 169, 170, 176
    simple, 106–108, 128, 129, 176
Diffusion, 5, 91, 109–111, 130–132, 141, 148, 162, 163, 166–169, 212, 221, 222, 226
    modeling, 130–132, 141, 148, 162, 163, 166–169
Diffusion coefficient, 111, 167
Dissociation, 134, 162, 170
    modeling, 134, 170
Dividing surface, 30
Dynamic correction factor, 90

**E**
Event
    event list, 42
    null event, 48
Event list, 42

**F**
First Reaction Method, 53–55, 219, 237
    algorithm, 53
    scaling with system size, 54, 66

A.P.J. Jansen, *An Introduction to Kinetic Monte Carlo Simulations of Surface Reactions*, Lecture Notes in Physics 856,
DOI 10.1007/978-3-642-29488-4, © Springer-Verlag Berlin Heidelberg 2012

time-dependent rate constant, with, 56
Floppy molecule, 84, 85
Fluctuation, 112, 113, 193, 195
Formation, 141, 142, 175
modeling, 141, 175
FRM, *see* First Reaction Method

**H**
Heat transfer, 233

**I**
Immediate process, 128, 142–153
modeling, 142–153, 156, 157, 161, 175, 178
priority, 142, 175
structure
counting, 146–148
flagging, 143–146
superstructure, 178
Interaction model, 95, 98, 99
Inverted list, 48

**L**
Label, 18–20, 126
adsorption site, 18, 126, 171
bimetallic surface, 140
bookkeeping site, 141
coordination number, 172
defect, 140
lattice point, 18, 126
modeling, 126, 171
occupation, 18, 128–132
process, 18, 127
reconstruction, surface, 140
step, 137–140
sublattice, 133–137
Lateral interaction, 5, 8, 34, 51, 94–104, 181–185, 194–196
Bayesian model selection, 98–102
calculating, 94–104
cluster expansion, 94–96, 161
cross validation, 97, 98
interaction model, 95, 98, 99
linear regression, 96, 97
long-range order, 183, 199
modeling, 159–162
pair interaction, 160
phase transition, 183, 197
rate equation, 184, 185
short-range order, 183
site preference, and, 185
systematic error, 95
transition state, effect on a, 103
Lattice, 13–22, 176
Bravais lattice, 15
coarse-graining, 226–231
composite lattice, 15
lattice vector, 15
primitive vector, 15
simple lattice, 15
sublattice, 15, 176
translational symmetry, 14
unit cell, 15
Lattice gas, 13–22
configuration, 18
label, 18–20
shortcomings, 20, 21
Lattice point, 15, 126
adsorption site, 14, 126
label, 18, 126
number of adsorbates, 226
occupation, 18
sublattice, 135–137

**M**
Macroscopic equation, 106
Macroscopic rate equation, 1, 106–115
lateral interaction, 6, 7, 184, 185
substrate, 189
Markov chain, 23
Markov process, 24
Mass transport, 232, 233
Master equation, 22–35, 171
absorbing state, 24
class structure, 24
configuration, 22
decomposable, 24
derivation, 26–32
integral form, 38, 55
irreducible, 25
lateral interaction, and, 34
lattice-gas model, for, 32–35
Markov chain, 23
Markov process, 24
rate constant, 23, 31, 75
reduced, 213–215
reducible, 24
splitting, 25
transition matrix, 216
transition probability, 22
Mean Field Approximation, 7, 112, 113, 228
MFA, *see* Mean Field Approximation
Microkinetics, 1, 8, 9
Modeling
adsorption, 136, 149, 169
dissociative, 165, 174
precursor-mediated, 167–170
simple, 128, 129, 156, 170, 174, 176

Modeling (*cont.*)
adsorption site, 133–141, 171
averaging over simulations, 122
averaging over system size, 123
averaging over time, 124
bimetallic surface, 140, 178, 190
bimolecular reaction, 130–132
bookkeeping site, 141, 144, 146, 160, 161
defect, 140
desorption, 156, 169, 170
associative, 130, 131, 165
simple, 128, 129, 176
diffusion, 130–132, 141, 148, 162, 163, 166–169
dissociation, 134, 170
Eley–Rideal reaction, 128
flow reactor, 233
formation, 141, 142, 175
immediate process, 142–153, 156, 157, 161, 175, 178
island, 178
isotope experiment, 165, 166
label, 126, 171
lateral interaction, 159–162
nanoparticle, 170–177
noise reduction, 122–125
occupation, 128–132
process, very fast, 142
reconstruction, surface, 140, 191–193
site blocking, 149, 150, 156–158, 191, 197
step, 137–140, 191, 199
sublattice, 133–137, 168, 170, 172, 174
superstructure, 178
unimolecular reaction, 128, 129

**N**
n-fold way, *see* Variable Step Size Method
Net, *see* lattice
Noise reduction, 122–125
Normal mode, 80
Null event, 48

**O**
Occupation
adsorption site, 14, 18
label, 18, 128–132
lattice point, 18
modeling, 128–132
Order
long-range, 183, 199
short-range, 183
Overfitting, 95, 98
Oversampling, 45, 49, 50, 52

**P**
Parallel algorithm, 61–64
approximate algorithm, 64
causality error, 62–64
control parallelism, 62
data parallelism, 62–64
conservative algorithm, 63
deadlock, 63
optimistic algorithm, 63
safe process, 63
noise reduction, 62
Partition function, 78–85
classical, 78
hindered rotation, 83, 84
normal mode, 80
quantum, 78
rotation, 81–83
translation, 84, 87
vibration, 80, 81
PES, *see* potential-energy surface
Phase transition, 5, 183, 193–203
equilibrium, 194
kinetic, 4, 194, 200
non-equilibrium, 194, 200
Phenomenological equation, 106
Poisoning, 5, 24, 202
Potential-energy surface, 11, 14, 27, 31, 32, 75, 79, 80, 83, 85, 87, 88, 235
Pre-exponential factor, *see* prefactor
Prefactor, 76, 86
Primitive vector, 15
Process
adsorption, 87–92, 136, 147, 149, 167–170, 212, 218, 219
dissociative, 114, 115, 165, 174
precursor-mediated, 167–170
simple, 108–110, 128, 129, 156, 170, 174, 176
sticking coefficient, 89–91
decomposition, 148
desorption, 86, 87, 89, 92, 122, 156, 169, 170, 212, 218, 219
associative, 130, 131, 165
simple, 106–108, 128, 129, 176
diffusion, 5, 91, 109–111, 130–132, 141, 148, 162, 163, 166–169, 212, 221, 222, 226
disabled, 42, 46–49, 65
dissociation, 134, 162, 170
Eley–Rideal reaction, 91
enabled, 41
flicker, 215
formation, 141, 175
immediate, *see* immediate process

Process (*cont.*)
label, 18, 127
Langmuir–Hinshelwood reaction, 85, 86
multistep, 216, 224

**R**
Random Selection Method, 51–53
algorithm, 52
oversampling, 52
scaling with system size, 66
Rate constant, 23, 79
activation barrier, 75, 78
activation energy, 76, 78
adsorption, 87–91, 108–110
dissociative, 114, 115
sticking coefficient, 89–91, 109, 115
Arrhenius form, 76, 77
bimolecular reactions, 111–114
calculating, 85–94
desorption, 86, 87, 89, 106–108
diffusion, 91, 110, 111
dynamic correction factor, 90
Eley–Rideal reaction, 91
experiment, from, 104–115
general expression, 31, 75
Langmuir–Hinshelwood reaction, 85, 86
partition function, 78–85
prefactor, 76, 86
time-dependent, 55–58
transition state, 31, 32, 75
Transition-State Theory, 75
tunneling, 78
unimolecular reactions, 110
Rate equation, 1, 106–115
lateral interaction, 6, 7, 184, 185
substrate, 189
Reaction, *see* process
Reaction-diffusion equation, 11, 203
Rejection-free algorithm, 48
RSM, *see* Random Selection Method

**S**
Selection, 43–46, 217, 222
hierarchical, 44
oversampling, 45, 49
Schulze's method, 46
uniform, 44
weighted, 43
Site, *see* adsorption site
Step, 9, 137–140
Sticking coefficient, 89–91, 109, 115
Sublattice, 15
label, 133–137
lattice point, 135–137
layer, 168
modeling, 133–137, 168, 170, 172, 174
Superbasin, 213
Supertype, 50, 51
Symmetry breaking, 5, 193–203

**T**
Temperature-Programmed Desorption, 5, 194–196
time that a process occurs, 57
Temperature-Programmed Reaction, 186–189
TPD, *see* Temperature-Programmed Desorption
TPR, *see* Temperature-Programmed Reaction
Transition matrix, 216
Transition probability, 22
rate constant, 23
Transition state, 31, 32, 75
early barrier, 103
late barrier, 103
Translational symmetry, 14
Tunneling, 78

**U**
Unit cell, 15

**V**
Variable Step Size Method, 38–51, 235, 237
algorithm, concept, 40
algorithm, improved version of the concept, 42
algorithm with approximate list of processes, 47
algorithm with random search, 49
lateral interactions, and, 51
scaling with system size, 41, 43–46, 49, 66
supertype, 50, 51
time-dependent rate constant, with, 55
Voltammetry, 190, 197–199
time that a process occurs, 57
VSSM, *see* Variable Step Size Method

**Z**
Zero-point energy, 78–80
ZGB, *see* Ziff–Gulari–Barshad model
Ziff–Gulari–Barshad model, 4, 24, 67, 122, 142, 163, 165, 172–176, 200–203